GW00871024

Institution of Civil Engineers

International Conference on

Coastal Management 2003

Proceedings of the International Conference on Coastal Management, organised by the Institution of Civil Engineers and held in Brighton, UK, on 15–17 October 2003

Edited by R. G. McInnes

Supported by:

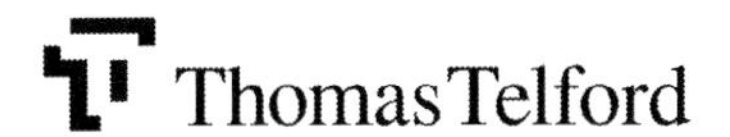

Organising committee

Eur Geol Robin G. McInnes (Chair)	*Isle of Wight Centre for the Coastal Environment*
Mat Cork	*Maritime Team, English Nature*
John Horne	*Defra*
Dr Keith Powell	*HR Wallingford Ltd*
Professor Dominic Reeve	*University of Nottingham*
Heidi Roberts	*Atkins*
Charlotte Street	*Severn Estuary Partnership*
Marianne Jones	*ICE Conferences*

Scientific advisory committee

Dr David Brook	*Office of the Deputy Prime Minister*
Stephen Cork	*PIANC (British Section)*
Dr Nick Dodd	*University of Nottingham*
Adam Hosking	*Halcrow Group*
Mark Lee	*Consultant*
Ian Meadowcroft	*Environment Agency*
Sam Rowbury	*Defra*

Overseas representatives

Dr Paolo Ciavola	*University of Ferrara, Italy*
Dr Jordi Serra	*University of Barcelona, Spain*
Professor Filomena Martins	*Universidade de Aviero, Portugal*

Published for the Organising Committee by Thomas Telford Publishing, Thomas Telford Ltd, 1 Heron Quay, London E14 4JD. www.thomastelford.com

Distributors for Thomas Telford books are
USA: ASCE Press, 1801 Alexander Bell Drive, Reston, VA 20191-4400, USA
Japan: Maruzen Co. Ltd, Book Department, 3–10 Nihonbashi 2-chome, Chuo-ku, Tokyo 103
Australia: DA Books and Journals, 648 Whitehorse Road, Mitcham 3132, Victoria

First Published 2003

A catalogue record for this book is available from the British Library
ISBN 0 7277 3255 2

Printed and bound in Great Britain by MPG Books, Bodmin, Cornwall

Editor's preface

The purpose of this conference is to translate policy into practice in terms of the management of coastal zones within the overall goal of achieving sustainable development. The objectives of the conference are fully supported by the European Commission and Defra, the lead government department for coastal issues in the UK.

Integrated coastal zone management (ICZM) is recognised as the most effective approach in terms of addressing the diverse problems and conflicts arising in coastal areas. The concept of ICZM has been strengthened following the completion of the European Union's 'Demonstration Programme on the integrated management of coastal zones' (1996-1999).

The dynamic coastal zones of the European Union encompass a range of physical problems including rapid coastal erosion, flooding and instability. In addition, there are considerable coastal development and tourism pressures. The coastal and nearshore waters are now recognised as being natural environments of considerable importance.

The European Commission regards it as essential to implement environmentally sustainable, economically equitable, socially responsible and culturally sensitive management of coastal zones. It also recognises the need for improved co-ordination of the actions taken by all the authorities concerned both at sea and on land in terms of managing the land / sea interface. It is anticipated that the wide-ranging and topical papers that are being presented at this conference will make a significant contribution towards this process.

The papers being presented at the conference relate to the following key areas:

- Coastal policy and management
- Conflicts, risks and uncertainty
- Estuarine and coastal engineering
- Shoreline management
- The coastal environment
- Ports, harbours, development and tourism
- Achieving better coastal management

The standard of papers submitted for this conference has been extremely high and this volume will prove to be a valuable contribution to the debate on integrated coastal zone management.

Eur Geol Robin G McInnes, FICE, FGS, FRSA
Editor, Chairman of the organising committee

Contents

7. Achieving better coastal management

Making Coastal Zone Management Work – Experiences from the Implementation Process on the Isle of Wight, UK

Robin G McInnes, Jenny Jakeways, Claire Marriott, Charlotte Street and Helen Houghton, Isle of Wight Centre for the Coastal Environment, Isle of Wight Council, Ventnor, Isle of Wight, UK

Introduction

For a number of years integrated coastal zone management (ICZM) has been recognised as the most effective approach in terms of addressing the diverse problems and conflicts arising in coastal areas. The concept of ICZM has been strengthened following the completion of the European Union's 'Demonstration Programme on the Integrated Management of Coastal Zones' (1996-1999).

The dynamic coastal zone of the Isle of Wight encompasses a range of physical problems including rapid coastal erosion and landsliding. In addition there are considerable coastal development and tourism pressures which have increased steadily since the middle of the 19th Century. The coastal and nearshore waters comprise an environment that is now recognised as being of European importance for both nature and earth science conservation reasons. The Isle of Wight Council is one of England's smallest local authorities, with limited financial resources. This paper describes both how the concept of ICZM has been introduced and applied on the Island and how the Council has made efforts to assist this process through participation in a range of coastal networks and groups. The process has been aided further by the establishment of the Isle of Wight Centre for the Coastal Environment in 1997, as well as through involvement in a wide range of research initiatives and demonstration projects supported by European Union and other funding programmes.

The paper describes the challenge of implementing sustainable coastal management policies on the Isle of Wight, taking full account of coastal evolution and the resulting risks to coastal development and infrastructure.

The gradual implementation of integrated coastal policies has required research into sedimentary processes, coastal erosion and the management of ground instability, as well as an understanding of existing climatic impacts and future scenarios. This work has been greatly assisted through the Isle of Wight Centre for the Coastal Environment's active participation in coastal initiatives at a sub-regional, national and international level. The paper describes the benefits perceived by the Isle of Wight Council from involvement in relevant networks and research programmes and illustrates examples of how best practice in the fields of both ICZM and management of natural hazards has been shared and disseminated. The Isle of Wight, which is managed by a Unitary Authority, provides a unique opportunity to implement ICZM and forms an excellent location for case studies; this aspect is also explained.

Coastal Management 2003, Thomas Telford, London, 2003.

Finally, this paper contemplates future coastal management and how the EU Recommendation may help local authorities such as the Isle of Wight fulfil the aims and objectives of sustainable development.

Research in recent years has demonstrated that integrated coastal zone management (ICZM) is an effective way of addressing problems that result from increasing pressures and conflicts of use within the coastal zone. Coastal management includes the full realm of activities that take place in the coastal zone and seeks to achieve an integrated approach to planning in order to address the management problems that arise in these often complex areas. It is important to recognise that ICZM includes both the physical and natural environments as well as the full range of human activities.

European Initiatives

The European Commission, which has played a leading role in encouraging the development of ICZM, regards it as essential to implement environmentally sustainable, economically equitable, socially responsible and culturally sensitive management of coastal zones. This process will maintain the integrity of this important resource whilst considering local activities and customs that do not present a threat to either sensitive natural areas or to the maintenance status of the wild species of coastal flora and fauna. This approach takes into account the Sustainable Development Strategy and Decision of the European Parliament and of the Council laid down in the Sixth Community Environmental Action Programme. It takes a strategic approach to the management of coastal zones based on the protection of the coastal environment, the recognition of the threats to coastal zones posed by climate change including sea level rise, the need for appropriate and responsible protection measures to protect coastal settlements and their heritage, as well as sustainable economic opportunities. In addition the Commission recognised the need for improved co-ordination of the actions taken by all the authorities concerned both at sea and on-land in terms of managing the land/sea interface (European Commission, 2002).

The Council of Europe began promoting integrated coastal planning in the early 1970s but the concept of coastal management has developed slowly since then. Organisations such as the Conference of Peripheral Maritime Regions and the European Union for Coastal Conservation fulfilled an important role in encouraging this approach, which led the European Commission to adopt a communication on ICZM in 1995 which reviewed the main features and the state of the European coastal zones. In order to expedite this process the Commission launched a Demonstration Programme in 1996, which aimed to fulfil two main purposes:

1. To test co-operative models for the integrated management of coastal zones and to provide the technical results needed to set ICZM in progress; and

2. To establish a structural dialogue between the European Institutions and all those stakeholders involved in the development of coastal zones.

During the period 1996-99 the Isle of Wight was one of thirty-five study areas selected for detailed review (European Commission, 1996) a process which resulted in the publication of two documents entitled 'Better management of coastal resources' (European Commission, 1997) and 'Lessons learnt from the EU Demonstration Programme' (European Commission, 1999). After considering these documents and other information arising from the EU Demonstration Programme, the European Parliament and the Council approved a Recommendation in May 2002 concerning the implementation of ICZM in Europe. It is hoped that in response to the Recommendation significant improvements will result in the way that the coastline of the European Union is managed.

Isle of Wight Issues

Coastal management is of great importance to vulnerable coastal zones such as the Isle of Wight. The Isle of Wight is situated off the coast of Hampshire on the south coast of England, separated from the mainland by the Solent channel, which varies in width from 3-6km. With a width of 35km from east to west and 22km from north to south the varied topography of the Isle of Wight is a result of its geological history. The rocks of the Island have been uplifted, compressed, folded, faulted and then subsequently weathered and eroded to provide a wide variety of scenery within a relatively small area. These natural processes, together with the effects of fluctuating sea levels over the last 10,000 years, have created a diverse landscape with a range of habitats including high sea cliffs, coastal landslides, salt marshes, estuaries and sand dunes. This diversity of habitats has led to over 90% of the Isle of Wight's coastline being designated as of European environmental importance.

Despite its special qualities, the coastal zone of the Isle of Wight does pose significant challenges in terms of its management. Contained within the Isle of Wight, for example, is the Undercliff, a 12km coastal landslide system along the southern coast which is inhabited by over 7,000 residents, the largest urban landslide complex in north-western Europe.

Situated in an exposed location facing the Channel, and taking account of the impact of storm waves from the prevailing Atlantic south-westerly direction, the Isle of Wight experiences some of the fastest rates of coastal erosion in Great Britain. Even in the more sheltered waters of the Solent, coastal erosion and instability are significant problems. Taking account of the predicted impacts of climate change including sea-level rise, drier summers and much wetter winters, new challenges must be faced by those with an interest in coastal management.

With a coastline of 110km the Isle of Wight has the longest length of coast of any Coast Protection Authority in England and Wales, but it is also one of the smallest local authorities. This presents a significant problem in terms of funding a range of necessary civil engineering measures as part of an overall management strategy. However, it is also recognised that, in terms of coastal management, the presence of one unitary authority, based upon an Island, presents particular opportunities as a scientific study area in terms of addressing these particular problems.

The Development of Coastal Networks and Plans

Local and Regional Networks and Plans

In 1985 the Isle of Wight County Council arranged a two-day conference entitled 'Problems Associated with the Coastline'. The event was organised to fulfil the need to discuss issues such as coastal erosion, ground instability, the potential impacts of aggregate dredging around the coast and the concern that the activities of some coast protection authorities may be having an adverse impact on adjacent coastlines. Following this meeting the Standing Conference on Problems Associated with the Coastline (SCOPAC) was established as the first Regional Coastal Group in England and Wales. SCOPAC brings together coastal engineers, planners and others with an interest in the shoreline (coastal defence) along a 400km length of the central south coast of England in order to exchange information and to encourage research (see figure 1). SCOPAC has fulfilled this role effectively since that time.

SCOPAC was actively involved as a consultee when the House of Commons Select Committee considered the subject of 'Coastal zone protection and planning' in 1992. The government confirmed the view of SCOPAC and others that one of the best ways of dealing with the many often conflicting issues that affect the coastal zone was through a non-statutory management plan process rather than more formal mechanisms within the planning system.

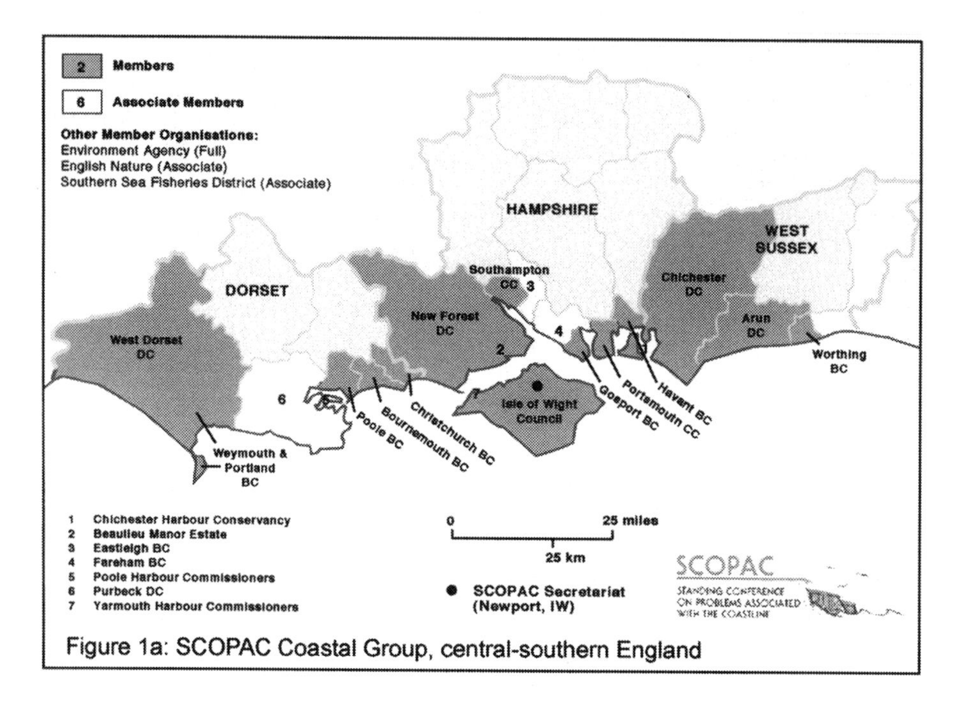

Figure 1a: SCOPAC Coastal Group, central-southern England

SCOPAC: TERMS OF REFERENCE

To promote sustainable shoreline management, and to facilitate the duties and responsibilities of local authorities and other organisations managing the coastal zone of central southern England by:

- Securing a consensus view on regional strategic and sustainability objectives and their means of delivery.
- Effecting consultation, co-operation and co-ordination between member organisations on issues of common concern.
- Promoting, co-ordinating and undertaking research to increase understanding and improve management of the regional shoreline environment.
- Actively representing the interests of member organisations in the development of national and supranational policies and initiatives relating to shoreline management.
- Increasing awareness of the need to deliver national and international statutory obligations.
- Providing a forum for communication of experience and best practice.
- Acting as a formal consultee on central government policy and guidance.

Figure 1b: Terms of reference of the SCOPAC Regional Coastal Group

Following this review, the former South Wight Borough Council on the Isle of Wight prepared 'A management strategy for the coastal zone' (McInnes, 1994), which was published after extensive consultation with stakeholders. As one of the first strategy documents in the UK relating to ICZM, and without the benefit of government guidance for documents of this kind, the strategy sought to consider all aspects of coastal zone management with recommendations for action. The document was important in that it first highlighted to Island residents the particular qualities and significance of the coastal zone, as well as the need for a more co-ordinated approach in terms of it's management.

Within a few years other coastal management plans were starting to appear (eg. New Forest District Council, (1997); and 'Strategic guidance for the Solent' (Badman and Sisman, 1997)) which provided improved formats for the development of coastal zone management plans.

Following local government re-organisation on the Isle of Wight in 1995 the replacement of two former Borough Councils and the Isle of Wight County Council by a new unitary authority created an excellent opportunity for the development of sustainable policies for the whole of the coastline, together with improved communication and integration between Council departments and stakeholders. This can be illustrated in many aspects of the work of the Isle of Wight Centre for the Coastal Environment, including the management process currently underway for the South Wight Maritime Special Area of Conservation.

Over the intervening period the Isle of Wight Council has led the development of a number of key strategy documents including a Shoreline Management Plan (coastal defence), Coastal Defence Strategy Studies, a Landslide Management Strategy, strategies for estuary management as well as a local Biodiversity Action Plan. The hierarchy of the most commonly prepared coastal plans is illustrated in figure 2.

The recognition of the need to update the South Wight's 'Management strategy for the coastal zone' coincided with the involvement of the Isle of Wight in the EU Demonstration Programme on ICZM in 1996. It seemed sensible to participate in, and learn from this project as it could assist the development of a better coastal management plan for the Island, taking advantage of the lessons learnt.

Whilst playing a leading role in SCOPAC has proved particularly rewarding and informative for the Isle of Wight, the contribution to ICZM implementation has also been partially fulfilled through the Council's membership of the Solent Forum. The Solent Forum was established in 1992 and sought to fulfil the need for improved understanding among local authorities and other agencies involved in the wider field of coastal planning and management in the Solent area (the estuary between the Isle of Wight and mainland England), to assist and encourage them in the execution of their functions. The aims and objectives of the Solent Forum are provided in figure 3. In 1997 the Solent Forum published 'Strategic guidance for the Solent' (Badman and Sisman, 1997) which provided advice and guidance on future management of the Solent, suggesting a series of priority actions which would be developed as 'flagship projects'.

The Solent Forum fulfils the role of an umbrella organisation for ICZM covering the central south coast of England. To the west, a similar organisation called the Dorset Coastal Forum was established in 1995. Elsewhere Fora were also developing such as that for the Forth Estuary based in Edinburgh, Scotland.

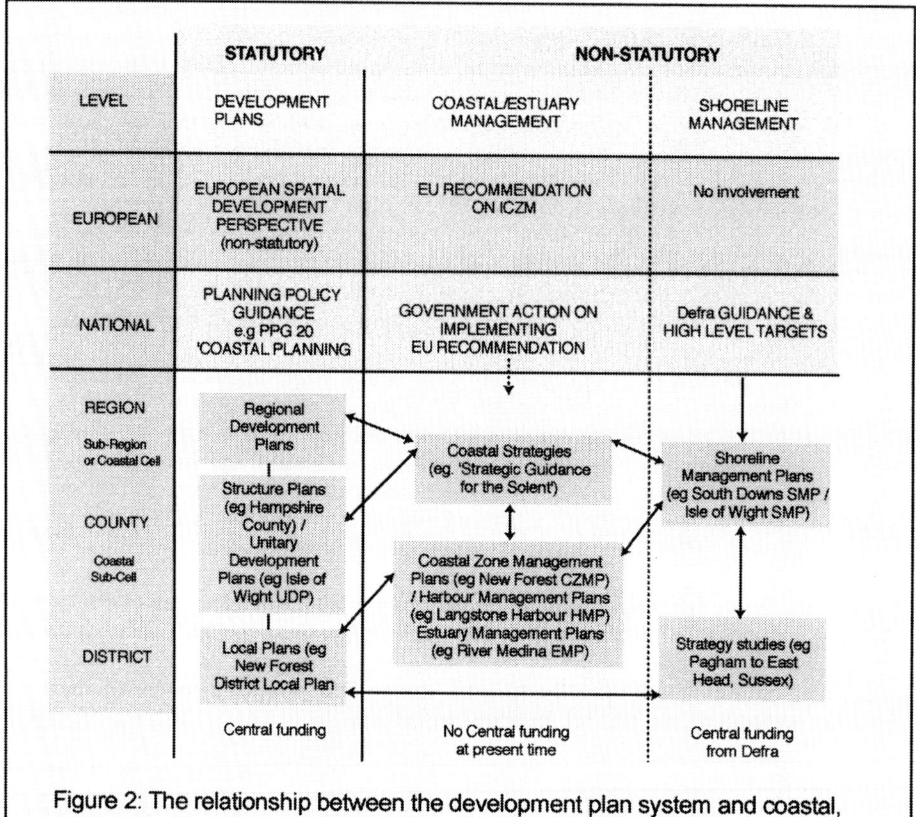

Figure 2: The relationship between the development plan system and coastal, estuarine and shoreline management. Adapted from Fleming C.A ed. (1996)

The value of organisations such as SCOPAC and the Solent Forum to their constituent members has proved to be the opportunities for sharing ideas and information as well as providing a way of keeping in touch with all the individual organisations interested in coastal management. These two partnerships have made a significant contribution towards ICZM in the south of England.

National Networks and Plans

In addition to the local and sub-regional coastal management networks and groupings such as SCOPAC and the Solent Forum, networks also exist at national and European levels. An increased awareness of the importance of coastal management and concerns about the continuity of delivery of the coast protection function through local authorities resulted in the establishment of the Local Government Association Coastal Special Interests Group (LGA SIG). Representing a large number of coastal authorities at County, Unitary and District Council levels the LGA SIG commissioned the preparation of a national coastal strategy which was published under the title 'On the edge' and launched in the House of Commons in the spring of 2001 (Shipman *et al*, 2001). The SIG provides a forum for national discussion of coastal issues of importance to local authorities. Subscriptions by its members enable the SIG to commission research on topics of particular interest or concern.

SOLENT FORUM

THE SOLENT FORUM (ESTABLISHED 1992) OBJECTIVES

1. TO PROVIDE A BROAD-BASED CONSULTATIVE FORUM
2. TO RAISE AWARENESS AND UNDERSTANDING
3. TO CONTRIBUTE TO POLICY DEVELOPMENT PLAN - MAKING AND STRATEGIES
4. TO IMPROVE THE INFORMATION BASE
5. TO IMPROVE COMMUNICATION, CONSULTATION AND LIAISON
6. TO COMMENT ON MAJOR DEVELOPMENT PROPOSALS AND OTHER CHANGES
7. TO PROMOTE THE IMPORTANCE OF THE SOLENT

Figure 3: Objectives of the Solent Forum

At a national level the Chairmen of the eighteen Regional Coastal Groups of England and Wales, which have a special interest in management of the shoreline (coastal defence), meet twice a year. These meetings, which are held at the Institution of Civil Engineers (ICE) in London, coincide with meetings of the ICE's Maritime Board, where the opportunity is taken for these two organisations to meet and exchange information at the national level. The Chairmen also meet Defra, English Nature and the Environment Agency twice a year at the Coastal Defence Forum. Defra (the Department for Environment, Food and Rural Affairs) is the national government department responsible for flood and coastal defence in England. The network of leading organisations and groups that are involved in coastal management issues is illustrated in figure 4.

European Networks and Plans

At a European level the majority of regional and local authorities are members of the Conference of Peripheral Maritime Regions (CPMR). Based in Rennes in France, the CPMR carries out its work through a number of Commissions including the Islands' Commission, the Atlantic Arc, the North Sea Commission, the Baltic Commission and the Inter-Mediterranean Commission. This well-structured arrangement provides a forum for debate on issues relating to the coastal zone and also the isolation and peripherality experienced by some coastal regions, including islands. The CPMR often fulfils the role of consultee on European initiatives and the European Commission values its structured approach, agendas and reports. The Isle of Wight is one of twenty-two European islands within the Islands' Commission. The Isle of Wight became an active member of the Islands' Commission after hosting the annual meeting on the Isle of Wight in 1993. It became apparent that there were many areas of common interest in relation to coastal management on islands and work commenced shortly afterwards on the preparation of 'A coastal strategy for the Islands of the European Union' (McInnes, 1995). This strategy for European islands was ratified as a framework document for development of coastal management plans for all islands within the group at the meeting held on Gozo in 1996.

The concept of 'regional seas' has become increasingly recognised, with examples such as those included within CPMR (e.g. North Sea Commission, Inter-Mediterranean Commission). However, the central part of the Channel (La Manche) does not benefit from inclusion within one of the current 'regional sea' groupings. The recognition of common aims and objectives on coastal and environmental issues in relation to the Channel led to the establishment of Arc

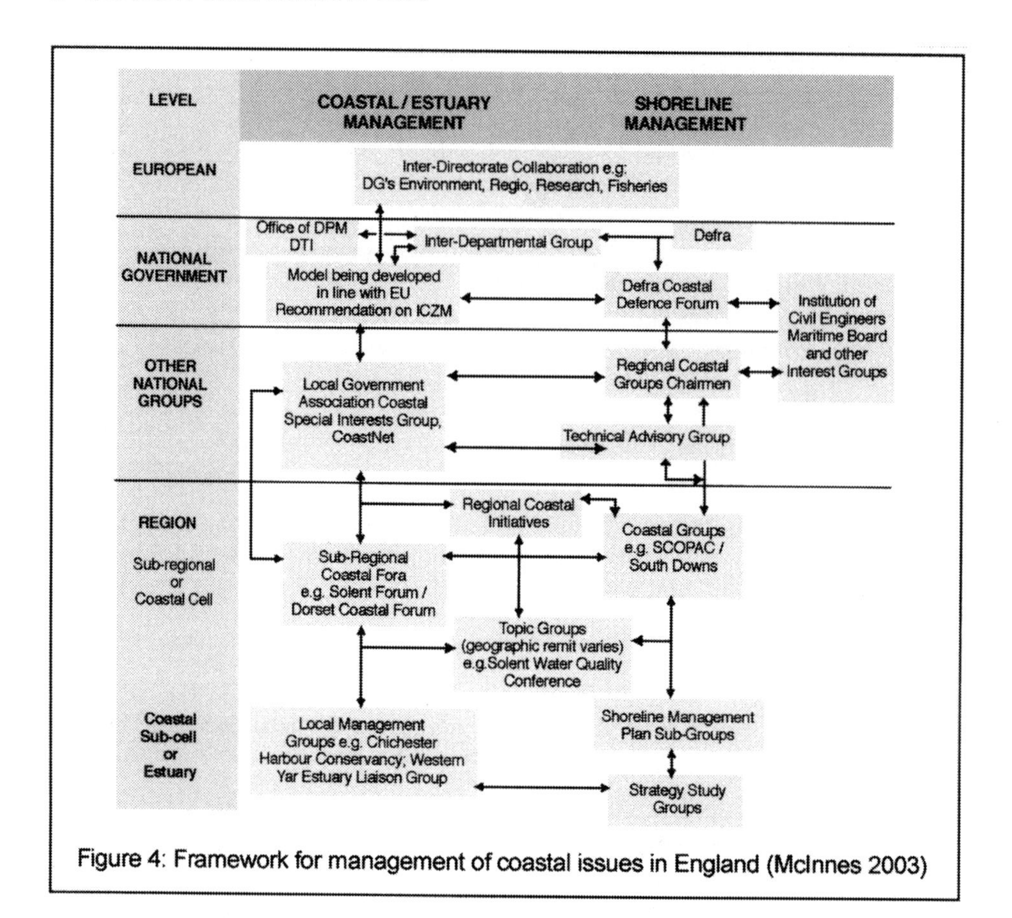

Figure 4: Framework for management of coastal issues in England (McInnes 2003)

Manche in 1996. Arc Manche, an alliance between the south coast Counties of England and the French Regions bordering the Channel (extending from the Belgian border to southern Brittany), developed a coastal and environmental agenda that involves information exchange and collaboration on European projects. In particular, projects have been undertaken through such programmes as LIFE Environment, Interreg and the Vth Framework Programme. A re-launch of the Arc Manche initiative took place at a meeting held in Rouen in March 2003. The purpose of the event was to re-define the remit of the partnership, to stimulate greater interest in coastal and marine-focused Arc Manche activities, to reinforce political support for the initiative and to set an updated agenda for future actions.

Networking and information exchange involving the Isle of Wight at a European level is being enhanced through projects such as 'EUrosion' (a DG Environment contract led by RIKZ in partnership with other leading research institutions), through the PROTECT Fifth Framework Programme Project (led by the British Geological Survey) which is studying chalk cliff instability problems, as well as through networks of excellence being developed for participation in the European Union Sixth Framework Programme.

Information Exchange

The European Commission and others in recent years have highlighted the vital need to ensure that coastal information and research is properly disseminated. For too long many researchers have been reluctant to properly disseminate the results of their studies and investigations which has proved to be a huge waste of resources and has resulted in overlap and duplication in research due to a lack of knowledge of the activities being undertaken by research organisations. A key objective of the European Commission's Vth Framework Programme for research was to ensure that dissemination of results was properly addressed, usually through the involvement of end-users in the research proposals.

European Projects

The Isle of Wight, as a small local authority with limited resources, was anxious to obtain the best possible information that could assist it in addressing its range of coastal and geotechnical problems. To engender this process the Isle of Wight Centre for the Coastal Environment was established in 1997 (see figure 5). The purpose of the Centre was to provide a focus for the Council's growing activities in the coastal, geotechnical and environmental fields. A team of coastal scientists, geologists and geographers was built up to manage and undertake its research programme. Development of coastal management on the Isle of Wight was significantly assisted following the award of European Union LIFE-Environment funding support for the Isle of Wight as a study area within the Integrated Management of Coastal Zones project (1996-99). In 1997 a further major EU LIFE-Environment project entitled 'Coastal change, climate and instability' (McInnes, Jakeways and Tomalin, 2000) was approved, with the Isle of Wight leading an international team of scientists and academics examining issues relating to long-term coastal change as well as risk management in relation to ground instability problems in coastal and mountainous areas. A wealth of information and experiences were gained from these collaborative projects with due consideration being given to dissemination of results. Consequently the Centre for the Coastal Environment established the Coastal Wight website (www.coastalwight.gov.uk) which disseminates information about the range of services provided as well as its research findings. The work of the IW Centre for the Coastal Environment and its partners in the coastal and geotechnical fields was highlighted in the European Commission's publication entitled 'LIFE Success Stories' (European Commission, 2001). The Centre now has a team of eleven post-graduate researchers who undertake a range of work programmes in order to assist addressing the Island's coastal management needs.

The Isle of Wight Coastal Visitors' Centre

In 1998 the Centre for the Coastal Environment, with the kind support of the Isle of Wight Council, managed to obtain occupancy of a former office building located in an outstanding position on the coastline at Ventnor in the south of the Isle of Wight. A 'Coastal Visitors' Centre' was gradually established over the next year at this site, comprising of a series of visual and interactive displays which illustrate the importance of the Isle of Wight coastal zone, the physical and natural environmental issues and how they were being addressed in terms of coastal zone management, as well as providing a mechanism for education for students aged 4-18 and beyond. A comprehensive technical library was also developed at the Coastal Visitors' Centre, one of the largest within a local authority in the UK, and it has been made available freely to researchers of all levels, including under-graduate and post-graduate.

Conferences and Workshops

A particular strength of the Centre for the Coastal Environment has proved to be the organisation of international conferences and workshops. A second Isle of Wight international geotechnical conference was held at Ventnor in May 2002 on the subject of 'Instability – Planning and Management' (the first being held on 'Slope Stability Engineering' in 1991). The event provided the mechanism for disseminating the results of the LIFE project on 'Coastal change, climate and instability' as well as allowing the presentation of some sixty papers by scientists and practitioners from across the world. Attended by representatives from over twenty countries, the Proceedings of the conference were published by Thomas Telford (McInnes and Jakeways (Ed.), 2002) to coincide with the conference event.

Other events arranged by the Centre have included a range of workshops on topics such as 'Coastal Engineering and Planning –Improving opportunities for collaboration and information exchange' (which included dissemination of the LGA Coastal SIG report 'Managing coastal risk: making the shared coastal responsibility work' by Ballinger, Taussik and Potts, 2002); the workshop was held in February 2003 for coastal engineers and planners in the SCOPAC region.

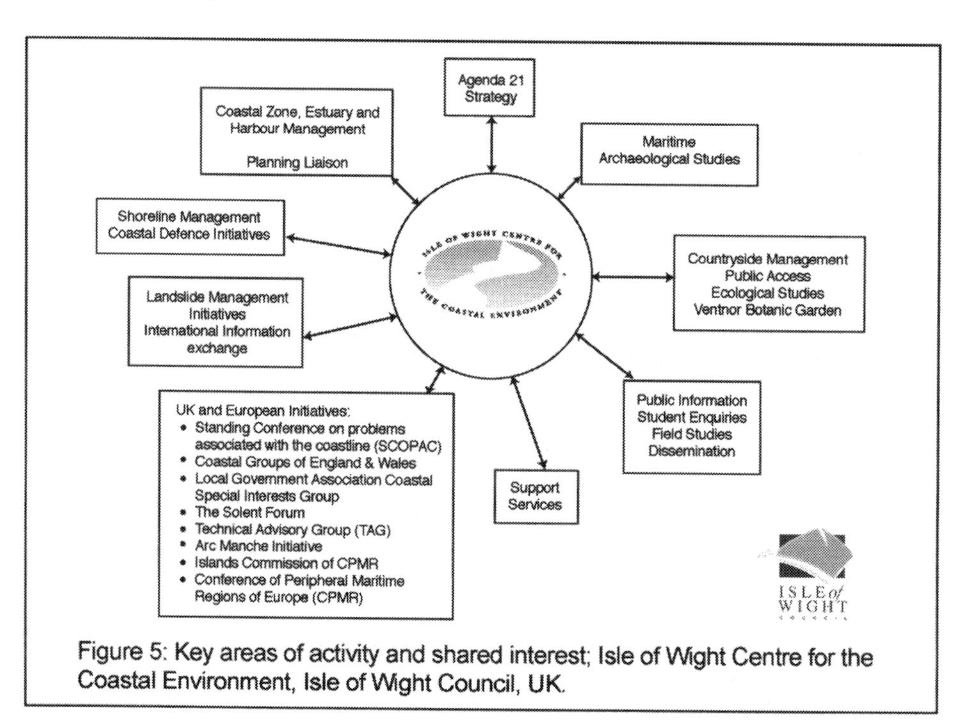

Figure 5: Key areas of activity and shared interest; Isle of Wight Centre for the Coastal Environment, Isle of Wight Council, UK.

Publications

The Centre for the Coastal Environment has recognised that it is vital for scientific information relating to the coastal zone to be disseminated to both technical and non-technical audiences. With this aim in mind efforts have been made to produce high quality, visually attractive publications, videos, cd-roms and brochures that will raise interest and awareness with those who do not necessarily have a detailed technical knowledge of the subject. A good example is the non-technical guide 'Managing ground instability in urban areas –A guide to best practice' (McInnes, 2000), an eighty-page book published as part of the European Union LIFE-Environment project on ground instability. This publication has attracted considerable interest including that of the United Nations ISDR Unit (International Strategy for Disaster

Reduction) which is now considering publishing a global version in a number of other European languages. A similar publication entitled 'Coastal defence –a non-technical guide' (McInnes, in press) will be published in June 2003 by SCOPAC.

Adopting a Strategic Approach

One of the particular successes in developing coastal policies for the Isle of Wight and central southern England has been recognition of the need for those involved in coastal management to work effectively together. In the field of coastal defence, for example, an uncoordinated approach by adjacent local authorities has in the past resulted in significant impacts down-drift, including cutting-off sediment supply and increased levels of coastal erosion. SCOPAC was influential in recognising this approach and commissioning research into the division of the southern England coastline into sediment 'sub-cells', discrete lengths of coast where the processes of erosion, sediment transport and deposition are relatively self-contained.

The complex coastal issues in central southern England have resulted in a range of sub-regional studies that have aided those involved in coastal planning in the area. These have included the commissioning of a study by SCOPAC entitled 'Preparing for the impacts of climate change' (Hosking and Moore, 2001), developing coastal evolution and risk mapping for the Isle of Wight, and participation in a pilot study for a new kind of ntegrated coastal zone map – an initiative of the Ordnance Survey in partnership with the British Geological Survey and the Hydrographic Office.

South-east England has also benefited from collaboration between local authorities and the Environment Agency taking advantage of new technology. The recognition of the need for strategic coastal monitoring and a collaboration between the coastal groups in the south-east led to a successful bid for an £8.1 million monitoring programme for the whole of the south-east of England which has been grant-aided by Defra, initially for five years (see figure 6). This strategic study will provide a uniform standard of monitoring along a varied section of coastline and will assist in providing baseline information against which, for example, the impacts of climate change can be assessed. It will help to ensure that inconsistencies within individual local authorities' monitoring programmes will be ironed out and this in turn will assist wise decision-making for coastal and shoreline management.

In England the developing Regional agenda has included interest in issues relating to the coastline. Regional working groups have been established examining such topics as climate change, flooding and biodiversity. In addition, the important role that the coast plays within the south-east of England Region has now been more fully recognised following the preparation of a report by the Isle of Wight Centre for the Coastal Environment (Roberts and McInnes, 2000) which highlighted the contribution the coast makes to the economy of the Region as well as identifying the risks to assets from erosion, flooding and climate change.

Best Value in Coastal Management

The principle of 'Best Value' is one which establishes a requirement for local authorities in the UK to deliver services to clear standards, covering both cost and quality, by a balance of the most effective, efficient and economic means available. A recent local government White Paper focussed on a wider improvement agenda and as part of this the Comprehensive Performance Assessment (CPA) was highlighted. The CPA (an assessment of local authority performance) is helping Councils to deliver better services such as coastal management and will lead to a greater focus on outputs and outcomes.

Figure 6: Strategic coastal defence planning in south-east England

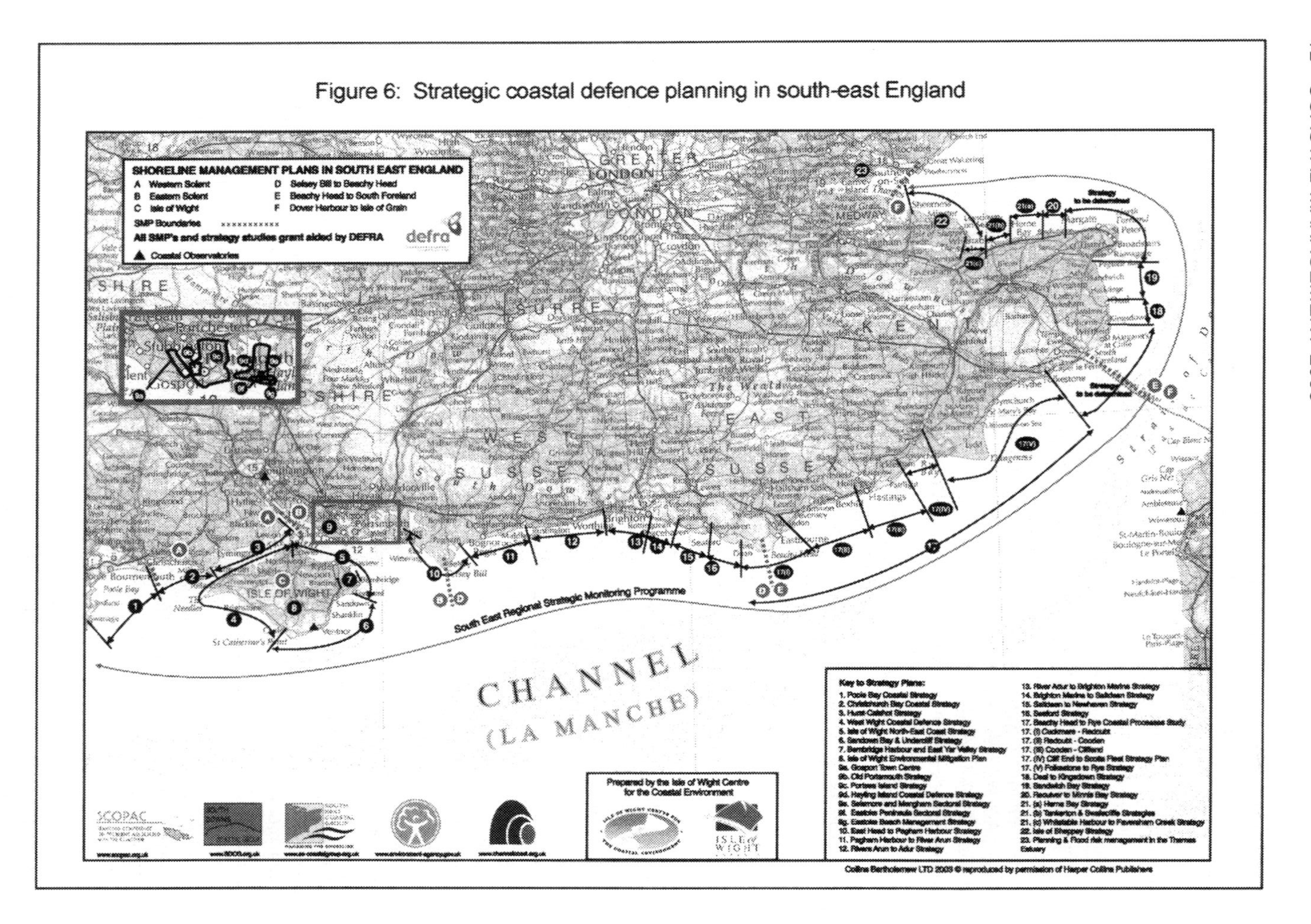

Within the broad field of coastal management there is a range of legislation that provides local authorities with responsibilities or powers to deliver services such as planning, coastal protection and environmental health. Furthermore, a wide range of guidance has been provided by central government and this is continuing to be updated. Through 'best value' local authorities must demonstrate that their method of service delivery is both effective and efficient. On behalf of the Local Government Association the Isle of Wight Centre for the Coastal Environment has established a matrix of proposed 'best value indicators' for the coast. The framework is capable of being applied at the county, unitary and district levels of local government.

Currently the subject of consultation within local authorities, it is intended that 'best value' should be addressed through the preparation of a short policy statement which sets out the physical, socio-economic and environmental attributes of each local authority's coastline and it should be accompanied by a map illustrating these aspects. The policy statement will explain how the coastal management objectives of the authority assist in developing the strategic objectives of the authority (which will be set out already in the authority's Community and Corporate Plans).

The Community Plan is of particular importance because it summarises the aspirations of the local community in terms of services it wishes to see provided by the local authority. The method by which these services will be delivered is set out in the authority's Corporate Plan. For most policy areas a purpose statement will be established which states the overall aim and objective for the service and describes particular service targets within its various fields of activity that it wishes to aspire to. An example of a typical local authority performance strategy for coastal management is illustrated as figure 7. The overall aim of the strategy is to assist achievement of the goal of sustainable development.

Working with National Government on the European Union Recommendation on ICZM

The implementation of the European Union 'Recommendation on ICZM' presents an exciting and challenging opportunity for EU Member States to implement effective coastal zone management. The European Commission has spelt out in detail how this should be progressed taking advantage of the results of the Demonstration Programme, the views of national experts and the conclusions from its publications 'Better management of coastal resources' (European Commission, 1997) and 'Lessons learnt from the EU Demonstration Programme' (European Commission, 1999). Networks and groups involving local authorities in the UK are anxious to work with the government in terms of assisting this process. The lead role to be played by Defra is particularly welcomed as it will assist in the integration of coastal and shoreline management policies and planning.

The important role played by Regional Coastal Groups and Fora such as SCOPAC and the Solent Forum has been recognised by national government in the past, which has supported a 'bottom up' approach towards the management of both the open coast and for estuaries. However, whilst Defra has provided sufficient funding in order to allow the delivery of sustainable policies for coastal defence, historically there has not been a similar funding arrangement in place for delivering ICZM through coastal and estuary partnerships. The result has been that the effectiveness of service delivery has been much reduced. It is very much hoped that the government will be able to allocate sufficient resources in order to ensure that sustainable management of the coastal zone can be implemented.

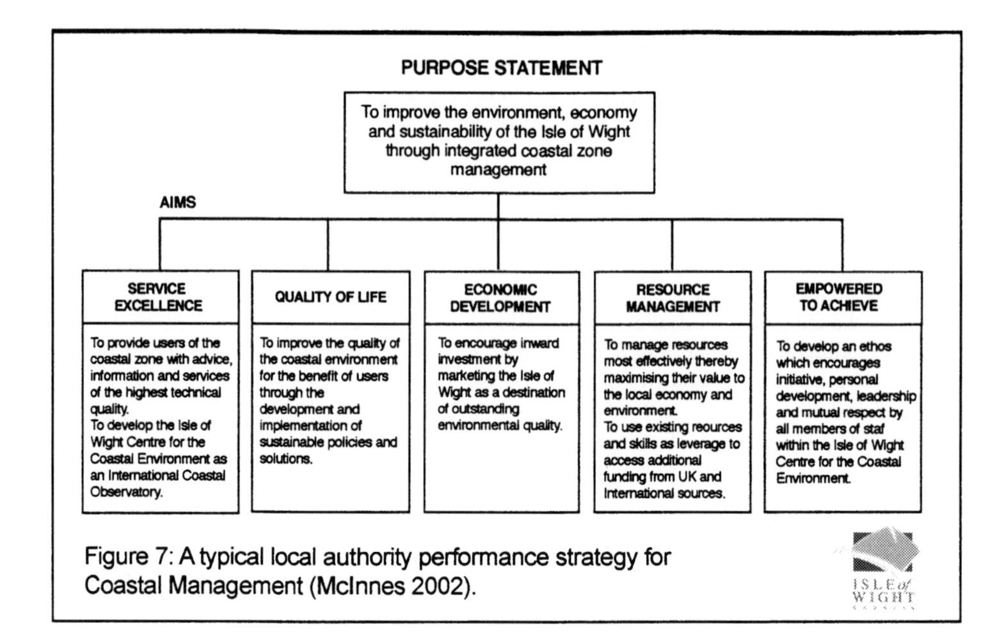

Figure 7: A typical local authority performance strategy for Coastal Management (McInnes 2002).

Conclusions

The Isle of Wight Council is a small local authority with limited resources. It has a long and complex coastline, parts of which were heavily developed in the 19th and early 20th Centuries. The Isle of Wight Council established a Centre for the Coastal Environment in 1997 and a Coastal Visitors' Centre in 1998 to help address its coastal problems.

Although the Isle of Wight Council has still to finalise a coastal zone management plan for the whole of its coastline, a great deal has been learnt since its 'Management strategy for the coastal zone' was written (McInnes, 1994). The Council's experience in the development of 'A coastal strategy for the Islands of the European Union' (McInnes, 1995) and work on the 'EU Demonstration Programme for the integrated management of coastal zones' (Jewell, McInnes and Roberts, 1999) has particularly helped this process.

A recent review of different types of plans and strategies suggests that a coastal zone management plan can comprise a relatively short, well-illustrated, concisely written and jargon-free document which sets out policies for different aspects of the coastal zone with key recommendations. A good example of this approach is the plan prepared recently by the Region of Aquitaine in France (Region Aquitaine, 2002); this is the approach that is likely to be followed by the Isle of Wight.

Successful coastal management involves a thorough understanding of natural coastal processes and the environment as well as a grasp of the socio-economic and human issues. Local authorities and their plans must be forward-looking, bear in mind the principles of sustainable development, and take full account of factors such as the predicted impacts of climate change. It is vital that the principles of ICZM are maintained and the Isle of Wight intends to ensure that successes achieved in the past are continued by employing innovative techniques and undertaking comprehensive research.

A particular success is believed to be the participatory approach that has been adopted at the various administrative levels in England and Wales in recent years. It is hoped that this success can be built upon through the implementation of the EU Recommendation to the benefit of all coastal regions in the United Kingdom and elsewhere.

References

Badman, T. & Sisman, K. 1997; Strategic guidance for the Solent. Southampton: Solent Forum.

Ballinger, R.C., Taussik J. & Potts, J. 2002; Managing coastal risk: making the shared coastal responsibility work. Summary Report to the LGA Coastal Special Interest Group. London: Local Government Association.

European Commission, 1996; Demonstration Programme on integrated management of coastal zones. Brussels: European Commission.

European Commission, 1997; Better management of coastal resources. Brussels: European Commission.

European Commission, 1999; Lessons learnt from the EU Demonstration Programme. Brussels: European Commission.

European Commission, 2001; LIFE-Environment in action – 56 new success stories for Europe's environment. Luxembourg: European Commission.

European Commission, 2002; Recommendation on the integrated management of coastal zones. Brussels: European Commission.

Gizard, X. 1997; The Conference of the Peripheral Maritime Regions of Europe (CPMR). Rennes, France: CPMR.

Hosking, A. & Moore, R. 2001; Preparing for the impacts of climate change'; report for SCOPAC by Halcrow Ltd. Swindon: SCOPAC

House of Commons Select Committee, 1992; Coastal zone protection and planning. London: HMSO.

Jewell, S., McInnes, R.G. & Roberts, H. 1999; IW LIFE Project – Integrated management of coastal zones. Newport, Isle of Wight: Isle of Wight Council.

McInnes, R.G. 1994; A management strategy for the coastal zone. Ventnor, Isle of Wight: South Wight Borough Council

McInnes, R.G, 1995; A coastal strategy for the Islands of the European Union. Report prepared by the IW Council for the Islands Commission of the Conference of Peripheral Maritime Regions of Europe. Newport, Isle of Wight: Isle of Wight Council.

McInnes, R.G., Jakeways, J. & Tomalin, D. 2000; Coastal change, climate and instability; LIFE-Environment Project. Ventnor, Isle of Wight: IW Centre for the Coastal Environment.

McInnes, R.G. 2000; Managing ground instability in urban areas –A guide to best practice. Newport, Isle of Wight: Isle of Wight Centre for the Coastal Environment.

McInnes, R.G. & Jakeways, J. (Ed.) 2002; Instability –Planning and Management: International Conference Proceedings, Ventnor, Isle of Wight, May 2002. London: Thomas Telford.

McInnes, R.G. (in press); Coastal Defence -a non-technical guide. Ventnor, Isle of Wight: SCOPAC.

New Forest District Council, 1997; Coastal Management Plan. Lyndhurst, Hampshire: New Forest District Council.

Region Aquitaine, 2002; Littoral Aquitain –état des lieux. Mission Littoral, Region Aquitaine, Bordeaux: Conseil Regional Aquitaine et République Française.

Roberts, H. & McInnes, R.G. 2000; Managing the coast of central-southern England –Final Report. Newport, Isle of Wight: Isle of Wight Council.

Shipman, B., Roberts, H. *et al* (member authority officers) 2001; On the edge: the coastal strategy. LGA Coastal Special Interest Group. London: Local Government Association.

Integrated Coastal Zone – Data Research Project (ICZMap®)

Martin Whitfield, UK Hydrographic Office
John Pepper, UK Hydrographic Office
David Overton, Ordnance Survey
Matthew Harrison, British Geological Survey

Background

The coastal zone has, through its social, economic and environmental significance, become a focus of growing attention in the UK. It is a dynamic environment under constant pressure from human and economic activity as well as physical and environmental forces, some of which are associated with climate change and rising sea levels.

Currently, the UK coastline has formed the boundary of both Ordnance Survey & Ordnance Survey Northern Ireland (OSNI) terrestrial mapping down to mean low water (mean low water springs in Scotland) and United Kingdom Hydrographic Office (UKHO) chart data (up to mean high water).

Coastal zone geographic data in UK is only available in diverse terrestrial, geological, and marine geographic datasets and this is a major problem for policy makers. Until now, it has been the remit of the user to resolve any integration issues resulting from differing projections, datum, scale of capture and other specification issues. This is a time consuming and laborious process and may produce inconsistent data, which are difficult to maintain and hence a major barrier to the management of a particularly sensitive environmental zone.

About the Project

The Integrated Coastal Zone Mapping Research Project (ICZMap®) is a two-year applied research venture, started in May 2001, between Ordnance Survey, UKHO and British Geological Survey (BGS) to resolve projection, datum and specification issues and produce a consistent, useable dataset. The project has been supported by HM Treasury through it's Invest to Save Budget (ISB) funding initiative.

The key stages of the project are to:

- establish user needs, priorities and service provision;
- draw up a Specification of Requirements for data integration;
- evaluate options for resolving technical differences;
- develop a Technical Specification for integrated coastal zone mapping;
- capture and integrate data for the defined "test" pilot areas;
- use the pilot test data to support innovative coastal management projects;
- present findings to the ICZMap® Steering Group and user community.
- derive post project options for electronic delivery of data & services

Coastal Management 2003, Thomas Telford, London, 2003.

The overall aim of the ICZMap® project was to investigate the creation of an integrated and accessible coastal dataset built from existing data, and how this might be used to satisfy the diverse range of coastal applications and services. In doing so, incompatibilities between horizontal and vertical datums, as used by the three organisations and the resultant problems, are being investigated by producing a methodology for datum transformation. Ultimately, this methodology would produce an integrated height model with a consistent height datum. In the case of this research we specifically looked at the transformation of UKHO bathymetry (at Chart Datum) to Ordnance Datum Newlyn (ODN).

Three pilot areas around the UK coast were selected as test cases over a wide enough geographic area and that could be delivered within a timescale to be practical for potential pilot users. Nominal limits of 5km inland and 20km out to sea were set and vector datasets from each of the Agencies produced and amalgamated. The main test area for the project covers a stretch of coastline between Shoreham and Lyme Regis on the South Coast including the Isle of Wight. Additionally, smaller test areas have been defined in the Firth of Forth in Scotland and Milford Haven in South West Wales.

The project has involved staff from all three Agencies covering specialists in areas such as Geographic Information Systems, Geodesy, Research & Development as well as Business Management.

Horizontal datum and projections

Both Ordnance Survey and BGS data are issued on OSGB36 horizontal datum with British National Grid on the Transverse Mercator Projection, whereas UKHO use WGS84 compatible datum on the normal Mercator Projection. In order to create integrated datasets the UKHO, using SafeSoft's Feature Manipulation Engine (FME), created parameter files containing ten 7-point parameter horizontal shifts for regions of the UK enabling UKHO to confidently transform its data to OSGB36. This transformation was applied to UKHO test data prior to its dispatch.

The Vertical Datum integration problem

Within the ICZMap® project there is a requirement to create a three dimensional height model integrating both terrestrial and marine geographic data held by Ordnance Survey and the United Kingdom Hydrographic Office (UKHO). This model is intended for use by scientific, research and coastal management and planning organisations and will not be used for navigation.

There is a fundamental factor currently preventing the creation of an accurate height model, that of the differing height datum's used by Ordnance Survey and UKHO. On mainland Great Britain, Ordnance Survey use Newlyn Datum (ODN), which is a land based fixed datum, and UKHO use Chart Datum, which is based on tidal data and varies, depending on location, with respect to Newlyn. Specific islands around the coast (e.g. Scilly Isles, Isle of Man, Shetland) use different vertical datums to Newlyn which also have to be tied in to the relevant chart datums. The UKHO publishes tidal data in Admiralty Tide Tables, which include data detailing the 'Height in metres of Chart Datum relative to Ordnance Survey Datum in the United Kingdom'. This provides, at coastal locations, the height difference between Chart and the relevant Ordnance Survey vertical datum. For the pilot test areas there are 49 published coastal locations where this height difference is provided, and over 800 around the entire UK (only 339 of them have all of the tidal information needed to add to the model). However, a problem arises when this coastal data is extrapolated seaward, as the effects of the tidal variation need to be considered. Charts are produced by the UKHO with co-tidal lines drawn through points of equal high water interval. These provide the necessary information to extrapolate the seaward component.

In order to create an integrated height model, the differences between the two datums throughout the test areas needed to be defined. Once known, the difference can be applied as a transformation from one datum to another. UKHO supply information regarding the difference between the two datums at points around the coastline (usually at ports). Unfortunately, this information is lacking offshore.

Work is now underway to establish the relationship between Ordnance Survey's vertical datums and the WGS84 spheroid. A checking exercise at a sample of tidal base stations was undertaken by the Ordnance Survey to confirm the validity of the historical heighting link between the tidal points and the related land based bench marks to aid this process.

See Fig 1. for the relationship between the different datums.

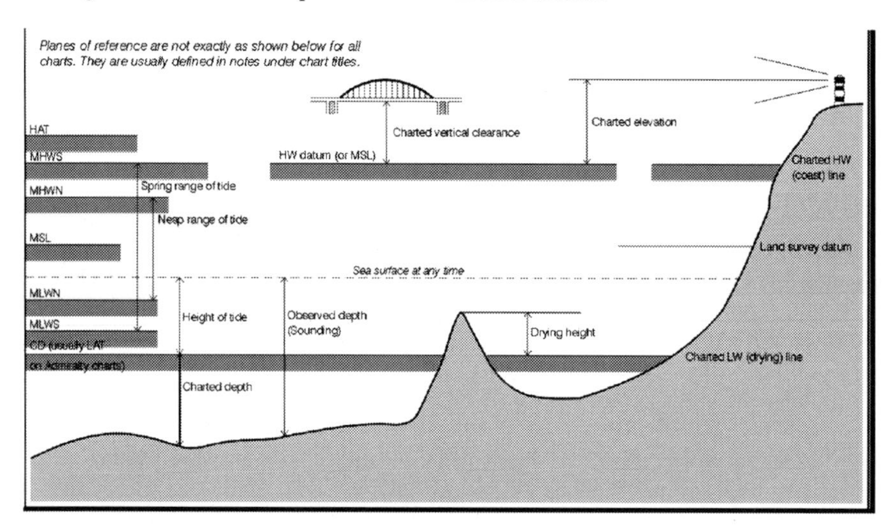

Fig 1. Relationship between the different datums.

As a result of this non-interoperability of data, this research had two main themes to it:
1. the calculation of the difference between the two datums at points of up to 20km offshore, and
2. the assessment of different interpolation algorithms as to their efficacy in producing a continuous interpolated surface covering the offshore sector from which the UKHO bathymetric data may be transformed to ODN.

The results of the first section directly feed into the second and ultimately a methodology can be produced that could be applied on a national scale.

Assumptions

The final methodology was based upon a number of assumptions, these are:-

- Due to the inconsistencies between OS data and UKHO data (variable depending on survey methodology), a rigorous correction method would not be beneficial.
- The connection between Mean Sea Level (MSL) and Chart Datum (CD) varies in proportion to the tidal range from the coast to seaward, thus the connection will vary across the pilot area, from the coast to 20km offshore.
- For the purpose of this research, in pilot areas the difference between MSL and ODN remains constant seaward, although it is acknowledged that in reality this is not so.
- At the coast Mean High Water Springs (MHWS) is recorded above MSL and Mean Low Water Springs (MLWS) is below MSL. The difference between MHWS and MLWS is known as the Mean Spring Range (MSR) and reduces seaward.

Integration of onshore and offshore height data

By relating the co-tidal value offshore to the nearest appropriate Port and the use of an equation developed at the UKHO, an offshore value for the height correction relative to the OS height datum can be obtained. Interpolation techniques can then be employed to grid the derived values around the UK. The production of the geo-referenced grid provides a means for applying a correction to the observed UKHO soundings relative to OS datum. The resulting data set can be used to re-contour the bathymetric depths and hence produce on the same vertical datum, an integrated terrestrial and marine geographic model for the UK.

A series of points 20km offshore where the difference between ODN and Chart Datum is known was produced. This was achieved by locating positions on the 20km offshore line where a co-range line (lines through points of equal MSR) cut. These positions were then used in conjunction with the co-range MSR values offshore, MSR values at the nearest ports and differences between the datums at the nearest ports to calculate the differences in datums. These points 20km offshore were then used in conjunction with the published differences in datums at locations along the coast to interpolate a 'transformation surface', which when applied to bathymetric data resulted in a consistent height model both onshore and offshore. Various interpolation algorithms were assessed as to their efficacy in producing an accurately modelled 'transformation surface'. Those algorithms assessed were, Inverse Distance Weighted (IDW), Spline Interpolator and Natural Neighbours.

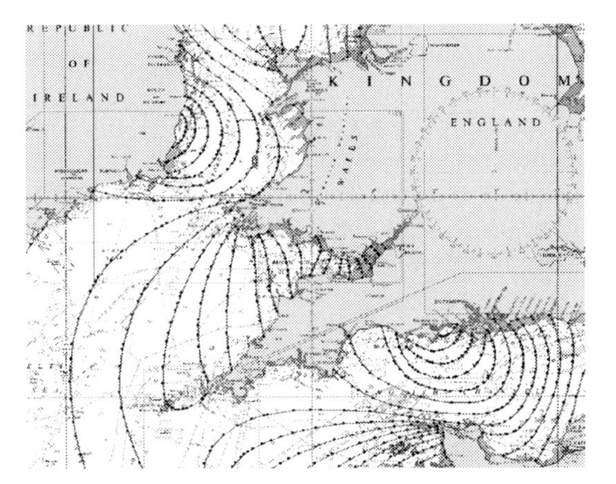

Fig 2 Co-tidal points, lines and ports

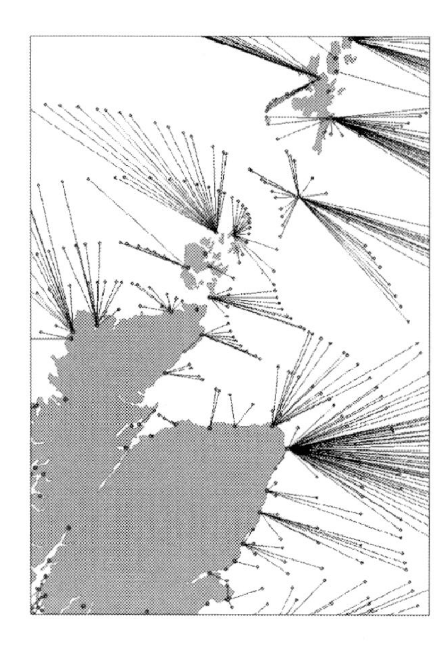

Fig 3:Co-tidal point related to nearest port *Fig 4: Contoured grid*

Of those interpolators assessed it was found that the Natural Neighbours algorithm produced the best results. This algorithm and the method of calculating datum differences offshore was then incorporated in to the methodology for datum transformation.

3D soundings into attributed 2D soundings

UKHO Sounding data is supplied as x,y,z. In order to enable height (depth) modelling attributed 2D Sounding data were required, with the depth being assigned to the feature as an attribute (2D+1). FME functionality (ElevationExtractor) was used to obtain attributed 2D soundings

Fig 5 3D Spatial model to common vertical and horizontal datum

Expanding integer values into meaningful text

UKHO data is in S57 format, whereby attribute values are coded (i.e. colour 1,6,1 equates to Red, Black, Red) in order that the data can be utilised across a range of ECDIS viewers, however this is meaningless to non-navigational end users. In order to add value to the s57 data the attribute codes were decoded and converted to meaningful text The user will view the text rather than the integer values.

Foreshore representation

This project investigated options available for the representation of the foreshore at the interface between these different datasets. At present Mean Low Water (MLW) marks the extent of Ordnance Survey data while the Mean High Water (MHW) marks the extent of the UK Hydrographic Office (UKHO) data. This creates an overlap in datasets and complications arise in their depiction due to conflicts between different datasets.

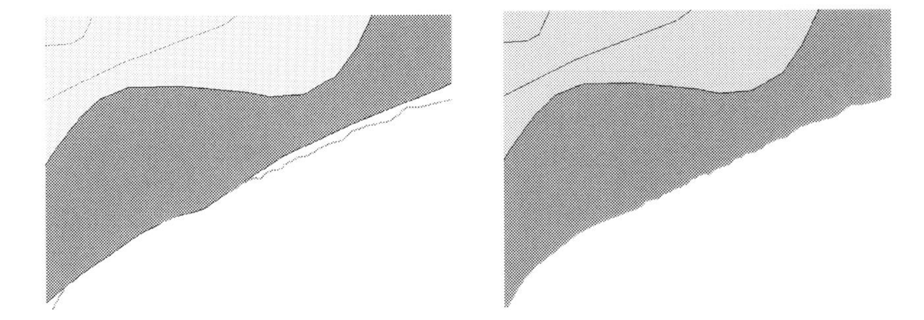

Figs 6 & 7 Incidence of Ordnance Survey and UKHO foreshore polygons before and after resolution

Due to their unique situation in that their data holdings traversed the seaward and landward divide, the BGS work was critical to the projects success. For this reason they led on the issue of creating a coast line to which all parties could align their data.

BGS updated the MLW boundary in the pilot areas to that which is currently depicted in OS MasterMap®. Although this does bring technical problems regarding the integration of a high-resolution (1:1250 and 1:2500) MLW with DiGMapGB-50 (1:50000) scale geology we believe that these are outweighed by the increased utility to users of BGS DiGMapGB-50 geology at the coast (figure 8).

For this process to be undertaken BGS required a clean and coherent MLW line, which was not available as standard in Ordnance Survey MasterMap®. After consultation with OS it was decided that BGS produce a new version of the MLW for use within the ICZMap pilot areas. This was then used to update the MLW of all the DiGMapGB-50 geology layers. The techniques used to create the new MLW are described in detail in the following section.

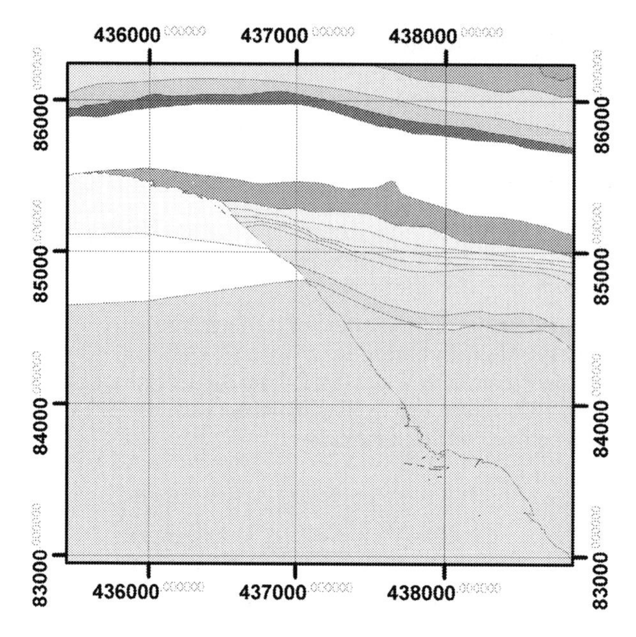

Fig 8. Example of the BGS ICZMap product. The geological classification, within the limitations of the resolution, meets seamlessly across the coastal zone. © NERC

The difficulties and complexities of offshore surveys results in an almost complete lack of geological data in shallow nearshore areas from MLW to approximately 10 m water depth. In practice this means that geological classifications and boundaries have not matched up across the coastal zone. Part of the pilot project for BGS was to re-assess this situation and attempt a more seamless classification across the coastal zone, within the limitations of data resolution and support. This work was done across the MLW boundary in each of the pilot areas

Linear re-processing using BGS generalised MLW and MHW

ArcView shapefiles of Ordnance Survey MasterMap® data were provided to the ICZMap project for the study areas and coatlines were converted to a "dgn" format.

On applying digital cleaning and validation processes to the linework, it was found that the basic extraction had left many loose ends. Primarily this was because lines such as jetties etc were not coded as MLW. If the missing parts were a simple straight line, the endpoints were joined by the extension of the available vertices with a straight line. If more complex, the original full dataset was referenced in and correct lines sourced from it.

In some cases many iterations of a cleaning process had to be run, due in some part to complex line strings with multiple vertices shattering into single length lines.

During the first phase of ICZMap work within the pilot areas, clean MLW linework was generated, but individual vertices were not edited. This gave relatively large file sizes and was time consuming. This would not be recommended for processing of the whole country beyond the ICZMap pilot areas.

For the second phase, 'weeded' (vertices thinned) linework was used which was extended up the estuaries/inlets with 'unweeded' Ordnance Survey MasterMap® data. After trials and validation it was decided to cut the geology layers against this MLW.

Linear processing was difficult in some areas, possibly due to the high resolution and physical size of Ordnance Survey MasterMap® vector data. Problems were again encountered, including shattering of linework into individual lines, which then had to be repaired. Once a clean MLW (or representation) had been achieved, it was processed using the GIS functionality of GeoGraphics.

For ICZMap the MLW in the DigMap50 was substituted with the new generalised MLW. Line cleaning of the MLW against the geology boundary lines was then effected to give clean junctions on all the vector components of the future polygons. On validation of the dataset it is obvious that some geology lines would overlap and some not reach the new MLW.

The software allows the flagging of all these points and with the assistance of an in-house macro, the affected vertices were sequentially edited.

Pilot Area Data – BGS – OS integration at MHW and MLW

The use of the new mean MLW allowed geology polygons to be revised either by extending or cutting them. Reseeding as necessary followed validation checks.

Cleaned and validated linework was checked for intelligent polygon seeding. The intelligence is supplied from a unique link on a text seed, provided through ODBC to rows in an Access database. All the values contained in the Access table are sourced from corporate Oracle tables. The software has automatic check routines which flag up any problems - usually 'no seeds' or 'duplicates'. This has become an automated function with the software's basic functionality having been improved through customisation by the BGS Cartographic Services System Developer. This flowline rapidly provides consistent and clean coded polygons.

Once onshore and offshore data met at MLW from a cartographic perspective, geological re-interpretation was undertaken to integrate the geological boundaries between the larger scale onshore and offshore geology.

The UKHO cut / extended its shoreward limit of the inter-tidal polygon to match the MHW as supplied by Ordnance Survey.

Symbology

A revised set of symbols for the UKHO element of ICZMap® has been created based on Chart 5011 (INT 1) symbol set so that features can be viewed in such a way so as to replicate those shown on UKHO navigational charts. The ENC s52 display standard (for use in ECDIS) is considered inappropriate for non navigational use.

Data formats

UKHO supply their data as S57 Ed 3 output format, Ordnance Survey supply their data in GML format and BGS in ESRI Shape or MapInfo TAB file formats. Two common data formats were agreed on for release in the test areas, namely ESRI Shape and MapInfo TAB files for the trial datasets.

Ongoing initiatives – Global datums

The task to integrate onshore and offshore height data for the UK will deliver a cartographically seamless vertical reference surface on which modelling of the coastal zone can be undertaken.

There is also a need to investigate new methodologies to improve the positional accuracy of vertical datum planes based on the WGS84 spheroid relative to the new Geoid Model for UK (OSGM02) and possibly the creation of a European or Global reference system based on alternative technologies (such as altimetry). Whilst UK has a vision of this as a long term goal, cooperation between world geospatial organisations will be needed.

User inputs

To date, over 500 organisations, ranging from nature conservation to mineral extraction companies, have expressed interest in ICZMap® through the website

www. iczmap.co.uk

The Inter-Agency Committee on Marine Science and Technology (IACMST) were represented as the chair of the ICZMap steering group. The rest of the steering group were made up of the participating organisations, the EA, DEFRA and ABP. Together with the 40 organisations pilot testing the data, this group forms the primary customers for the data. The membership of this group covers the wide range of users of coastal information within Central, Local Government and the commercial sectors.

The user group is assisting in defining the specification for the data and investigate practical use of the data in their applications and hence priorities. Additionally users will provide a key input to the project through the Coastal Groups involved in testing the datasets at present.

Success criteria

Specific criteria for measurement of project success include:

- users of ICZMap® are able to access, receive, integrate and use the data sets more easily
- users can develop their own applications using the coastal zone mapping
- data is widely used and integrated
- ICZMap® is recognised as an effective solution to user needs

ICZMap® trial data has been tested by members of The Standing Conference for Problems Associated with the Coast (SCOPAC), Forth Estuary Forum and the Pembrokeshire Coast Forum for trial and evaluation purposes.

Up to date

The project is now complete and professional user views have been gleaned for the 3 pilot areas. The data has been incorporated in projects and MSc studies and the vertical datum model is currently on trial with two key Agencies.

Some key findings from the process are outlined below:-

- 85% of the respondents would find the proposed joined up data of use, 70% would find it more useful than existing data.
- The Majority of Local Authority respondents would recommend ICZMap to DEFRA for Shoreline Management Plans.
- Research, planning and meeting legislative drivers represent the greatest foreseen use for the data
- Horizontal Datums, Vertical Datums and High Water lines are the most important elements to unify.
- The proposition to make three interoperable datasets available is preferable to a combined product.
- Purchase decisions are largely led by legislation , followed by spending plans and business objectives.
- The largest possible scale was requested by the majority (up to 1:1250 mapping scale).

Feedback on specification, update cycles and specific usage were also gleaned from this process. This will help the decision processes within the 3 organisations beyond the project phase.

Conclusions

- Through the project it has proved technically feasible to join datasets from OS, BGS and UKHO across the coastal zone

- Users found ICZMap data a major improvement in mapping of the coastal zone

- Feedback during the project highlighted the vital importance of the coastal zone to a wide and enormously diverse range of sectors

- The market appears genuinely interested in ICZMap data; 85% of those asked would find the combined data useful, and an indicative proportion of those who used the data were supportive in its application to Shoreline Management Plans.

- ICZMap would contribute to the delivery of a number of Government and European policies, initiatives and directives which impact on the coastal zone

- There is a key requirement for common vertical datum referencing systems. This project has created a cartographic solution fit for purpose in many uses. Beyond this an academic approach is approved by this project and will be undertaken by the UKHO.

- The perception and use of digital coastal datasets varies with the needs and expertise of the users. There is no universal requirement with regard to many issues, including, delivery format, scale, horizontal or vertical datums.

- Lack of data within the near-shore between MLW and approximately 10 m water depth is a major constraint on full integration across the coastal zone.

- Agreements need to be finalised between agencies on:-
  - an agreed common coastline
  - a process for creating a TIN traversing the coastal zone.
  - An approach to funding of the considerable amount of work needed to roll-out beyond pilot

  - Rollout of ICZMap on a UK wide scale will require substantial investment and commitment on the behalf of the three organisations. This could only be embarked upon with the approval and sponsorship of Government.

Recommendations

- The three organisations need to assess funding opportunities and consider government funds to roll out ICZMap for the nation.
- The three organisations need to agree a coastline to match their data to (eg, generalised or pure OS MasterMap®).
- Ordnance Survey need to create continuous joined up coastlines.
- Only once this has been agreed can a mechanism for joining the data together be progressed.

The Manhood Peninsula Partnership – Coastal Zone Management in Practice

C. Cobbold, BSc (Hons) Eng., Risk Management Journalist, Birdham UK
B. E. Waters, CEng., FICE, FCIWEM, Consulting Engineer, Eastbourne UK

Introduction

This paper describes the processes by which two local authorities decided to promote the Manhood Peninsula as a case study on sustainable integrated planning.

The peninsula lies south of Chichester, on the central south coast of England. It is low lying and under threat from coastal erosion, coastal flooding, rising ground water levels and overflowing water courses, although clearly not all these apply to all areas all the time. Climate change predictions indicate that these problems will increase.

The town of Selsey, at the southern tip of the peninsula, has grown from 2,000 population to over 10,000 during the last five decades. The other coastal settlements of Bracklesham and East Wittering also have experienced rapid growth in recent years, while the inland villages have grown fast but to a lesser extent. In total, the Manhood Peninsula has absorbed a significant proportion of all new housing in the Chichester District for the last forty years, despite flooding and infrastructure difficulties, notably transport.

It is reasonable to consider the whole peninsula as a coastal zone, not least because it is likely that it will be impacted greatly by rising sea levels, caused by climate change and the tilting of the UK. Indeed, as a distinctive area with natural boundaries, a unique landscape, under pressure from increasing population, its problems and qualities make it an ideal case study, which may suggest solutions to both its difficulties and those of other areas.

Perhaps the most pressing decision is with respect to the future of the coastline. This will have a profound effect on the increasing conflict between housing, industry, farming, the environment and tourism, the mainstay of the area's economy. Tourism, in particular, is centred on the coast, the rural environment, Chichester Harbour and a large caravan park, the largest such park in Europe, in Selsey. All these aspects will be affected by whatever decision is made about the coastline. The authorities believe that all aspects should be taken into consideration before any decision is made as to its future. They also believe that security of the coastline should be maintained, at least in the short term, to allow the environment, community and economy to adapt to its future.

Coastal Management 2003, Thomas Telford, London, 2003.

The Peninsula

The triangular peninsula is bordered by the sea to the south east and south west, by Pagham Harbour to the east and Chichester Harbour to the west. To the north lies Chichester and the mainland, consisting of gravels over alluvium fronting the chalk downs. The gravel decreases south of Chichester, so the 15x20 km peninsula is formed largely of alluvium, with the Medmerry shingle bank, west of Selsey Bill, believed to be sited on an ancient shingle base.

Fig 1
Part of the Manhood Peninsula from the air, showing Selsey in the foreground and Pagham Harbour to the top right.

Fig 2
The Medmerry frontage – a shingle bank protecting a site of some 3000 caravans on low-lying land.

Selsey contains some 10,000 of the peninsula's total population of 24,700 with East Wittering and Bracklesham being the next largest communities. Most of peninsula's interior is relatively undeveloped and contains substantial Grade I and Grade II agricultural land. The built up area of Selsey is shown at Figure 1. Much of the land behind and to the west is under 5m above Ordnance Datum (A.O.D).

Coastal erosion has been a problem since recorded history, although not consistent around the coast. By 1950 East Beach in Selsey was the second fastest eroding coastline in the U.K. with a rate exemplified by the loss of a house a year adjacent to the lifeboat station. The present station was built in the 1960's to replace one far seaward which had been built on the beach in the 1920's. Other beach movement and deposition had closed off the sea passage at Medmerry, which had, in medieval times, made Selsey an island. Indeed at the end of the causeway carrying the main road to Selsey, the farmhouse is still known at Ferry House.

The coast protection authority, Chichester Rural District Council (since 1974 Chichester District Council) instituted coastal protection works starting in 1950 under powers conferred by the 1949 Coast Protection Act. Government financing was through MAFF (now Defra) and over a period of some 35 years the built-up coastline was largely secured. Areas threatened by flooding, notably in the Medmerry area, were the responsibility of the Sussex Rivers Board, later Southern Water, National Rivers Authority, and now the Environment Agency. The tests for economic benefits and possibly these management changes have meant that large sums have been expended on emergency works and maintenance, which might otherwise have funded an engineered scheme. Even now, the Medmerry bank at its weakest points has a 1 in 1 year standard of defence against a breach occurring.

The Medmerry shingle beach is regularly re-profiled to a height of three to four metres. However, it is often overtopped and during storms plant is on standby to maintain the cross section. Behind it lies one of Europe's largest static caravan parks catering to a summer population of up to 15,000 and contributing, according to a recent study, some £25 million per annum to the local economy. (Fig 2).

As well as high quality agricultural land, the peninsula supports a horticultural industry, much of it under glass, as the area boasts particularly good quality light.

The peninsula contains several important nature conservation areas including Chichester and Pagham Harbours, both designated as Special Areas for Conservation under the EU Habitats Directive; as Special Protection Areas under the EU Birds Directive; and as Sites of Special Scientific Interest under UK legislation. This gives wildlife in these internationally important areas the highest level of legal protection possible in the European Union.

Pagham Harbour, designated a Ramsar site under the Convention on Wetlands of International Importance, has historically had a mobile entrance but is now controlled by retaining walls to the north and east and a shingle bank to the south west. Both require maintenance and the proposed diversion of the River Lavant flood water to the harbour may introduce further problems.

Chichester Harbour also has been designated an Area of Outstanding Natural Beauty (AONB) and is an important contributor to the local economy, both as visitor attraction and because of the substantial recreational marine industry it supports. A study commissioned by

Chichester Harbour Conservancy estimates that the Harbour generates some £60 million a year to the local economy. [3]

However, the value of the Harbour and its environs is greater than the sum of its parts, not least because of the difficulty in pricing the value of its designation as an AONB. Like the entire peninsula itself, its worth is made up of both tangible and intangible benefits. In any consideration of coastal zone management for the peninsula, the harbour is a highly important component of the peninsula's content.

The entrance to the Harbour is protected by a naturally formed mobile sand spit known as East Head. The harbour has a complex regime, maintaining a deep entry channel on the Hayling Island side. As a result of the 'double dip' Solent Tides the harbour fills in about five hours and empties in three hours. Studies have demonstrated the complexity of the tidal and current movements in the harbour mouth and it is very difficult to predict the likely impact of any change to the spit. However, the spit, which is owned by the National Trust, is attached to the mainland by a small 'hinge' which is subject to coastal erosion. A long term solution for the future of the East Head has yet to be agreed, with many differing views expressed by local agencies and groups.

Climate change, predicted to result in rising sea levels, increased storminess and increased winter rainfall, as well as the isostatic re-adjustment of the UK landmass since the last ice age causing a subsiding effect in the south east, will exacerbate the coastal and inland flooding problems suffered by the peninsula.

More work needs to be done to understand the coastal and groundwater dynamics in the area; what impact climate change will have; what possible solutions are available; which are the most sustainable options; and how the local community, environment and economy needs to adapt.

These are the issues that the Manhood Peninsula Partnership is addressing on a medium to longer term co-ordinated manner, involving the agencies responsible working together and closely with the community.

The Manhood Peninsula Partnership

The Manhood Peninsula Partnership (MPP) arose from an initiative by two local residents to encourage a more integrated and long-term approach to planning and coastal and water management. Renee Santema, a Dutch spatial planner and Carolyn Cobbold, a risk management journalist, were concerned that the apparent lack of a co-ordinated approach and of a long term view was, at best, ignoring some of the opportunities the area could exploit and, at worst, producing a situation where local communities and local and national authorities would find their options constrained dramatically in the future and their interests ultimately put at risk physically, economically and socially.

For example, large housing estates were being built adjacent to a vulnerable coastline whose defence in the future was, and still is, a matter of considerable debate between Defra, the Environment Agency and the local authorities. Expansion of the coastal settlements had given many local people a greater sense of security than reality. Other residents were, and are, concerned that the local authorities have allowed new building to continue without being prepared to spend more money on coastal defence.

As a result of these concerns, Santema approached NIROV, the Dutch Institute of Spatial Planning, suggesting that it should hold one of its brain-storming, problem-tackling workshops on the Manhood Peninsula. These workshops, to which experts from different fields are invited to share their ideas and experience, are held by NIROV every few years to tackle particular planning issues in the Netherlands.

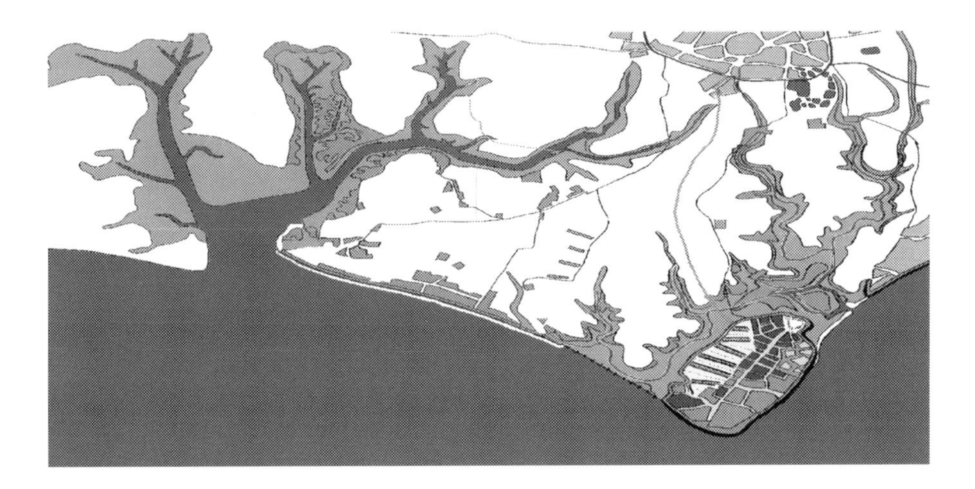

Fig 3

One of the NIROV workshop proposals: Selsey as "Stone Island", a heavily protected development able to withstand future flooding.

NIROV officials believed that the diversity of the issues facing the Manhood, and the likely impact of climate change on the area, would make the Peninsula a fascinating subject for its first such workshop overseas.

West Sussex County Council, Chichester District Council and the Environment Agency offered to co-host the five-day workshop on the Manhood Peninsula and provided a total of £10,000. A further £20,000 was raised from various local and national organisations (both Dutch and British) and 28 Dutch and British practitioners and academics in planning, the environment, infrastructure and water and coastal management were invited to attend the five-day workshop.

Before arriving in the country, the participants were briefed about the area, from documentation and background papers provided by a variety of sources, including local authorities, parishes, schools, community groups, farmers and businesses.

The first day of the workshop was spent on a tour of the area, during which participants met with representatives of the local authorities, Chichester Harbour Conservancy, businesses, landowners and others to get a better feel of the issues, coastline and environment.

The participants then were divided into mixed teams, with each team containing one or more coastal geomorphologist, coastal engineer or water management specialist; town and country planner, landscape architect or infrastructure specialist; environmental or climate change specialist. During the workshop people with specific knowledge, members of the public and elected councillors were invited to sit with the teams to inform them in more detail and discuss proposed ideas.

The workshop participants had to come up with ideas for a long-term integrated planning approach, based on an overall vision for the whole Manhood. In this approach, all planning problems had to be addressed, including coastal management, flooding, housing development, infrastructure, agriculture and horticulture, employment, tourism, recreation, nature conservation and environmental issues, all in the light of climate change.

In addition, a separate group of Dutch and British coastal experts was formed to discuss the costs and benefits of alternative coastal protection.

By combining different professions and different nationalities, and removing the political and administrative constraints normally imposed on coastal management and planning decisions, the groups were able to take a new approach to the problems and solutions. Working and living together for five days, in an unconstrained, creative and dynamic manner – with teams working into early hours of the morning – produced some innovative and thought-provoking options for the future. (Fig 3)

The results of each group's vision for the future of the Peninsula were presented to representatives of the public and local authorities and organisations at County Hall, Chichester on the final day of the workshop and a book outlining the results 'Going Dutch on the Manhood Peninsula' was published by NIROV.[(2)]

Not meant as blue-prints for the future, the scenarios devised by the workshop participants have prompted lively and diverse discussions about the area's options, both amongst local people and local authorities, and have attracted interest from national and international organisations.

Interestingly, despite total freedom given to the groups, there were several striking similarities between their responses. Most importantly, all the groups recognised that climate change and the so-called 'problems' the area faced, such as its increased threat of coastal flooding, could provide opportunities for the area, environmentally and economically. By starting with that positive premise, the groups looked at future options that acknowledged the inevitability of change but focused on how change might benefit the local community and economy.

All groups agreed that a mixture of hard and soft defence was the best way of protecting the peninsula; all groups allowed more space for both coastal and fluvial flooding in their visions; all groups stressed that the area was unsuitable for increased housing development, unless hard coastal defences were massively improved; and all groups identified the environmental enhancement opportunities that could be created through different coastal and water management techniques.

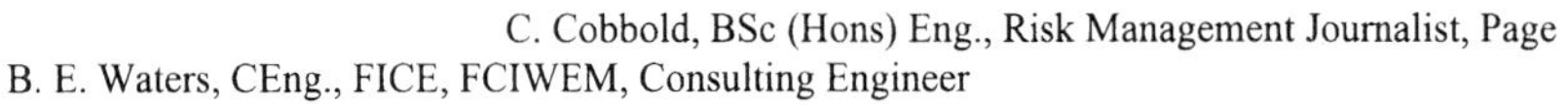
C. Cobbold, BSc (Hons) Eng., Risk Management Journalist, Page
B. E. Waters, CEng., FICE, FCIWEM, Consulting Engineer

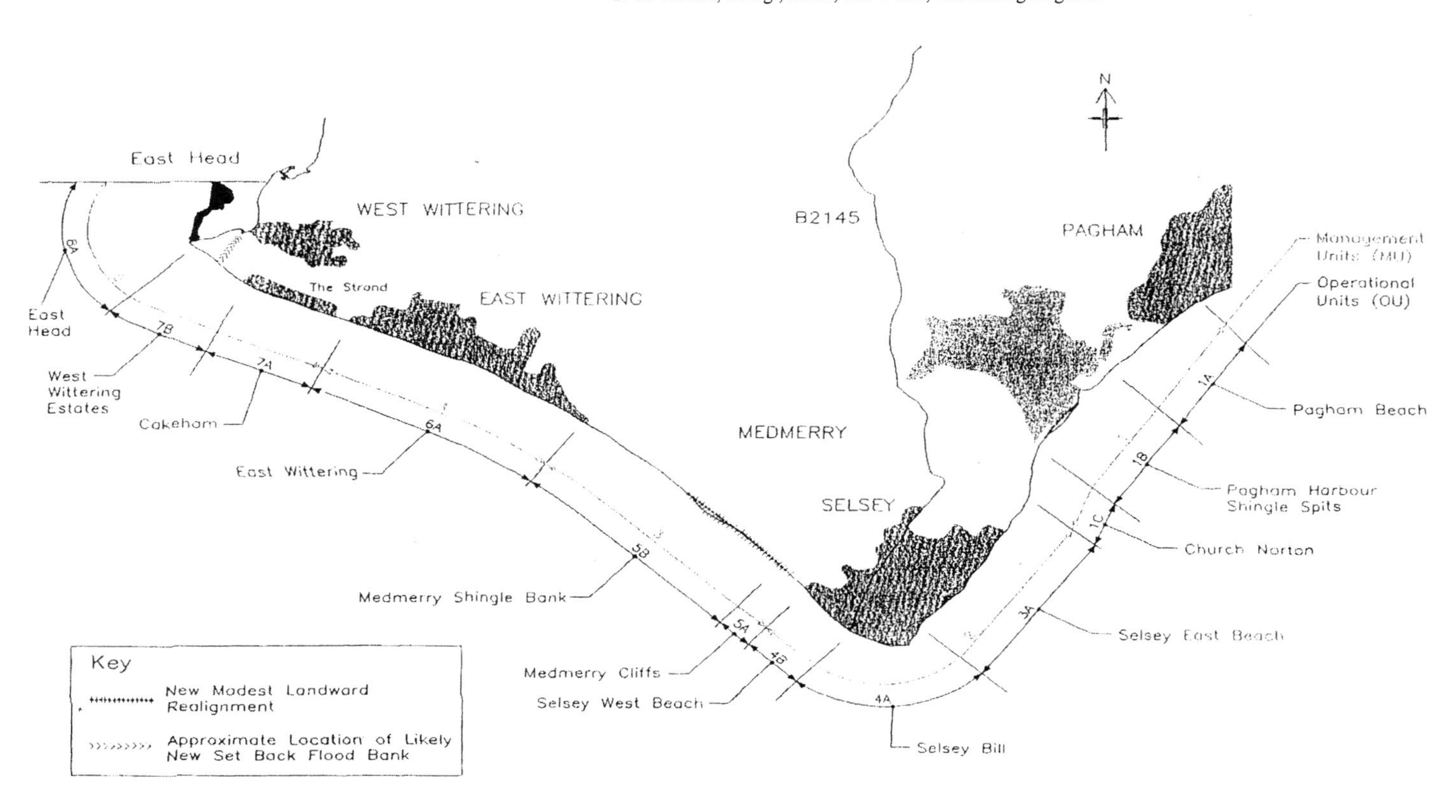

Figure 5 Arrangement of Management Units

Several months after the workshop, a follow-up meeting was held at which West Sussex County Council and Chichester District Council decided to form the Manhood Peninsula Partnership, to allow local authorities, agencies and the community to work together to formulate a long-term, sustainable plan for the peninsula.

The MPP has a steering group which consists of a chairman (both a County and District Council Councillor), the two project leaders, and representatives of West Sussex County Council, Chichester District Council, English Nature, Environment Agency, Chichester Harbour Conservancy, Sussex Association of Local Councils, Selsey Town Council and Sussex Wildlife Trust.

Current Work

The partnership's standing committee resolved to examine the peninsula's present physical position to determine how it might evolve. Certain basic parameters existed. The coastal protection authorities (Chichester District Council – CDC - and the Environment Agency – EA) had commissioned a coastal strategy study which indicated management units (Fig 4) and that the coastline, with some exceptions, should be held on its present position. [I]

However the proposed realignment on the EA (Medmerry) frontage had implications for future maintenance, and the absence of defence at the harbour entrance suggested problems in maintaining the status quo.

Currently, sea defences works at Medmerry cost £300,000 annually, which is not funded by central government.

Realigning the Medmerry frontage between Selsey and Bracklesham would allow a shallower, more sustainable beach to be formed, providing greater protection. However, the realigned beach will need to be carefully monitored and may prove to be unsustainable in the long term, as the foreshore steepens.

The Environment Agency and English Nature, both members of the MPP, believe that the Medmerry front ultimately should be left undefended, allowing salt-marshes to form and provide a natural sea defence. Both organisations believe that the area provides an opportunity to create important coastal habitat which is being lost elsewhere because of coastal squeeze. Replacing this lost habitat is a requirement under the European Union Coastal Habitat Directive.

Other members of the MPP, however, believe that the re-aligned coastline should be maintained, at least on a short-term basis.

A commitment to maintain a re-aligned frontage, they argue, would provide several benefits, including:

- Allowing for further monitoring and studies to be made as to the likely impact of abandoning sea defences.

- Allowing more time for alternative defences to be investigated. For example, Selsey Town and Chichester District Council currently are sponsoring research into the use of scrapped tyres as cores for offshore breakwaters or below armouring on softer land defences.

- Allowing the community and economy to adapt to a changed environment. For example, the creation of wetland areas behind the Medmerry Frontage could provide a drainage storage function for the Peninsula as well as provide valuable natural habitat. This would enhance the Peninsula's environment and could encourage more long-stay tourists and allow some of the more vulnerable caravans to be re-sited to an equally attractive but less exposed plot.

At the end of the expected 'life-time' of the realigned coastline, a better understanding of the physical process may be possible; a better understanding of the consequences of a "do-nothing" approach; or even the development of alternative coastal defences, may result in a decision to continue to defend the frontage. If, however, further defence is not sustainable, the community (and hopefully the economy) will be in a better position to understand and accept a managed retreat option.

It may be that a policy achieving a wet-land progression to salt marsh would be seen as less dramatic than defended agricultural land to salt marsh. Indeed the community may well embrace a decision to retreat if it has seen the benefits (socially, environmentally, and economically) from enhanced environmental habitat of the wetlands.[(4)]

The regime at Chichester Harbour mouth is known to be extremely complex and a failure to maintain the existing deep water channel (by, for instance, a breach at East Head and a secondary channel forming) could jeopardise a major economic generator for the area.

Public consultation suggested strong support for hard defences for Selsey and for the Bracklesham – East Wittering frontages. Protecting the low lying caravan park area at Medmerry was seen to be more problematical but again the contribution of the facility to the local economy was thought to be important.

One of the difficulties of public consultation is in explaining what can appear to be semantics. The Treasury, in financing Defra, consider caravan parks to be transferable, that is to say there would be no "national" loss in caravans being moved from a threatened site to a safer one. Thus the full economic value of the installation is not used. In fact, of course, the installation costs are only one part of the sum: the other is the economic contribution to the area.

Some partnership members argue that the site is unique: that the transfer of 3000 living sized vans to a site within the same mileage of London is, to say, the least, unlikely: and that the full economics involved in providing the site should be recognised. They also argue that a case could be made for revising the benefit cost calculation for the peninsula such that instead of viewing the coastal strip only, the economic activity of all the land should be taken into account. The logic of this is that a major breach could have far reaching effects, even to making Selsey a tidal island, thus affecting transportation, horticulture, tourism and services. Should the argument be successful, and the principle of "hold the line" be accepted it is believed that even on current point scoring, government funds would be available both for updating defences which are themselves now mainly to a design 50 years old, and to finance more permanent defence on the EA frontage.

This, in theory, is standard practice under the project appraisal guidelines for economic appraisal in the 1999 Flood and Coastal Defence document but it would require the acceptance of different criteria for this particular caravan park.

The partnership, however, are receptive to ideas of creating wetland areas, perhaps as part of agricultural land set-aside, providing the integrity of the coastal process can be safeguarded. Any larger scale set back or breach might well act as a sink for littoral drift material (which on the western beaches runs conversely west to east) and thus, again, exacerbate problems at the harbour entrance.

Much of the peninsula coastline is protected by one or more environmental criteria. In general, English Nature prefer to see coastal zones develop naturally and are keen to see development of sites for environmental use to counter losses elsewhere, such as by coastal squeeze. It is also of note that certain sites may be required under EU legislation to be protected, and, again, in community level consultations, it can be difficult to explain that the protection is mandatory, but coast protection powers are otherwise only permissive, and then only if demonstrating compliance with economic criteria.

The partnership propose that both existing and future environmental sites offer considerable scope for environmental tourism.

The partnership's project managers – the authors – are aware of the research being carried out by Hydraulics Research at Wallingford on the use of scrapped tyres as cores for offshore breakwaters or below armouring on softer land defences. A further more nebulous idea has been for major offshore breakwaters carrying wind turbines. Much research would be needed of the effect of such breakwaters, particularly as it would appear they might inhibit littoral drift, or indeed the periodic onshore movements of offshore banks. However there is a clear case for investigating lower cost solutions because of the effect they would have on the economics for defending the coastline on a present or modified line.

Aside from coastal defence issues, the MPP has been working closely with the Chichester Harbour Conservancy, which is trying to establish a cycle route from Chichester to West Wittering.

Support also is being given to the community-based Manhood Cycle Network, to help establish a network of cycle routes throughout the Manhood Peninsula, to link Selsey and Chichester, and to promote the use of cycling as an alternative form of transport and to promote 'cycle' tourism.

The harbour conservancy, formed by Act of Parliament in 1971 to manage the harbour for recreational activity and protect its natural beauty, already has a programme of self-guided walks and activities designed to encourage the public to explore and enjoy the Area of Outstanding Natural Beauty. This is in addition to "afloat" activities.

The MPPs major current work concerns ground water. In recent years and especially in the autumn of 2000 there has been considerable surface water flooding to homes and highways largely due to overflowing ditches.

An initial investigation by Chichester District Council found that much of the peninsula ditch system had no flow and presumably was intended to work as a horizontal land drain.

Anecdotal evidence suggested the problems might be due to alleged interference by developers and/or riparian owners unaware of their responsibilities. It was thus resolved that the partnership's first independent study would be to determine the reasons for flooding and to suggest measures to improve the situation. Tender documents were sent to five consultants with a brief designed to effect added value as well as economic price. Funded by the District Council, the successful firm started work in February 2003. Included in the brief was a requirement to consult Parish Councils and other public and business bodies, to emphasise community involvement.

Mention has been made elsewhere on the matter of agricultural land set-aside. The partnership is seeking funding to research how compensation might be arranged for agricultural land becoming a wetland or other environmental site possibly with cycle way access, the idea being, in due course, to attract to the peninsula a form of "green" tourism with a longer-than-one-day clientele.

Meanwhile, the MPP is encouraging Manhood based Parish Councils to work together and to promote active community involvement in the area's future plans. For example, the MPP is helping a group of parish councils put a joint bid together for the Countryside Agency's Parish Plans initiative which would entail running community workshops to discuss local issues such as coastal management, transport and other infrastructure and other social, economic and environmental issues to produce a strategic plan for the peninsula.

The MPP was chosen to participate in the Examination in Public debate of WSCC's Structure Plan before the Government Planning Inspector who noted the MPP's views in his report published February 2003.

Transportation

The road layout (Fig 5) on the peninsula has not changed materially since it supported purely rural and fishing communities one hundred years ago. The marked increase in population, referred to earlier, together with lack of employment within the peninsula and growth of car ownership has placed an enormous strain on exit roads at peak times on weekdays and over longer periods on holiday weekends.

These problems are exacerbated by the A27 east-west trunk road, which runs south of Chichester and thus intersects all the peninsula exit roads to the north. Introducing development south of the Chichester By-Pass to fund improved highway design is probably not an option since available sites are themselves liable to possible future flooding.

The MPP response is to look at cycle based tourism but this will introduce a further problem of providing a "park-and-ride" facility. In the case of Selsey, if the town is to be allowed to develop further, with all that implies in coastal and flood protection, another access road with limited access may be needed on the east side of the peninsula. This could connect with a grade-separated junction on the A27 between Chichester and Tangmere and might even service a new Chichester Parkway station on the coastal railway line. These are, however just ideas for the future. A recent Government sponsored study recommends Grade – separated junctions at the Fishbourne and Stockbridge roundabouts but with no such improvements at the Bognor Regis roundabout to the east of Chichester. This could result in more transport impacting Chichester Harbour and the rural hinterland of the peninsula.

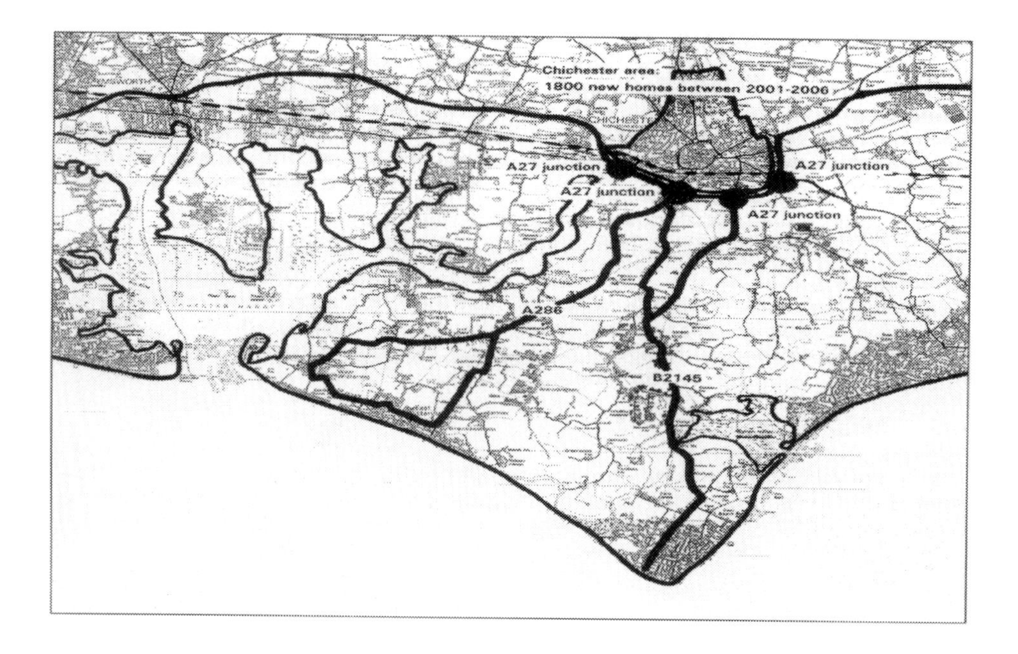

Fig 5
The Manhood Peninsula showing the main transport links and how they are intercepted by the A27 Chichester By-Pass. The B2145 is the only direct Selsey – Chichester route.

Funding

The MPP exists at present largely on sufferance of its members and their employers. The District Council provides technical and administration back up, plus with the Conservancy, meeting rooms. West Sussex County Council provides staff input and has been handling funding bids to Europe under Intereg IIIB, in association with Hampshire County Council.

The County also supports the Manhood Parish Partnership bid with SEERA (South East England Regional Assembly).

The Manhood Peninsula Partnership's work also is a key part of a joint European – funded project examining climate change and planning. The project, which was conditionally given the go-ahead by the European Commission in June, involves West Sussex County Council, SEERA, Hampshire County Council, the Environment Agency and agencies from the Netherlands and Germany.

Commentary

Involving the community in developing a strategy for coastal zone management produces diametrically opposed views and even among professionally qualified bodies considerable differences of opinion. Most of these have their roots in economics. To the people of Selsey, a community which has out-grown its infrastructure and transport links, the answer is in further development, much improved links to Chichester and the north and guaranteed hard defences.

Residents of smaller villages on the peninsula would prefer, in general, to see less change, providing their physical security was assured.

A further tourism boost would be the refurbishment of Chichester Canal, the residual length of what was the Arundel to Chichester Harbour route onward to Portsmouth. After a substantial basin in Chichester, much of the waterway is now reed bed.

The County Council, who own the Chichester Canal, have stated that it is their long term intention that it should be re opened. One of the authors produced a costed scheme whereby this might be achieved, only to be told by a then senior planning officer that whilst there might be an apparent public consensus on the desirability of re opening the canal, on the day it was announced that such action was to be taken, objectors would flood out of its banks.

It is the view of the authors that the Manhood Peninsula could provide a valuable environmental and recreational link between the Downs and the sea, bringing additional visitor spend to Chichester and its environs, and a revitalised canal, with attendant cycle routes along its banks, would be an inherent part of such a scheme.

While a managed retreat at the Medmerry frontage as recommended by English Nature and the Environment Agency may provide an opportunity for the area to increase its eco-tourism potential, it is important for the community to accept and be prepared for such a radical change in their environment. Sufficient research needs to be done to ensure the physical, social and economic safety of all the communities on the peninsula before such an option can be chosen. Research should also be undertaken to assess the impact of a managed retreat at Medmerry on adjacent coastlines.

Studies need to be done to compare the economic social and environmental benefits to the local community and, indeed the wider community, of all the options still available, and which may be available in the future.

As a result, there is a valid argument for maintaining the current coastline, with perhaps some slight realignment to create a shallower, more sustainable beach, as long as possible. During this period, further research can be done both on the consequences of undertaking a managed retreat and into possible alternative defences. Also, during this period the community and local agencies can consider changing some land use to adapt to climate change and also to prepare the local economy for such a fundamental change if that is deemed to be the best alternative.

For the past fifty years the Manhood Peninsula, particularly the town of Selsey, has been regarded as a dormitory settlement area for Chichester and the larger towns along the coast. A radical change in this policy may have to be adopted by local planners if the economy and community is to adapt to climate change and changes in the management of its coastline.

The authors conclude that some local relocation of, say, the caravan facility may be required on the basis of zone management and in particular to allow part of the Medmerry hinterland to be used initially as additional drainage capacity (to cope with wetter winters in the future) and to encourage the creation of a flora and fauna habitat.

Again, such an approach needs a compensation mechanism and an appreciation that investment now could bring into being an environmental amenity, which would have economic value in the future.

An added value might be that in educating the general public to accept such changes, there would be better understanding of the processes that climate change could bring. Certainly, one concern that has to be addressed is that increased eco-tourism to coastal areas does not impose a yet higher traffic burden on small villages or access roads, but the economics, let alone environmental objections, for new road building do not need repetition here. To quote a Dutch contributor to the NIROV study "Think first about the future of the area, then re-think about its problems, instead of thinking about its problems first.".

Conclusions

The authors' intention in producing this paper was to "flesh out" the concept of coastal zone planning in a discrete area, especially where the communities are directly involved in the consultation process. The authors suggest that considerably more needs to be done in educating the public, and also to ensure decision taking is both transparent and, as far as possible, fair. To this end the financing of schemes and compensation for those affected needs to be re-visited. Finally, whilst the remit of the partnership is to look at scenaria up to thirty years on, clearly proposals must incorporate flexibility in case there is evidence that current views on climate change become invalid.[5]

There is a vision, worth pursuing: there are deliverables in the medium term: involving the community has proved exciting, frustrating, and ultimately satisfying: an integrated approach to planning, engineering and the environment is clearly the best option: and the authors hope their account will be an example to others.

Acknowledgements

The authors would wish to thank their chairman, and the local authorities involved, for permission to present this paper, which represents their views and which are not necessarily those which will become policy in the future. They acknowledge generous assistance from Posford Haskoning, engineering and environmental consultants, and ready advice from their colleagues and the standing committee. Their particular thanks go to Mrs Tessa Garnham for her secretarial expertise and Mr Mark Casey who transferred the figures.

References

1. The Pagham Harbour to East Head Coastal Defence Strategy" – Posford Duvivier. Jan 2001

2. "Going Dutch on the Manhood Peninsula" – Carolyn Cobbold & Renee Santema. May 2001

3. "Valuing Chichester Harbour" – Chichester Harbour Conservancy. June 2001

4. "Climate Change & Some Implications for Shoreline management in the UK" – John Andrews & Simon Howard (Posford Haskoning). May 2002

5. "Climate Change Scenarios for the UK" – UKCIP Briefing Report – Defra. April 2002

Interventions for Environmental Management of the Coastal Zone of Sri Lanka

Dr. S.S.L.Hettiarachchi, Associate Professor, Department of Civil Engineering, University of Moratuwa, Sri Lanka

Dr. S.P.Samarawickrama, Senior Lecturer, Department of Civil Engineering, University of Moratuwa, Sri Lanka

---

Introduction

Coastal problems of Sri Lanka were recognised from the turn of the last century. However, it was in the early seventies that growing attention was focused on these problems because of their serious aggravation. When the initial response of seeking engineering solutions to control immediate coastal erosion problems- by the construction of ad-hoc protective structures – proved to be non-effective it was realised that there was a need to adopt a coordinated approach to the problem within a wider coastal zone management framework. This need combined with increased environmental awareness resulted in the acceptance of the requirement for some measures of regulation of development activity within the coastal zone. It was also realised that the legislative and administrative framework which was then in existence was insufficient to meet the demands of effective coastal zone management.

Sri Lanka is one of the earliest island states to have developed and operated fully, a Coastal Zone Management Plan on a national scale. The Coast Conservation Department, by Act of Parliament, has the full responsibility for the implementation of the plan. It is therefore necessary that all issues related to coastal zone management are addressed via the Coastal Zone Management Plan. This plan is periodically revised to incorporate the emerging challenges giving due consideration to current and projected development trends in the country in refining policies and guidelines.

This paper covers four broad areas of interest. Firstly it identifies the significance of the coastal zone of Sri Lanka and describes its present status of management. Secondly it discusses critical issues relating to actions and measures adopted. Thirdly it identifies the challenges and constraints and finally it focuses attention on the way forward in the light of lessons learnt.

Coastal Management 2003, Thomas Telford, London, 2003.

Significance of the coastal zone of Sri Lanka and its present status of management

Aligned with the UN convention on the Law of the Sea which was ratified by Sri Lanka in July, 1994, the country enjoys a total extent of approximately 489,000 square kilometers of maritime waters. The maritime zones consist of Internal waters, Historic waters, Territorial Sea, Contiguous Zone and an Exclusive Economic Zone (EEZ) as shown in **Figure 1**. The majority area (437,000 square kilometers) belongs to the EEZ. The island on the other hand, has a relatively small land area of 65,000 square kilometers which gives a land to ocean area ratio of 1 to 7.5. The coastal zone is therefore of strategic significance to its populace due to accessibility to the vast resource base of the marine environment surrounding the island, in principle, from any point of the 1585 km long coastline.

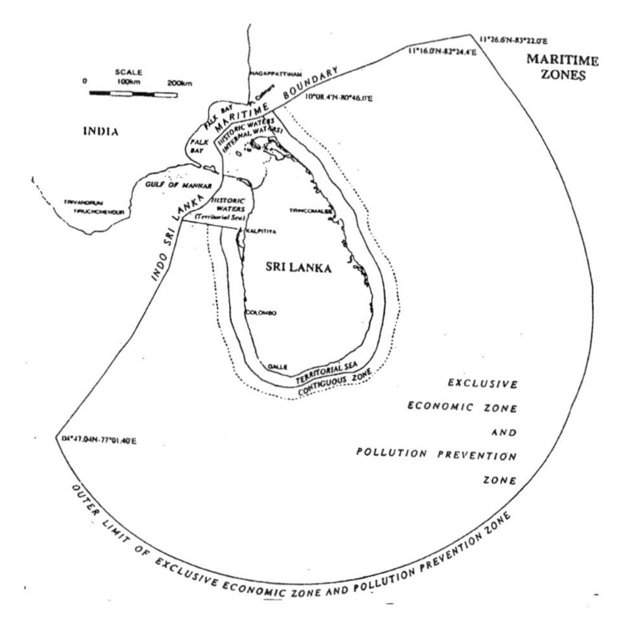

Figure 1 : Maritime Zones

The coastal zone includes both the area of land subject to marine influence and the area of sea subject to land influence. The coastal zone can be defined based on either one or a combination of geo-physical, ecosystem and human development considerations. There are 67 coastal administrative units in the country and if the coastal zone is identified landward by these administrative units and seawards by the narrow continental shelf, the coastal zone will include:

a) Approximately 24% of the land area and 32% of the population (according to 1981 census)
b) 65% of the urbanized land area, 65% of the industrial output, and 80% of fish production
c) Principal road and rail transport infrastructure
d) Principal commercial ports, fishery harbours and anchorages
e) 80% of tourism related infrastructure with the majority being located in the western and south-western coastal regions and most of them within close proximity to the shoreline

f) A significant extent of agricultural land, sizeable areas of usable land which remain undeveloped, substantial reserves of valuable minerals
g) Some of the richest areas of biodiversity including coral reefs, seagrass beds, mangroves, lagoons, estuaries, wetlands and sanctuaries covering an extent over 160,000 hectares
h) Areas subject to extensive water pollution associated with industrial pollution sources, domestic waste water and sewage disposal and garbage disposal
i) Many areas of cultural, historical and religious significance and scenic beauty

Coastal regions of Sri Lanka produce a large number of goods and services of economical benefit. This not only led to the development of urban centres along the coastline but also attracted human settlements and major investments to areas that are susceptible to hazards such as storm surges and coastal erosion. During the last three decades there has been increasing pressure for development in the coastal zone and communities have exploited the use of natural resources such as sand and coral on a commercial basis. Severe impacts of coastal erosion and flooding and the need to discharge inland flood water imposed increasing demands for coastal and flood protection and effective drainage of inland waters. The impacts of coastal erosion had also increased at alarming rates due to unplanned human intervention in the coastal zone. Environmental degradation of the coastal zone is a major hazard faced by Sri Lanka.

In order to correct this situation the Government of Sri Lanka established a Coast Conservation Division in the Ministry of Fisheries in January 1978 and this Division was upgraded to a Department in January 1984. The Government also introduced the Coast Conservation Act **(1)** which came into operation in October 1983. This act decreed the appointment of a Director of Coast Conservation with specific responsibilities for implementing the Act. It includes the administration and implementation of the provisions of the Act and the execution of schemes of work for coast conservation within the coastal zone as defined in the Act. It is important to note that the Coastal Zone as presented in the act has a very narrow geographic definition as illustrated in **Figure 2**.

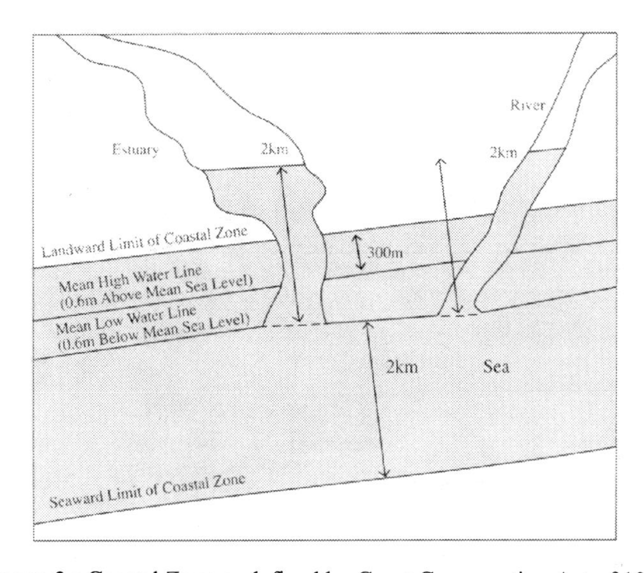

Figure 2 : Coastal Zone as defined by Coast Conservation Act of 1981

The Act also required the Director of Coast Conservation to conduct a survey of the coastal zone as defined in the Act, and on the basis of the results of the survey to prepare a comprehensive Coastal Zone Management Plan **(2)**. This plan received the approval of the Cabinet of Ministers in April 1990.

The objectives of the Coastal Zone Management Plan were as follows;

- to identify and prioritise coastal problems that need to be addressed;
- to present the management programme adopted by the Coast Conservation Department to address these problems;
- to identify the measures which should be adopted by the Government and all stakeholders to reduce the scope and magnitude of coastal problems; and
- to identify research activities of immediate importance to enhance the management of coastal resources.

Within the context of the above objectives the Coastal Zone Management Plan focuses attention in three major concerns, namely,

1. Coastal erosion management;
2. Conservation of natural coastal habitats; and
3. Conservation of cultural, religious and historic sites and areas of scenic and recreational value.

The plan describes the nature, scope, severity and causes associated with each of these problems. Objectives and policies for the management of each problem are identified along with specific management techniques.

The Coast Conservation Department also prepared a Master Plan for Coastal Erosion Management **(3)** which was urgently required for the planning, design and construction of coast protection works in 1986. This plan has been subsequently updated.

The Coastal Zone Management Plan was revised in 1997 **(4)** to take into account the experience gained through the first plan and to address in greater detail issues which had not received such attention. The second generation plan has also given due consideration to the development trends in the country in refining policies and guidelines.

Critical Issues relating to Actions and Measures adopted

The Government authorities duly recognized the magnitude of the coastal problems and undertook remedial measures by formulating a well planned programme of action for the management of the coastal zone. These management strategies, which accept the need for some measure of regulation of development activity within the coastal zone, have been implemented since the early eighties and they have proved to be successful in controlling the degradation of the coastal zone and introducing planned development. The government authorities also monitored closely all problems and issues related to the implementation of the Coastal Zone Management Plan in order to revise the same at a later stage. This section identifies the issues relating to the implementation of the Coastal Zone Management Plan and reviews recent developments in coastal zone management.

Implementation of the Coastal Zone Management Plan

Regulation of development activity and the Issue of permits

The control and management of development activities (including natural resources utilization) constitute a major area in the implementation of the Coastal Zone Management Plan. 'Development Activity' is defined in the Coast Conservation Act as an activity likely to alter the physical nature of the coastal zone in any way and includes the construction of buildings and other structures, the deposition of waste or other materials from outfalls, vessels or by other means, the removal of sand, seashells, natural vegetation, sea grass or other substances, dredging, filling, land reclamation and mining or drilling for minerals.

The principal means of regulation is via the appraisal of proposed development activities in the designated coastal zone by the Coast Conservation Department prior to the issue or refusal of a permit to proceed. Although a permit is required for development activities that are likely to alter the physical nature of the coastal zone, it is important to note that fishing, cultivation of crops and planting of trees or other forms of vegetation may be carried out in the coastal zone without a permit. The procedure for the issue of permits is laid down in the Act and is handled by the Department.

The Coast Conservation Department has decentralized several of its functions to the Divisional Secretaries under the terms of Public Administration Circular No. 21/92 of May 1992. The delegation of administrative authority has been made under Section 5 of the Coast Conservation Act No. 57 of 1981. This delegation of authority has been formulated to improve the efficiency of coastal management programmes by permitting the local authority to issue permits for restricted sand mining and for construction of small buildings.

Prohibition of Coral Mining

The Coast Conservation (Amendment) Act No: 64 of 1988 prohibits engaging in mining, collecting, possessing, processing, storing, burning and transporting in any form whatsoever of coral within the coastal zone. Although considerable progress has been made in this respect, the ban on coral mining remains a difficult issue to implement without identifying and making available alternative means of livelihood for the coral miners who have been engaged in this activity for decades. Active governmental support is required to resolve the socio-economic aspects of this problem by generating attractive employment opportunities in the same locality. If such opportunities are created elsewhere due consideration has to be given to problems arising from re-settlement issues. In November 2000, a decision was taken by the authorities to restrict the usage of coral based lime in the construction work of the government sector and it is now implemented through the construction agencies in the country.

Control of sand mining

Sand mining has been controlled to a satisfactory degree by the Department and this activity is allowed only in selected areas of the coastal zone on certain days of the week and to an extent supervised by the staff. In other areas it is a prohibited activity. Sand mining is still extensively carried out in river estuaries and along reaches upstream of estuaries. This has certainly affected the sediment budget along the coastline leading to coastal erosion. The total estimated annual sand mining of the four main rivers in the north-western and western provinces of the island is in the order of 4 million cubic meters. Violations and infringements of regulation relating to sand mining do occur and due consideration has to be given to resolving the socio-economic aspects of the problem.

Setback and Variance

With respect to physical infrastructure development in the coastal zone due consideration has been given to setback guidelines which have been identified in the Coastal Zone Management Plan. A setback is defined as an area left free of any physical modification to allow for dynamics of seasonal and long term fluctuations of the coastline and to ensure public access to the water front and visual access to it. Setback is identified based on coastal erosion rates, beach dynamics, shoreline ecology and related activities. Setbacks are an effective and inexpensive approach towards coast protection and conservation. In 'no build' areas, depending on site conditions, the developer can apply for a variance from the Coast Conservation Advisory Council for setback reduction to build in that area. The setback guidelines introduced have been successful in ensuring the protection and security of assets which otherwise may be in grave danger due to unfavourable environment conditions from coastal erosion degradation.

Environmental Impact Assessment and Monitoring

In the case of larger development projects which could have significant impacts on the environment it is necessary for the developer to prepare and submit an Environmental Impact Assessment (EIA) report prior to obtaining approval for the implementation of the project.

The Coast Conservation Department has the responsibility to ensure compliance with the conditions stipulated in the permit through a monitoring system. Monitoring is achieved via periodic site visits or, on some occasions, by direct supervision undertaken by the staff of the Department. The Department may also nominate a state authority for this purpose. Further, permit compliance surveys are carried out through the research teams of the universities. In the case of large development projects, well-formulated monitoring programmes are incorporated in the EIA report and usually a monitoring committee is appointed to supervise all specified surveys, tests and field investigations.

Coastal Erosion Management

Coastal erosion was a severe problem faced by Sri Lanka. Unfavourable natural conditions and unplanned human intervention had increased erosion rates around the island. This led to the damage or loss of houses, hotels and other infrastructure, undermining of roads and the rail track, loss or degradation of valuable land and caused disruption to fishing, navigation, recreation and other activities. In economic terms the public and private costs of coastal erosion are enormous.

During the period 1983-84 the southwest coast of Sri Lanka was subjected to very severe erosion. It was recognised that any major coast protection work should form an integral part of an overall Coastal Erosion Management Plan and/or a Coastal Protection Plan and that if this approach was not adopted it would only facilitate the transfer of the problem from one location to another. With Danish assistance the Coastal Conservation Department prepared a Coastal Erosion Master Plan and undertook the construction of coast protection works in two stages. The sites were selected on a priority basis after careful screening. In view of the severity of the erosion problem and the receding coastline endangering infrastructure and assets of the urban community, most of the protection works were rock armoured revetments which provided a frontline defence. For one area which offered substantial tourist facilities and had a strong local fishing industry, offshore breakwaters were adopted for beach development.

Although the construction of revetments may not be the best solution to meet the user requirements and the development of a healthy beach, the authorities had no option but to adopt this form of frontline defence to control large scale erosion. With limited funding available the Department also undertook emergency protection works at several locations to combat cases of severe erosion arising from heavy monsoonal wave attack. The shoreline was continuously monitored thus enabling the identification of other sites and areas to be protected. These sites were prioritised for the purpose of implementation of coastal works.

Second generation initiatives in coastal zone management

The first generation efforts in coastal management in Sri Lanka has had many follow up activities of national interest. The Coast Conservation Department and the Central Environmental Authority have implemented special management projects in order to investigate critical problem areas. The following projects are examples of such positive initiatives.

- Special Area Management (SAM) Projects to study in detail the problems relating to specific areas which are under severe development pressure. Strategies for management have been developed giving due consideration to all critical issues as well as impacts arising from their interactions.

- National Sand Study **(5)** and the follow up Interim Sand Study **(6)** to investigate the ways and means of preventing and mitigating environmental hazards and degradation caused by present practices of sand mining and investigating practical alternatives for river sand. The government authorities are now investigating the option of using cleaned sea sand pumped from deep offshore waters as a viable alternative.

- Environmental Quality Standards for surface water and air quality. These standards are being used to evaluate the present environmental quality, to assess environmental impacts due to discharges and to prepare environmental management plans.

- Wetland Conservation Project to assist in the conservation and management of Sri Lanka's wetlands, including several coastal wetlands which are under severe pressure due to rapid urbanization process.

Another important follow up activity was the preparation of Coastal 2000: A Resource Management Strategy for Sri Lanka's Coastal Region **(7)**. This document focuses attention on specific problems relating to the implementation of the Coast Conservation Act and recommendations are made. These include the following.

- Single agency and sectoral approaches to solving coastal resources management problems to be replaced by a more comprehensive perspective approach.
- The reduction on the emphasis of regulation needs.
- The recognition of the interrelationships among important resource management concerns such as water quality, habitat degradation, use of natural resources and institutional weaknesses and the need to adopt effective strategies involving more than one agency and a range of management techniques.
- The inadequacy of the narrow geographic definition of the coastal zone in the Coast Conservation Act which does not reflect in actual terms the interconnections between coastal ecosystems and resources. This definition has proved inadequate for even the basic tasks of effectively managing coastal erosion and shorefront construction. It is

totally inadequate for integrated land–use and water–use plans for coastal ecosystems or habitat management.
- The increase of participation by local and provincial officials and coastal communities in the formulation of plans and strategies for managing coastal resources.

In practical terms, the Coastal 2000 strategy enables planners to formulate individual coastal resources management efforts against the background of the entire coastal region that includes all critical coastal habitats and other valuable resources.

The Coast Conservation Department reviewed the Coastal Zone Management Plan and revised the main objectives and policies which will led to the preparation of a revised plan in 1997 **(4)**. The revised plan has taken into account the wide experience gained through the first plan and addresses in greater detail issues which had not received such attention earlier. The second generation plan has also given due consideration to the current and projected development trends in the country in refining policies and guidelines. This plan like its predecessor of 1990, outlines interventions to reduce coastal erosion which may also increase by sea level rise, to minimize depletion and degradation of coastal habitats, and to minimize loss and degradation of sites of archaeological, historical, cultural, recreational and scenic interests. Due recognition has also been given to Coastal Pollution Control and Special Area Management.

Current initiatives in coastal zone management

The government authorities are currently implementing the Coastal Resources Management Project with donor funded assistance. The project comprises four components namely, Coastal Stabilisation relating to coast protection, Fishery Management and Fishery Harbour Construction for improved management of the fishery industry, Coastal Environment and Resources Management relating to conservation of coastal resources and Institutional Strengthening relating to enhancing the organisational capabilities of governmental and other stakeholder agencies. The project has a wide ranging portfolio of activities and it is expected that it will contribute very positively towards improved coastal zone management with considerable benefits to the coastal community at large and in particular to the fishing community. One of the project activities is the revision of the Coastal Zone Management Plan which is being currently implemented.

Challenges and Constraints

The implementation of the Coastal Zone Management Plan imposed major challenges and constraints to the government authorities. Environmental degradation of a high magnitude, arising due to a number of reasons had already resulted by the time the implementation of the management plan commenced leading to severe adverse impacts on the coastal environment. There existed strong and often conflicting pressures for exploitation of the coastal region with crucial economic and social implications.

The government authorities had to educate and convince the coastal population on the reasons leading to a high degree of environmental degradation and the need for them to accept some measure of regulation of development activity within the coastal zone. The authorities were only too aware of the socio-economic implications of introducing such regulation in the coastal zone, which has provided and will continue to provide opportunities for economic development in a wide field of activities. In this context the authorities have been successful in controlling environmental degradation while sustaining multiple uses of the coastal zone. However increased settlements and development activities have imposed further stresses on this fragile zone and their consequences have had adverse impacts in particular on coastal ecosystems.

Coastal erosion

Coastal erosion has apparent and immediate consequences to man and society and is therefore often a principal concern and challenge in the administration of coastal districts. The Coast Conservation Department has adopted the construction of coast protection structures, use of setback lines and directed development away from eroding areas through the implementation of a regulatory system. It was identified that there is a strong need to obtain funding for the implementation of coastal stabilization scheme to combat increasing erosion. This has been achieved to a considerable extent through the Coastal Resources Management Project (CRMP). Seven sites have been identified along the northwestern, western and southwestern coastline.

One of the main objectives during the project planning phase was to change the prevailing erosion control methodology from a reactive/defensive approach to a proactive/preventive approach. The Coast Conservation Department has chosen to maintain the existing defence line where possible. Encroachment of beaches and lack of public access were concerns of the authorities. High prioritiy has been given for stakeholder consultations in formulating designs to achieve sustainable multiple uses of the coastline. Varying demands of the stakeholders, often conflicting in nature have been adequately met by adopting a 'hybrid' approach for coastal stabilization. Such an approach enables the consideration of both stakeholder demands and the hydraulic/structural function of the protection measures. The 'hybrid' approach comprising an appropriate combination of hard structures, sand fill and beach nourishment have been adopted for each site to satisfy the design objectives.

Environmental management of coastal ecosystems

Sri Lanka's coastal zone consists of diverse shoreline and nearshore habitats. These ecosystems in their natural state not only support marine life but also provide a buffer against the erosive forces of the hostile ocean climate. The physical and ecological characteristics of many of Sri Lanka's coastal ecosystems make them susceptible to degradation. They lack resilience and have a low threshold for irreversible damage. Once degradation exceeds the limit set by the low threshold, rehabilitation becomes prohibitively expensive or impossible. Increased human settlements and development pressures have enhanced the stresses on these unique ecosystems. The management policies of the Coast Conservation Department for these areas rely on regulatory and non-regulatory initiatives. Such initiatives have been enhanced by inter-governmental coordination, research studies oriented towards obtaining a greater understanding of the problems encountered leading to the improved knowledge of coastal resource users on the need for management. Special management objectives identified by the Coastal Zone Management Plan include the preservation of coastal ecosystems and the promotion of sustainable development of resources available within these systems. Management of coastal ecosystems for sustaining multiple uses remains a challenge and in this context the government authorities will have to further focus attention on Special Area Management (SAM) Projects as well as Wetland Conservation Projects.

Global warming and Sea level rise

Against the above background Sri Lanka also faces the challenges of possible impacts of global climate change and sea level rise over the next few decades leading to the aggravation of existing environmental pressures. Global warming is expected to lead to a rise in sea level, higher temperatures, more frequent and prolonged droughts, high intensity rainfall and increased thunder activity. When assessing the impacts of global sea level rise on coastal regions, it is important to identify and understand the connectivity of these inter-related issues, all emerging as a result of global warming. It is also important to recognize the global, regional and local scales of impacts as it would be the resultant that would finally affect a given environment.

Sea level rise on its own would lead to inundation and displacement of low lying coastal areas and wetlands, erosion and degradation of shorelines, salinisation of estuaries and freshwater aquifers and resulting varying impacts on coastal habitats.

A rise of the mean sea level will also lead to increased wave height thereby disturbing equilibrium beaches and making them more prone to erosion and interfering with existing longshore sediment transport rates and distribution. In addition, the combined effects of an increased mean sea level rise and increased wave heights would result in further impacts such as undermining the stability of coastal structures, and the altering of circulation patterns inside coastal embayments and estuaries. It is also recognized that a change in climate due to global warming would contribute to the changes in frequency and intensity of extreme events such as increased coastal flooding, thereby further complicating the analysis of overall impacts.

Some of the impacts of climate change and sea level rise on the coastal zone which have to be given consideration are **(8)**;

- Inundation of low lying areas, including coastal settlements and coastal wetlands;
- Coastal erosion;
- Flooding and storm damage;
- Quality of surface and groundwater leading to salinisation of estuaries and freshwater aquifers;
- Degradation of marine ecosystems – coral reefs; and
- Changes in the hydraulic force regimes of sea defence structures and breakwaters leading to greater vulnerability to impacts of increased erosion and extreme events.

Climate change and sea level rise will also impose severe impacts on the Fishery industry and the Tourist industry. Impacts of sea level rise will have to be considered in the planning and design of land based infrastructure and land reclamation in nearshore regions.

Disaster preparedness in the coastal zone

During the recent past the government authorities have focused attention on the potential disasters and their impacts on the marine environment and on the coastal zone. Until now the major concern was large scale erosion which could arise as result of storm attack. On many occasions coastal infrastructure such as road and rail network have been damaged by storm attack. Strategies for the adoption of emergency works have been prepared and implemented. Fishery and tourism industries and the conservation of natural resources demand pollution free coastal waters and clean beaches. In view of the close proximity of the principal shipping routes to the south-western coastlines, there was a need to investigate the possible impacts of oil spills and its dispersion on coastal waters and the coastline. With Norwegian assistance, the Marine Pollution Prevention Authority of Sri Lanka has implemented a successful project to model a range of scenarios relating to possible oil spills in order to investigate the impact of such spills on the coastline, coastal ecosystems and other assets **(9).** The outcome of this study is equally important for the environmental management of coastal ecosystems. **Figure 3** illustrates the dispersion of an oil spill in the vicinity of the Negombo Lagoon a unique wetland north of the Port of Colombo.

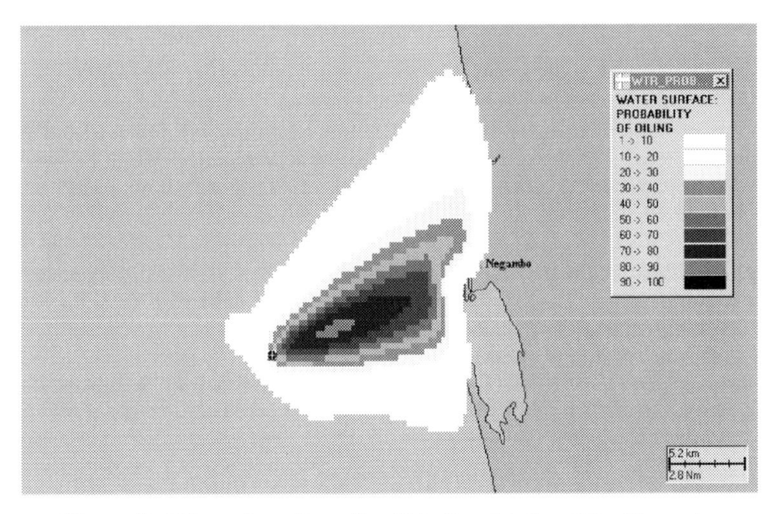

Figure 3 : Dispersion of an oil spill in the vicinity of the Negombo

Lessons learnt and the Way forward

The experience gained in coastal zone management in Sri Lanka in the last decade has shown that an approach to resource management that focuses on regulation alone tends to alienate the coastal residents affected. It indicates that a collaborative effort on the part of governmental agencies, non-governmental organizations and the local community is required to address the root causes of environmental degradation in the coastal zone. Experience has clearly illustrated that local communities can be organised to manage their natural resources only if they perceive that they will derive tangible benefits from such management and therefore there is need to adopt management policies which can accommodate such interaction.

There is a greater demand for the Coast Conservation Department to transform itself from a primarily regulatory agency to a service-oriented organization. The department should provide the leadership, the coordination, the technical assistance and the training that will be required for the successful implementation of a scientifically based coastal planning and management strategy. It needs to facilitate locally based planning and implementation efforts. Such an expanded agency must become proactive in its approach to coastal zone management and must cover a wider area and scope of coastal related activities.

Critical areas of adverse impacts

The Coast Conservation Department has been successful in planning and implementing well formulated programmes which have been successful in controlling environmental degradation and improving conservation in the coastal zone. However, increasing human pressure and unfavourable natural conditions will continue to impose adverse impacts on this fragile zone. The impacts can be broadly classified as arising from

- Natural hazards
- Pressure for unplanned development of coastal infrastructure
- Over exploitation of natural resources
- Non-conservation of coastal aquatic systems
- Overall coastal pollution
- Global warming and sea level rise.

The government authorities have focused attention on the above impacts and have been monitoring them to a limited extent. Of these the impacts arising from global warming and sea level rise have received less attention in view of the more recent origin of concerns related to the subject.

It has been identified that the three principal objectives of coastal zone management related to sea level rise are to,
1. avoid development in areas that are vulnerable to inundation
2. ensure that critical natural systems continue to function
3. protect human lives, essential properties, and economic activities against the adverse impacts of the sea.

It is identified that almost all the possible responses to sea level rise are best addressed within a broader context of coastal zone planning and management. If considered in isolation the effectiveness of such responses would be reduced mainly due to incompatible policies and / or actions taken by other coastal sectors. Many responses to sea level rise are very similar to those required to address existing coastal zone management problems, thus identifying a clear demand for the planning of sea level rise to be integrated with coastal zone management practices. Since Sri Lanka has a fully operative Coastal Zone Management Plan with a central organisation having the full responsibility to implement the same, the impacts of sea level rise are best addressed via the Coastal Zone Management Plan.

Application of an integrated coastal zone management plan

Coastal zone management involves management and decision-making, based on effective strategies, encompassing a range of activities. It also includes effective monitoring and control procedures to sustain multiple uses of the coastal zone while protecting its resources and without causing adverse impacts on the environment. In this context, the use of an appropriate management framework for its efficient management is justified. Of the several types of frameworks which have been adopted for such management, the use of an effective integrated coastal zone management framework seems most relevant for Sri Lanka in order to accommodate the widely varying and interrelated issues of its coastal zone.

The application of an integrated coastal management framework is a dynamic process by which decisions are made for the use, development and conservation of coastal areas. Provisions are also made to achieve goals established in association with relevant authorities, both national and local, and with other stakeholders who utilize coastal resources. Due attention should be given to strengthen the public participation process for formulating national policies and developing a coastal management plan leading to the establishment of a coastal governance system that applies these policies in an integrated manner.

To achieve success in the decision making process relating to engineering interventions, coastal engineers should have access to good quality data, simulation techniques and efficient data management systems. Therefore the proposed framework should make provisions for a continuing process to collect and disseminate the necessary scientific information and data resources, on coastal problems and issues, on functional uses and development, and on the needs of the private and public sector. It should also include information management systems which incorporate computational modelling tools for the analysis of data and simulation of aquatic systems. These models should have the ability to analyse the hydraulic regimes of coastal waters.

References

1. Coast Conservation Act of No. 57 of 1981. Parliament of the Democratic Socialist Republic of Sri Lanka.

2. Coastal Zone Management Plan. Coast Conservation Department of Sri Lanka, 1986.

3. Master Plan for Coastal Erosion Management. Coast Conservation Department of Sri Lanka, 1986.

4. Revised Coastal Zone Management Plan. Coast Conservation Department of Sri Lanka, 1997.

5. National Sand Study for Sri Lanka, Phase One, Volumes 1 & 2. Delft Hydraulics and Netherlands Economic Institute , 1992.

6. Interim Sand Study (Alternatives of River Sand), University of Moratuwa Department of Civil Engineering, 1997.

7. Coastal 2000: Recommendation for A Resource Management Strategy for Sri Lanka's Coastal Region, Volumes I and II. Coast Conservation Department and the Coastal Resources Management Project, Sri Lanka, 2000.

8. Initial National Communication under the UN Framework Convention on Climate Change for Sri Lanka. Ministry of Forestry and Environment, Sri Lanka, 2000.

9. Assessment of Risk of Oil spill Damage to Specific Coastal areas and Resources, Report of task 3 of the INSTCOM Project of the Marine Pollution Prevention Authority, Sri Lanka. Lanka Hydraulic Institute and University of Moratuwa, 2003.

Violent wave overtopping at the coast, when are we safe?

WILLIAM ALLSOP [1], TOM BRUCE [2], JONATHAN PEARSON [2],
JOHN ALDERSON [3] & TIM PULLEN [3]
[1] University of Southampton, c/o HR Wallingford, Wallingford, OX10 8BA UK
[2] Division of Engineering, University of Edinburgh, Edinburgh, EH9 3JL
[3] HR Wallingford, Wallingford, OX10 8BA UK

SUMMARY

Every year, people drown being swept from UK coastal paths, breakwaters and seawalls (at least 12 deaths in 1999 – 2002). It is likely that the people concerned had little true idea of the hazard to which they were exposed, yet most overtopping hazards are easily predicted using results of recent research. This paper uses the results of recent and current UK and European research to: improve understanding of overtopping, including effects of different types of wave breaking; improve prediction methods for wave overtopping discharges and velocities; and extend / validate suggested limits to acceptable overtopping. The paper re-states and extends advice in the EA overtopping manual by Besley (1999) applying new information / advice where appropriate.

1. WAVE OVERTOPPING AT THE COAST

Many kilometres of urban infrastructure around coasts of UK and Europe are protected against wave overtopping and/or erosion by steep sea walls, often vertical or near vertical, sometimes with a toe berm. Such walls (Fig. 1) are also widely used to protect railway lines or roads along the coast, cliff protection as seen along lengths of the southern England and Italian coasts, and around ports world-wide.

Fig. 1 Vertical or battered seawalls

Fig. 2 Overtopping watching at Oostende

It is generally appreciated that seawalls reduce wave overtopping, but a more sophisticated understanding is needed to be aware that seawalls cannot always stop overtopping. Under storm action, waves still overtop seawalls, sometimes frequently and sometimes violently. These processes may excite considerable public interest, see the example in Fig. 2 at Oostende where tourists gather during storms, and in Figs. 3 & 4 at Marine Drive, Scarborough, before the 2002 / 2003 improvements.

Coastal Management 2003, Thomas Telford, London, 2003.

In winter storms, wave overtopping may cause local flooding, and/or potential hazards to people close behind the seawall. The most severe hazards to pedestrians are probably:

- direct impact causing direct injury;
- direct impact causing the person to fall backwards against a hard object;
- impact or flow velocities causing the person to lose their footing;
- backwash flows carrying a person off the wall into the sea.

Less severe hazards are getting wet and cold (itself a potential hazard); or being frightened by the threat of inundation.

Fig. 3 Overtopping on Marine Drive, Scarborough, 2002

Fig. 4 Public attitude to overtopping hazard!

Other hazards affect drivers and passengers in road vehicles or trains, see particularly the example in Fig. 5 of a driver swerving to avoid an overtopping wave, and the discussion by Kimura et al (2000) on road accidents on a coastal highway in Japan. In the UK, a number of coastal railway lines suffer significant overtopping, although the consequent dangers are not well-established.

Fig. 5 Driving hazard from overtopping

It is generally understood that climate change may cause sea-level rise and perhaps more severe storms, so many involved in coastal management are aware that there will be more locations where overtopping hazards will increase, although most analysis hitherto has concentrated on the contribution to flooding. Many of the public are also aware of climate change and flooding, but are generally much less aware of the hazards or frequency of seawall overtopping. Indeed few are aware that at least 12 people have been killed in the UK by wave overtopping or related processes during 1999-2002, and approximately 60 killed in Italy over the last 20 years. Any gaps in understanding common hazards on the shoreline are further aggravated by media references to occurrence of "freak waves", often phenomena that could be predicted by an informed person.

This paper uses results from VOWS, CLASH and SHADOW research projects to identify overtopping hazards; and to describe overtopping performance of steep and composite seawalls which frequently defend urban infrastrusture. The paper analyses hazards at a monitored coastal site, and draws in initial results from research in the UK and Europe.

2. OVERTOPPING PROCESSES AND PREDICTIONS

2.1 Wave processes

Any discusion on wave processes requires standard terms. Of these, the most critical processes for overtopping and wave forces are the form and severity of wave breaking. Historically these may have been divided into "breaking" and "non-breaking", but those terms convey erroneous messages and are imprecise. Whilst universal definitions are not yet available, two sets will be used in this paper.

For sloping structures like embankment seawalls, the surf similarity parameter $\xi_{op} = \tan\alpha / \sqrt{s_{op}}$ (where α is the structure slope and $s_{op} = 2\pi H_s / g T_p^2$ is the wave steepness) is used to separate **"plunging"** ($\xi_{op} < 2$) and **"surging"** conditions ($\xi_{op} > 2$), see Fig. 6. These definitions are most commonly used in calculating armour stability for rubble mounds, see Meer (1984), also in the CIRIA / CUR Rock Manual (1991). The overtopping prediction method by Meer uses different relationships for "plunging" or "surging" waves, but the method by Owen does not apply such a distinction, see below.

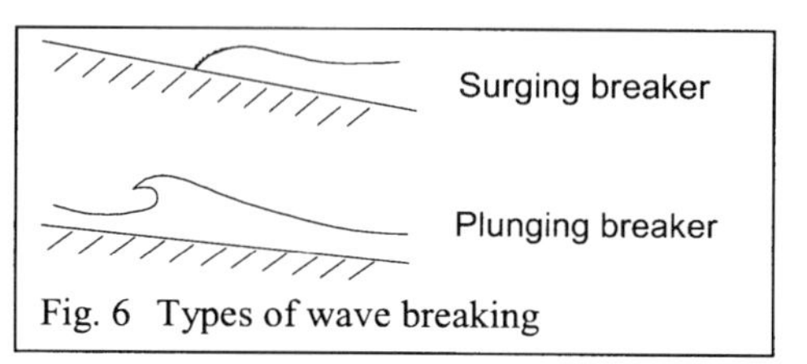

Fig. 6 Types of wave breaking

On steep walls (vertical, battered or composite), **"pulsating"** overtopping occurs when waves are relatively small in relation to the local water depth. These waves are not strongly influenced by the structure toe or approach slope. In contrast, **"impulsive"** breaking on steep walls occur when waves are larger in relation to local water depths, perhaps shoaling up over the approach bathymetry or structure toe itself. Under these conditions, some waves will break violently against the wall with (short-duration) forces reaching 10-40 times greater than for "pulsating" conditions, see McKenna (1995), Allsop et al (1996) and Allsop (2000).

2.2 Overtopping processes

Overtopping occurs when waves run up the beach, revetment, seawall or breakwater and pass over the crest of the defence. The frequencies, volumes and velocities of these overtopping events substantially influence the safety of the defence and of people living, working or travelling close behind the defence structure. Overtopping rates predicted by empirical formulae generally include "green water" discharges and splash, since both parameters were recorded during the scale model tests on which these prediction methods are based. Most laboratory studies on wave overtopping (research and site specific) have concentrated on measuring the mean overtopping discharge, Q_{bar}, usually derived from the total overtopping volume collected over 1000 waves, divided by the collection time. Those results have then been used to derive empirical prediction methods.

A second form of overtopping occurs when waves break on or seaward of the face of the structure and produce significant volumes of fine droplets. This "spray" can be carried over the wall under their own momentum and/or driven by wind. Spray overtopping may be generated directly by wind acting on wave crests, most noticeable when waves reflected from steep walls interact with incoming waves to give severe local 'clapotii'. Effects of wind on spray overtopping are seldom modelled. Tests by de Waal et al (1992, 1996) suggest that onshore winds will have relatively little effect on green water overtopping, but may increase discharges under Q_{bar} = 1 l/s.m where much of the overtopping may take the form of spray. Studies by Ward et al (1994, 1996) explored wind effects on waves and overtopping processes at laboratory scale, and noted changes to shoaling, breaking and up-rush processes, but did not lead to any firm guidance on wind effects. Spray is not therefore presently believed to contribute significantly to overtopping volumes, and generally causes little hazard except reducing visibility and extending the spatial extent of salt spray effects. An important exception is the effect of spray in reducing visibility on coastal highways where the sudden loss of visibility may cause significant driving hazard, see the example for a Japanese coastal highway discussed by Kimura et al (2000).

2.3 Overtopping prediction methods

2.3.1 Empirical methods

For sea defence structures, the mean overtopping discharge may be predicted by empirical or numerical models. Overtopping varies with wall shape, crest level, water level and wave conditions. Generally design procedures are expected to calculate the crest freeboard (R_c = height of crest above water level) that would limit overtopping to below a chosen discharge limit, Q_{crit}, see Besley (1999) and section 3 below. Empirical models or formulae use relatively simple equations to describe mean overtopping discharges, Q_{bar}, in relation to defined wave and structure parameters. Empirical equations and coefficients based on use of dimensionless discharge parameters (e.g. Q^*, $Q\#$, Q_b, Q_n and Q_h) and freeboards (e.g. R^*, R_b, R_n, R_h or simply R_c/H_{si}) are, however, limited to a relatively small number of simplified structure configurations. Use out of range, or for other structure types, may require extrapolation or may not be valid.

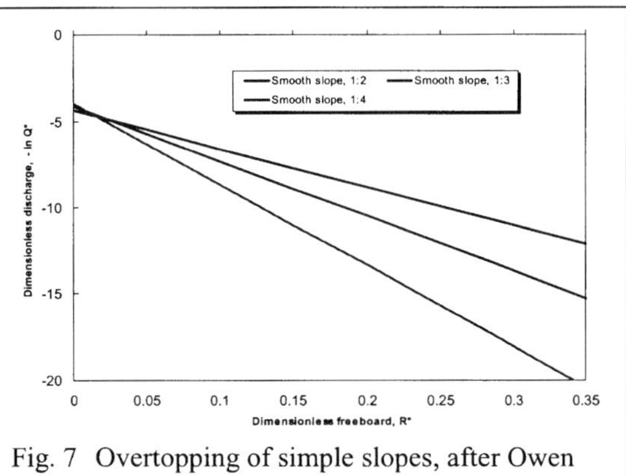

Fig. 7 Overtopping of simple slopes, after Owen

Simple slopes

Rural seawalls around UK are often of simple trapezoidal section, with slopes of 1:2 - 1:4. Overtopping of these slopes was related to freeboard R_c, and wave parameters H_s and T_m by Owen (1980, 1982). Dimensionless parameters Q^* and R^* are used in an exponential equation with roughness coefficient, r, and coefficients A and B for each slope:

$$Q^* = A \exp(-B\, R^*/r) \quad (2.1a)$$

where $$Q^* = q / (gT_mH_s) \quad (2.1b)$$

and $$R^* = R_c / T_m(gH_s)^{0.5} \quad (2.1c)$$

Coefficients A and B were initially given by Owen (1980) and revised by Besley (1999).

Owen's method is plotted as Q^* against R^* in Fig. 7. For embankments with small relative freeboards and/or large wave heights, the predictions come together, indicating that the slope angle no longer has much influence in controlling overtopping. At this point, the slope is said to be "drowned out". Over the normal range of freeboards, the discharge characteristics for slopes 1:1, 1:1.15 and 1:2 are similar, but overtopping reduces significantly for slopes shallower than 1:2. Owen's method was developed for smooth slopes, but use of the roughness factor, *r*, allowed it to be extended to rough, and even armoured slopes.

Since 1980, alternative prediction methods for armoured slopes have been explored. In the Netherlands, methods for estimating overtopping on sea dikes have been developed by de Waal & Meer (1992) and Meer & Janssen (1994). Their method distinguishes between plunging and surging conditions as identified by the surf similarity or breaker parameter:

For **plunging waves**, $\xi_{op} < 2$, overtopping is calculated from:

$$Q_b = 0.06 \cdot e^{-4.7 \cdot R_b} \qquad (2.2a)$$

$$Q_b = \frac{q}{\sqrt{gH_s^3}} \cdot \sqrt{\frac{s_{op}}{\tan\alpha}} \qquad (2.2b)$$

$$R_b = \frac{R_c}{H_s} \cdot \frac{\sqrt{s_{op}}}{\tan\alpha} \cdot \frac{1}{\gamma_b \cdot \gamma_h \cdot \gamma_f \cdot \gamma_\beta} \qquad (2.2c)$$

where Q_b = dimensionless overtopping discharge for breaking waves; R_b = dimensionless freeboard for breaking waves, and γ_b, γ_h, γ_f, and γ_β are reduction factors berm width, shallow depth, friction and angle of wave attack

Similar relationships are available for **surging waves**, $\xi_{op} > 2$, using different dimensionless parameters:

$$Q_n = 0.2 \cdot e^{-2.3 \cdot R_n} \qquad (2.3a)$$

$$Q_n = \frac{q}{\sqrt{gH_s^3}} \qquad (2.3b)$$

$$R_n = \frac{R_c}{H_s} \cdot \frac{1}{\gamma_b \cdot \gamma_h \cdot \gamma_f \cdot \gamma_\beta} \qquad (2.3c)$$

where Q_n = dimensionless overtopping discharge for non-breaking waves, and R_n = dimensionless freeboard for non-breaking waves

Vertical walls

Historically, predictions of overtopping for vertical walls used a single formula, the method developed by Franco et al (1994) was applicable to deeper water relative to wave height. Allsop *et al* (1995), later refined by Besley et al (1998), demonstrated that overtopping processes at vertical and composite walls are also strongly influenced by the form of the incident waves, not just H_s and T_p. When waves are small compared to depth, waves impinging on a vertical / composite wall are generally reflected back. If the waves at the wall are large relative to depth, then they can break onto the structure, leading to significantly more abrupt overtopping characteristics. These observations led to formulation of a wave breaking parameter, h_*, given by:

$$h_* = \frac{h}{H_s}\left(\frac{2\pi h}{gT^2}\right) \qquad (2.4)$$

Pulsating waves predominate when $h_* > 0.3$, for which the following is valid over $0.03 < R_c/H_s < 3.2$:

$$Q\# = 0.05 \exp(-2.78\, R_c/H_s) \quad (2.5a)$$
$$Q\# = Q/(gH_s^3)^{0.5} \quad (2.5b)$$

For **impulsive waves**, when $h_* \leq 0.3$, new dimensionless discharge, Q_h, and freeboard parameters, R_h, incorporated h_* to give a different prediction equation:

$$Q_h = 1.37 \times 10^{-4}\, R_h^{-3.24} \quad (2.6a)$$
$$Q_h = Q / (gh^3)^{0.5} / h_*^2 \quad (2.6b)$$
$$R_h = (R_c / H_s)\, h_* \quad (2.6d)$$

For composite structures, Besley *et al* (1998) re-defined the breaking parameter d_* based on h_*:

$$d_* = \frac{d}{H_s}\left(\frac{2\pi h}{gT^2}\right) \quad (2.7)$$

If $d_* \leq 0.3$, the mound was classified as small, and similar overtopping characteristics for vertical wall are predicted. When $d_* > 0.3$, the mound was classified as large and overtopping characteristics are corrected for the presence of the mound. These methods are described fully by Besley (1999).

Recently, Bruce *et al* (2001) extended the prediction method for impulsive waves to steep (nearly vertical) walls of 10:1 and 5:1 batter. Using the form of equation 2.5a, modifed coefficients were developed. For 10:1 batter:

$$Q_h = 1.78 \times 10^{-4}\, R_h^{-3.24} \quad (2.8a)$$

and for 5:1 batter:

$$Q_h = 1.92 \times 10^{-4}\, R_h^{-3.24} \quad (2.8b)$$

3. PERMISSIBLE OVERTOPPING LIMITS

In assessments of the effects of overtopping, most analysis has either evaluated flood volumes / areas, or has tried to estimate damage against suggested overtopping limits. Most advice on tolerable overtopping has used mean overtopping discharges, generally derived from total overtopping volumes collected over 500 to 1000 T_m. The mean discharge is then expressed as flow rate per metre run of seawall, typically m^3/s.m or l/s.m. Mean overtopping discharges are the responsive measure of hydraulic performance, and are much more stable than any peak measures.

3.1 Mean discharges

Limits to identify onset of damage to seawalls, buildings or infrastructure, or danger to pedestrians and vehicles have been defined relative to mean overtopping discharges. Guidelance on tolerable limits were developed by Owen (1980) based on work in Japan by Fukuda et al (1974) were cited in the CIRIA Rock Manual edited by Simm (1991). Owen's suggested limits for safety of vehicles and pedestrians are summarised below and in Fig. 8.

Pedestrians :-					
Wet, but not unsafe			Q_{bar}	<	0.003 l/s.m
Uncomfortable	0.003 l/s.m	<	Q_{bar}	<	0.03 l/s.m
Dangerous	0.03 l/s.m	<	Q_{bar}		
Vehicles :-					
Safe at moderate / higher speeds			Q_{bar}	<	0.001 l/s.m
Unsafe at moderate / higher speeds	0.001 l/s.m	<	Q_{bar}	<	0.02 l/s.m
Dangerous	0.02 l/s.m	<	Q_{bar}		

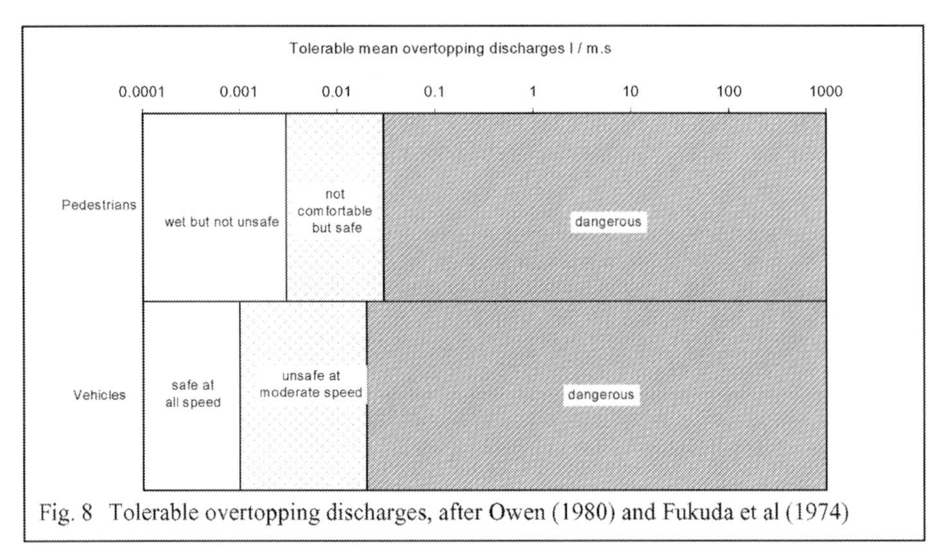

Fig. 8 Tolerable overtopping discharges, after Owen (1980) and Fukuda et al (1974)

Later revisions to these limits were suggested by Franco et al (1994), see Fig. 9. The changes were based on tests on vertical breakwaters with high walls, experiments with falling jets and studies in the Netherlands on the safety of "dyke masters" or inspectors on embankment seawalls. The limits suggested in Fig. 9 therefore apply to trained personnel, ready and equipped to get wet, but probably not to the general public!

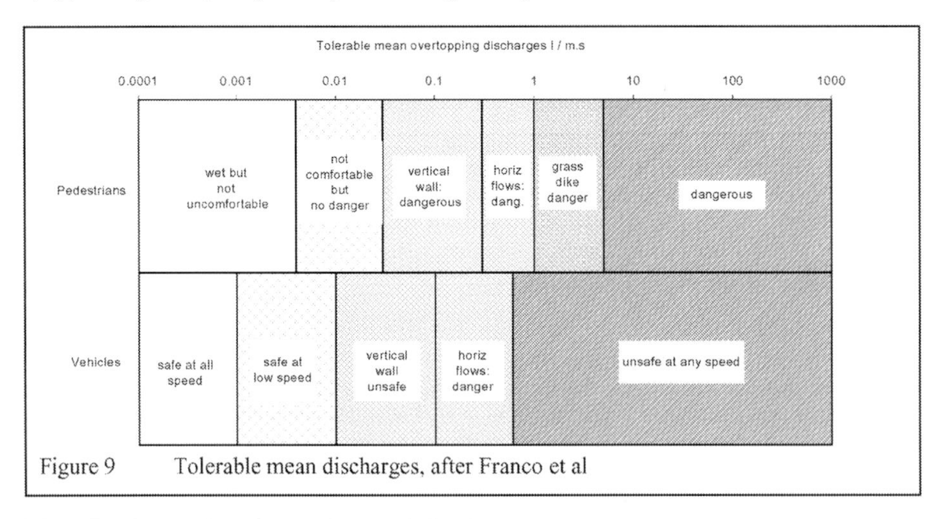

Figure 9 Tolerable mean discharges, after Franco et al

3.2 Peak overtopping volumes / discharges

Whilst common in practice, use of mean overtopping discharges in assessing safety levels without any other information is questionable. For many cases, it is probable that maximum individual volume (and velocity) are of much greater significance than mean discharge, both for damage to structures, and hazard to people. Franco et al (1994) and Besley (1999) have shown that, for a given level of mean discharge, the volume of the largest overtopping event can vary significantly with wave condition and structural type. It may therefore become inconsistent to specify safety levels with reference to mean discharge levels alone. At present however, predictions of peak overtopping volumes or velocities remain subject to large uncertainties. This paper is intended to make some reduction in those uncertainties in advance of full results in early 2005 from the CLASH project (see: http://www.clash-eu.org).

Individual overtopping events have not yet been firmly linked with hazard levels, although some useful suggestions have been made. Franco *et al* (1994) demonstrated that danger to people or vehicles from individual overtopping events could be related to their volumes. Franco *et al* (1994) suggested a "safe" limit for an individual overtopping volume for people operating behind a vertical wall as V_{max} = 100 l/m, but for a horizontally composite structure (and a trained Dutch dyke inspector) it might be increased to V_{max} = 750 l/m. Franco *et al* (1994) however observed that a volume as low as V_{max} = 50 l/m could unbalance an individual when striking their upper body without warning. These experimenters were prepared to be hit by water and will therefore have tolerated more severe conditions than would be reasonable for workers or the public who are hit by water without warning.

Franco et al (1994) also noted that any "safe" limit would vary with structural type. Any given volume overtopping a vertical structure was more dangerous than the same volume following a more horizontal trajectory. Different velocities will influence the danger caused by any particular volume, and the elevation at which a person is hit will alter the degree of danger. These effects will be influenced by the form of wave breaking onto the structure, and by the geometry of the structure's crest detail, in particular the height of any parapet wall, if present. This is illustrated by Smith *et al* (1994) who report full-scale tests conducted on grass dykes, to determine safe overtopping limits for "dyke masters" carrying out inspection and repair work. These tests concluded that work on the dyke was unsafe for trained staff when the mean discharge exceeded Q_{bar} = 0.01 m^3/s.m or 10 l/s.m, probably corresponding to approximately $V_{max} \approx$ 1 m^3/m, or 1000 l/m. This is considerably higher than the limits determined by Franco *et al* (1994) for work behind a tall wall, but accords with their observation that safe limit of V_{max} varies with structural type. In Smith *et al*'s tests, most of the flow acted on the observer's lower legs only. Again, safety limits for trained personnel working on a structure and anticipating overtopping will be higher than those for other users.

Herbert (1996) monitored overtopping behind a vertical seawall at Colwyn Bay. During installation and operation of measurement equipment, Herbert observed that personnel could work safely on the crest of the wall up to Q_{bar} = 0.1 l/s.m. Individual overtopping volumes were not measured, but methods by Besley (1999) give V_{max} = 40 l/m for Q_{bar} = 0.1 l/s.m, in close agreement with Franco's limit of V_{max} = 50 l/m. Herbert (1996) also noted that overtopping became dangerous to vehicles when the mean discharge exceeded Q_{bar} = 0.2 l/s.m, suggesting a limit of V_{max} = 50 l/m should be applied as the upper safe limit for pedestrians and vehicles.

In summary, best present guidance for areas accessed by the public are to limit overtopping to:

For pedestrians (unaware) $Q_{bar} < 0.03$ l/s.m, V_{max} = 40 l/m
For trained staff (aware) $Q_{bar} < 0.1$ l/s.m, V_{max} = 100 l/m

These suggested discharges / volumes may probably be revised upwards where the overtopping discharges are not at high velocities, or only relate to flows at low level.

4. OVERTOPPING CHARACTERISTICS

4.1 Wave-by-wave volumes

The main wave overtopping characteristics (chiefly mean overtopping discharge, but also the proportion of overtopping waves and peak overtopping volumes) can be predicted for most simple structure types using methods described in the EA overtopping manual by Besley (1999) with additional methods and explanation by Allsop & Besley (2000), Besley *et al* (1998), Bruce *et al* (2001) and Pearson *et al* (2002).

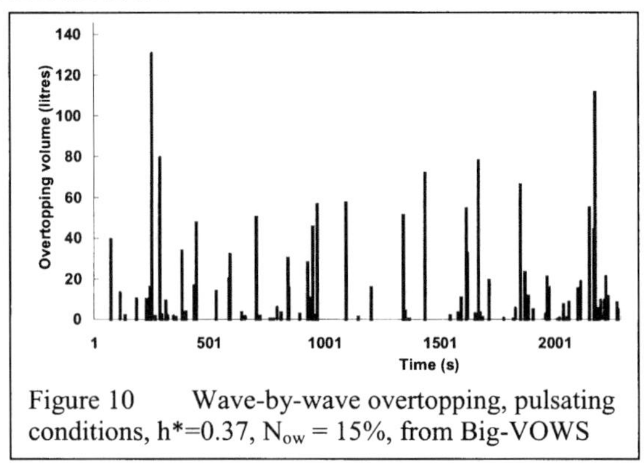

Figure 10 Wave-by-wave overtopping, pulsating conditions, h*=0.37, N_{ow} = 15%, from Big-VOWS

Relationships between peak and mean overtopping volumes can be illustrated in Figs. 10 and 11 by example results from Big-VOWS tests by Pearson *et al* (2002). Here the frequency of overtopping is quite high at $N_{ow\%}$ = 15%.

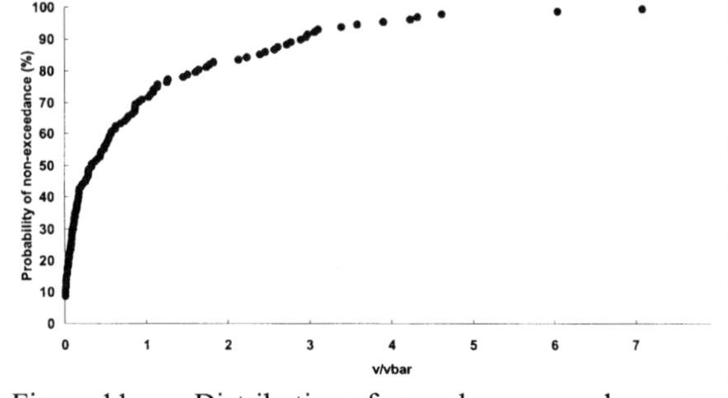

Figure 11 Distribution of wave-by-wave volumes, pulsating conditions, h*=0.37, N_{ow} = 15%, from Big-

The relationships between peak and mean overtopping volumes can be illustrated by considering the example results in Figs. 10 and 11. Here the frequency of overtopping is quite high at N_{ow}:N_w = 0.15 (= 15%). An average overtopping volume V_{bar} can be defined as

$$V_{\text{bar}} \equiv \frac{V_{\text{total}}}{N_{\text{ow}}} \tag{4.1}$$

The distribution of individual overtopping volumes shown in Figure 11, typical of tests under impulsive conditions, suggested that the highest volume in the sequence, $V_{\max}$ is about 7 - 8 times the average overtopping volume, V_{bar}. Current guidance on admissible overtopping is based upon the mean discharge, Q_{bar} (in m^3/ s / m run). We can use our approximate relation between V_{bar} and $V_{\max}$ to arrive at a relation between Q_{bar} and $V_{\max}$, and thus relate guidance based upon mean overtopping to the size of an individual, large event.

$$Q = \frac{\text{total volume}}{\text{total time}} \text{per metre run} = \frac{N_{ow} \times V_{\text{bar}}}{\text{total time}} \tag{4.2a}$$

If $V_{\text{bar}} \approx \frac{V_{\max}}{8}$ then $Q \approx \frac{N_{ow} \times V_{\max}}{8 \times \text{total time}}$ per metre run (4.2b)

The total time $\approx N_w \times T_m$ $\Rightarrow Q \approx \frac{N_{ow}}{N_W} \frac{V_{\max}}{8} \frac{1}{T_m}$ $\Rightarrow V_{\max} \approx 8QT_m \frac{N_w}{N_{ow}}$ (4.2c)

For a seawall with design waves of T_m = 10s designed for minimal overtopping, *eg* Q_{bar} < 0.05 l/s.m, with (*eg*) 10% of waves overtopping, then peak overtopping volume, $V_{max} \approx 40$ l per m. As an individual volume, this may not seem very large in terms of flooding, but it is clearly more hazardous if projected at any significant speed, see below.

4.2 Overtopping velocities

Pearson et al (2002) and Bruce et al (2002) analysed velocities of waves overtopping vertical walls at both small and large scales. The largest 20 individual overtopping events were selected. For each of these overtopping events, the upward velocity (u_z) of the leading edge of the water was determined, and was non-dimensionalised by the inshore wave celerity c_i, given by $c_i = (gh)^{1/2}$. These relative velocities are plotted in Fig 12 against the wave breaking parameter, h_*.

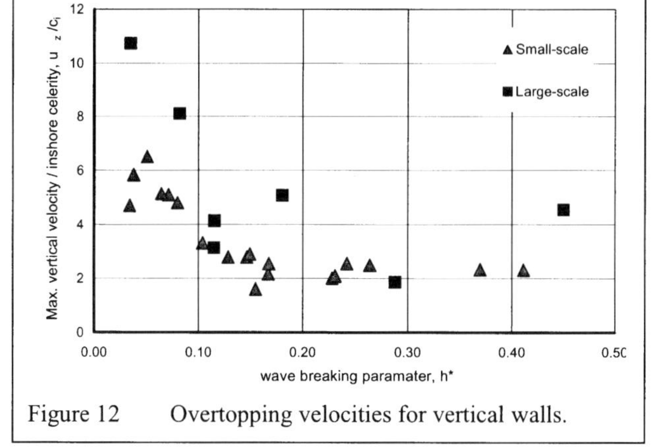

Figure 12 Overtopping velocities for vertical walls.

It is noticeable in Fig. 12 that, when $h_* > 0.15$, the non-dimensionalised throw velocity is roughly constant at $u_z/c_i \approx 2.5$, but these velocities increase very significantly when $h_* \leq 0.15$ reaching $u_z/c_i \approx 4$ - 10. This confirms that the hazard derived from overtopping discharges may vary dramatically with changes of wave breaking characteristics. The largest velocities measured here suggest prototype velocities equivalent to u_z = 40 m/s, at which speed an overtopping volume of v_{max} = 25 l per metre run becomes quite serious!

5. OVERTOPPING CASE STUDIES

Under the EC project CLASH (see: http://www.clash-eu.org) and de Rouck *et al* (2002), wave overtopping will be measured at full scale at 3 or 4 sites around Europe: the large breakwater at Zeebrugge; the rubble mound at Ostia; the vertical / composite wall at Samphire Hoe; and (possibly) part way up a shallow slope embankment dyke at Petten, see Pullen *et al* (2003). Those measurements will be presented at future conferences. At Samphire Hoe, as well as the current measurements, observations of overtopping hazards have also been made for the last seven years, and those observations are analysed here.

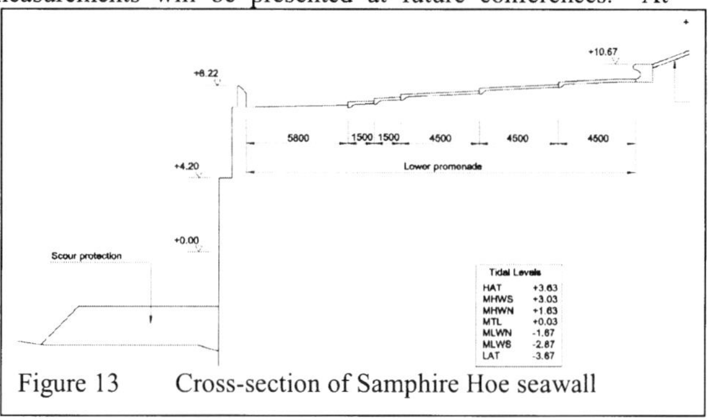

Figure 13 Cross-section of Samphire Hoe seawall

5.1 Samphire Hoe

Samphire Hoe reclamation (just west of Dover) was formed by 5 million m^3 of chalk spoil excavated from the Channel Tunnel, and designated for public recreation. The seawall in Fig. 13 is exposed to waves from south and south west, but is popular with walkers and anglers. Eurotunnel was concerned to ensure that access to Samphire Hoe was safe, so commissioned HRW in October 1995 to devise a hazard warning system.

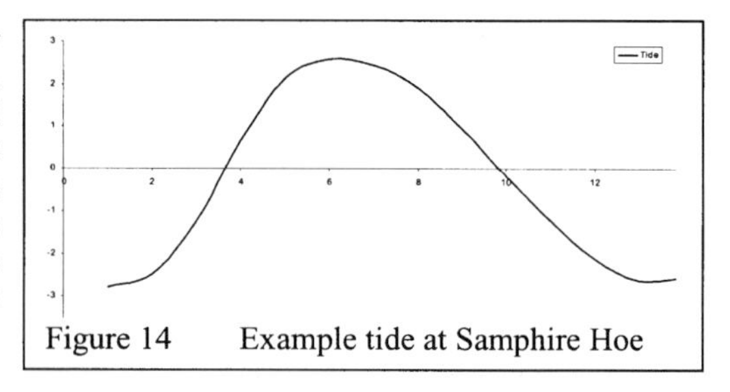

Figure 14 Example tide at Samphire Hoe

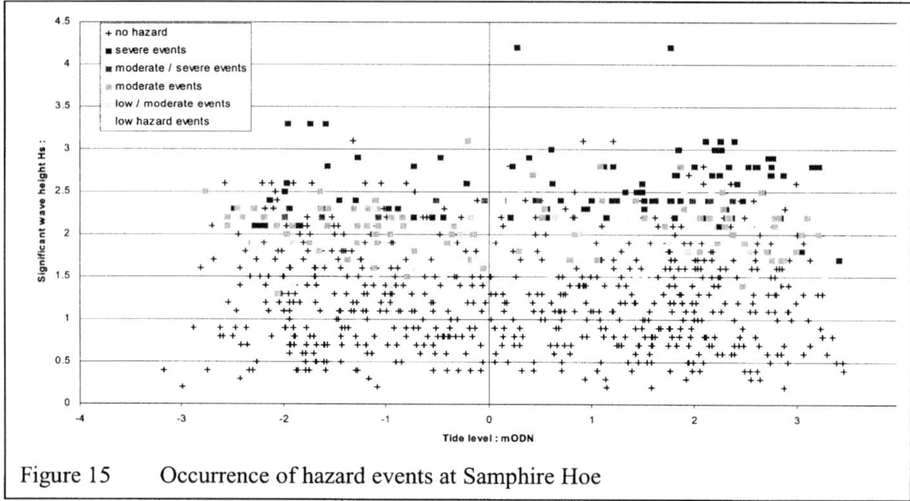

Figure 15 Occurrence of hazard events at Samphire Hoe

The project team argued then that guidance on acceptable levels of overtopping had not significantly improved since Owen's compilation of advice in 1980. Guidance on permissible overtopping was based on limited data, and was insufficient for estimation of safety limits for people. Herbert (1996) had conducted trial experiments for vertical walls with some success, but the range of comparison was very small and subject to uncertainties. Methods adopted by Sayers *et al* (1996) and Gouldby *et al* (1999) were therefore not related to overtopping discharges, but used direct observations to identify ranges of water level and wave conditions giving hazards. The UK Met Office's local area numerical weather model predicted hourly wind speeds 24 hrs in advance. Predicted winds, together with surge and tide levels, see example in Fig. 14, were then used to predict waves at Samphire Hoe using hindcasting and transformation models. On site four levels of hazard were used to record observations:

None	No observed overtopping.
Low	Occasional splash, white water (spray) only. A person may feel nervous, but no substantial danger.
Moderate	Occasional wave overtops the personnel barrier, momentary "green water" overtopping and some personal danger.
Severe	Consistent "green water" overtopping, causing substantial danger.

5.2 Overtopping hazard analysis, Samphire Hoe

Observations for October 2000 to March 2002 allow recorded overtopping hazards to be compared with predictions. At each observation time, the degree of hazard assessed locally, the tide level and the wave height are plotted as a point in Fig. 15. In general, the more severe hazards occur for higher wave heights, as expected. The surprise is that there is no similar correlation between hazards and high tide levels, despite a range of water levels over 6-7m. Hazardous effects occur at both high and low water levels. For simple monotonic overtopping responses, as in Fig. 7, high overtopping would be expected for high tide level and/or high waves, with low or no overtopping expected for low waves and/or low water levels. This can however be explained by careful use of overtopping prediction methods developed / improved by Defra / EA funded research at HRW and results from the VOWS project.

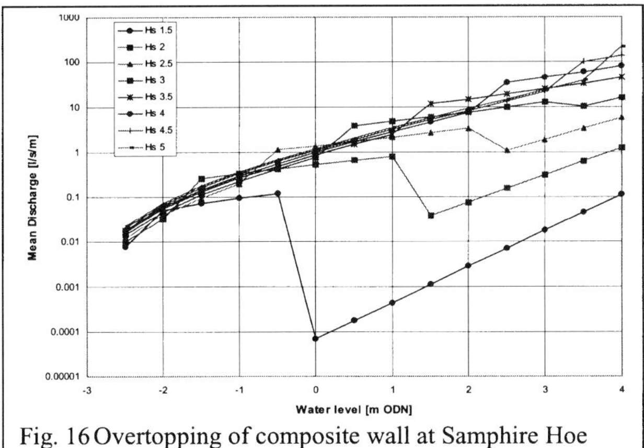

Fig. 16 Overtopping of composite wall at Samphire Hoe related to water level and nearshore wave height.

Three different methods have been used in the calculations in Fig. 16, corresponding to: **pulsating** conditions; **impulsive** breaking; and **broken** waves. The graph shows mean overtopping discharges against water level, for wave heights H_s = 1.5m to 5m. The mean wave steepness was taken as s_m = 0.05, typical of storms in the English Channel. The overtopping response functions are taken for composite walls under "pulsating" or "impacting" conditions after Bruce *et al* (2001), with a further part of the curves based on "broken" wave conditions, see Bruce *et al* (2003).

The straight lines towards bottom right of Fig. 16 predict overtopping disharges under "pulsating" wave conditions. Reducing water levels then lead to the onset of "impulsive" breaking onto the wall as waves shoal over the toe berm, at which point overtopping increases suddenly over a relatively small drop in water level, contrary to expectations given by any monotonic prediction method. At the lower water levels shown here, waves are broken before reaching the wall, but present guidance still suggests using the impulsive overtopping prediction method. The larger wave heights do not show the same behaviour as they are already too large to give "pulsating" conditions.

The methods illustrated above can then be applied to the data from Samphire Hoe shown previously in Fig. 15. The simple limit of Q_{bar} < 0.03 l/s.m is shown in Fig. 17 against observations from Samphire Hoe for October 2000 - March 2002 with tide levels and wave heights from the Met Office wave model. These comparisons reinforce the suggested upper limit for moderate hazard given by Q_{bar} < 0.03 l/s.m.

5.3 Overtopping hazard analysis, Colwyn Bay

Similar analysis for the A55 seawall at Colwyn Bay derived from observations made by Herbert (1996) are shown in Fig. 18. Here the transition zones between wave breaking types

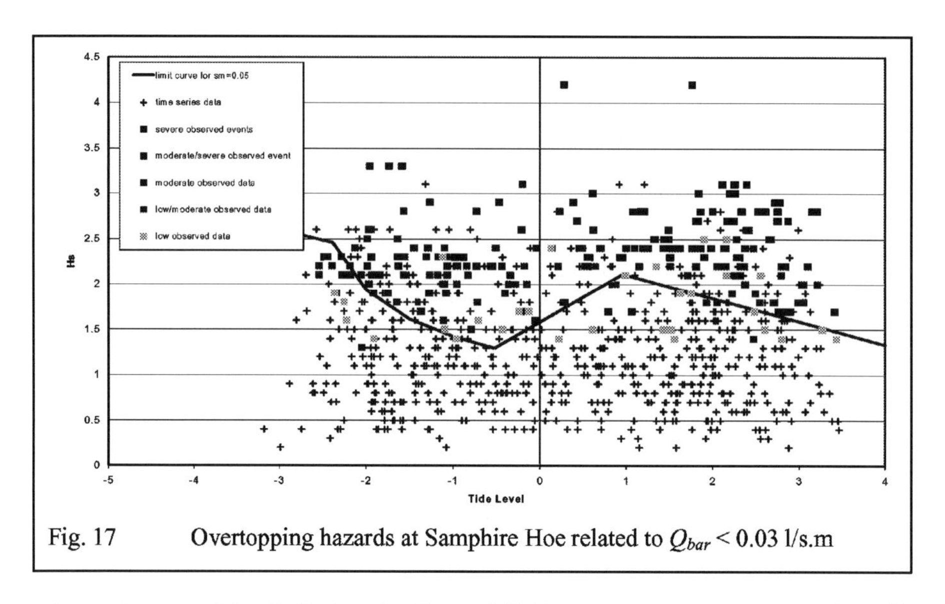

Fig. 17 Overtopping hazards at Samphire Hoe related to Q_{bar} < 0.03 l/s.m

are less extreme, and the limit given by Q_{bar} < 0.03 l/s.m. appears more conservative. The differences may however again relate to the observations at Colwyn Bay being made by trained personnel who were expecting to get wet.

5.4 Overtopping hazards elsewhere

During the second half of the CLASH project, July 2003-December 2004, observations of overtopping hazards at Oostende, Zeebrugge, Ostia, Samphire Hoe, and other sites will be used to generate a greater database from which to make firmer recommendations. In the meantime, it is worth noting occurrences of wave overtopping onto Japan national highway 336, Hokkaido discussed by Kimura *et al* (2000). A length of highway immediately after a tunnel is protected by a steep seawall (battered at 2.5:1) exposed to the Pacific. Analysis of video records of overtopping, together with waves measured up-coast and tides measured down-coast, were used to correlate damage to a passing vehicle with hind-cast calculations of overtopping. Kimura *et al* concluded that the car windscreen was damaged under overtopping of order $Q_{bar} = 10^{-5}$ m^3/s.m, despite being on the farther carriageway from the seawall. Overtopping was calculated using Goda's (1975) diagrams. This level of discharge is very low in relation to the hazard level in earlier guidance. It is possible that the simple method of deducing the overtopping discharge has missed impulsive breaking effects. This require further analysis.

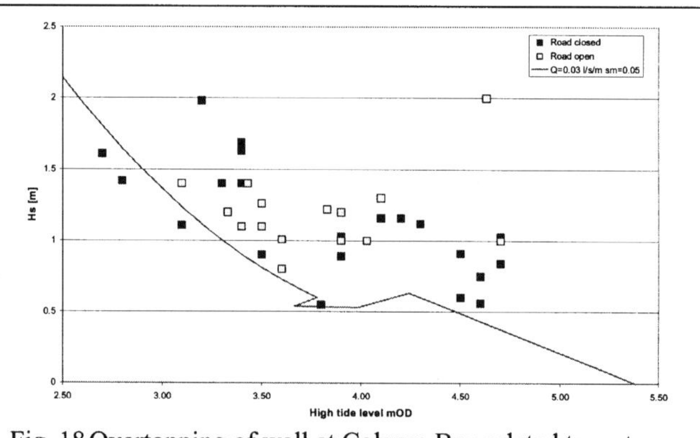

Fig. 18 Overtopping of wall at Colwyn Bay related to water level and nearshore wave height.

6. DISCUSSION AND RECOMMENDATIONS

Methods to predict wave overtopping at seawalls (primarily mean overtopping discharge, Q_{bar}) have improved in recent years, and will continue to do so under current research projects. Some recent extensions to empirical methods have been summarised here. New methods have been developed to predict overtopping characteristics for infrequent events, $N_{wo}\% < 5\%$, and to estimate overtopping velocities, u_z under impulsive conditions Further extensions and reductions in uncertainties are being developed in describing wave-by-wave and peak overtopping volumes, v_{max}, and overtopping behaviour under broken waves.

Guidance on overtopping hazards is still primarily based on mean overtopping discharge, although refinements to include peak volumes and velocities have been proposed. The best present guidance for areas accessed by the public is to limit overtopping to:

For pedestrians (unaware) $Q_{bar} < 0.03$ l/s.m, $V_{max} = 40$ l/m
For trained staff (aware) $Q_{bar} < 0.1$ l/s.m, $V_{max} = 100$ l/m

It is noted that these discharges / volumes may be revised upwards where overtopping velocities are low, or where flows only occur close to promenade / road level. It is anticipated that this guidance will be improved and extended when results from hazard measurements under the CLASH project become available.

Hazards from wave overtopping should be assessed using a staged approach. The stages will depend on the degree and probability of hazard, whether it is long-standing or is anticipated as a result of new-build, and the availability of site specific data. The key steps may be summarised:

1. Identify occurrence of overtopping hazards – For existing defences, local experience should be used to identify the occurrence of hazards (date/time and severity) and to relate each hazard event to key input parameters for the event: tide + surge; wave height, period and direction; beach level / slope, even if only be estimated within a range. Offshore wave conditions may be extracted for given dates / times from the most appropriate grid point on the UKMO wave model (archive copies held at HR Wallingford). Waves then need to be transformed inshore (shoaling, refraction, breaking) using suitable empirical methods. (Numerical models are probably not appropriate for this analysis). Data on hazards and input conditions might be presented in a form similar to that used here for Figs. 15, 17, 18, linking occurrence of hazard to key input parameters to identify the general trends of performance. For new defences, this step is taken after Stage 3 below.
2. Understand the overtopping characteristics of the defence – For any defence, new or existing, overtopping characteristics should be described against the likely range of the key input parameters including: tide + surge; wave height, period and direction; beach level / slope, to give Q_{bar}, and $N_{wo\%}$, u_z, and V_{max}, These characteristics might be presented in a form similar to Fig. 16 to identify sudden changes of performance.
3. Analyse experience to date – For an existing defence, incidents of overtopping hazards can be "back-analysed" using the data above to calculate overtopping characteristics, Q_{bar}, $N_{wo\%}$, u_z, and V_{max}, It may be worthwile to include "near miss" conditions when no hazard was observed despite the conditions, as long as this does not simply reflect the absence of any information. Each incident of hazard (or no hazard) can then be compared with calculations of overtopping characteristics, Q_{bar}, $N_{wo\%}$, u_z, and V_{max}. This process can be repeated covering the likely range for those parameters that can only be estimated. The result of this will be some "calibration" of the overtopping response characteristics, and design thresholds, for this defence.

4. Develop hazard warning system – The methods discussed above can be used to set up a simple warning system driven by wave conditions predicted by the UKMO wave model. This model computes at 1 hour intervals, and output can be configured to give tide + surge, wind and wave. Forecasts of waves / tide can be made up to 48 hours ahead, updated at, say, 4 hour intervals. Wave conditions for the offshore grid point can then be transformed to a nearshore point using empirical formulae set up for the chosen site and driven by the wave model output. The final stages are then to apply nearshore wave conditions / water levels to the overtopping model set-up and calibrated in Stages 2-3 above. This model can then calculate Q_{bar}, $N_{wo\%}$, u_z, and V_{max} to be contrasted with thresholds calibrated in Stage 3.

ACKNOWLEDGEMENTS

The VOWS research project was supported by EPSRC under GR/M42312 and GR/R42306, and built on earlier research at HR Wallingford supported by MAFF and EA. Work on hazard analysis from overtopping is being conducted under the CLASH project under EU contract EVK3-2001-0058 and supported by Defra / EA under FD2412. Further work by Universities of Edinburgh and Southampton is supported by an EPSRC project in Partnership in Public Awareness (GR/ S23827/01).

The White Cliffs Countryside Project and EuroTunnel are thanked for access to data from Samphire Hoe. Previous data and guidance from Phillip Besley at HR Wallingford is gratefully acknowledged, as is the helpful supervision of the collaborative VOWS work by the EPSRC CEWE Project Manager, Michael Owen.

REFERENCES

Allsop, N.W.H., Besley, P. & Madurini, L.. (1995), "Overtopping performance of vertical and composite breakwaters, seawalls and low reflection alternatives", Paper final MCS Project Workshop, Alderney.

Allsop, N.W.H., McKenna, J.E., Vicinanza, D. & Whittaker, T.T.J. (1996), "New design methods for wave impact loadings on vertical breakwaters and seawalls", Proc 25th Int. Conf. Coastal Eng., Orlando, ASCE, New York.

Besley P. (1999), "Overtopping of seawalls – design and assessment manual", R & D Technical Report W 178, ISBN 1 85705 069 X, Environment Agency, Bristol.

Besley, P., Stewart, T. & Allsop, N.W.H. (1998), "Overtopping of vertical structures: new prediction methods to account for shallow water conditions", Proc. ICE Conf. Coastlines, Structures and Breakwaters, Thomas Telford, London.

Bruce T., Allsop N.W.H. & Pearson J. (2001) "Violent overtopping of seawalls – extended prediction methods" Proc. ICE Conf. On Shorelines, Structures & Breakwaters, September 2001, pp 245-255, ICE, London

Bruce T., Allsop N.W.H. & Pearson J. (2002) "Hazards at coast and harbour seawalls - velocities and trajectories of violent overtopping jets" Proc. 28th Int. Conf. Coastal Eng. (ASCE), Cardiff

Bruce T, Pearson J & Allsop N.W.H. (2003) "Violent wave overtopping – extension of prediction method to broken waves" Proc. Conf. Coastal Structures '03, Portland, ASCE / COPRI, New York.

CIRIA / CUR (1991), "Manual on the use of rock in coastal and shoreline engineering", CIRIA special publication 83, Simm, J.D. (Editor), CIRIA, London.

Franco, L., de Gerloni, M. & van der Meer, J.W. (1994), "Wave overtopping on vertical and composite breakwaters", Proc 24th Int. Conf. Coastal Eng., Kobe, ASCE.

Fukuda N., Uno T. & Irie I (1974) "Field observations of wave overtopping of wave absorbing revetment" Coastal Engineering in Japan, Vol 17, pp 117-128, Japan Society of Civil Engineers, Tokyo.

Goda Y. (1971) "Expected rate of irregular wave overtopping of seawalls" Coastal engineering in Japan, Vol 14, pp 45-51, JSCE, Tokyo.

Goda, Y, Kishira, Y, & Kamiyama, Y. (1975) 'Laboratory investigation on the overtopping rates of seawalls by irregular waves'. Ports and Harbour Research Institute, Vol 14, No. 4, pp 3-44, PHRI, Yokosuka.

Gouldby B.P., Sayers P.B. & Johnson D (1999) "Real-time hazard forecasting: implementation and two years operation at Samphire Hoe, Dover" MAFF Conf. on River and Coastal Engineers, Keele.

Hedges, T.S. & Reis, M.T. (1998), "Random wave overtopping of simple sea walls: a new regression model", Proc. Instn. Civil Engrs. Water, Maritime & Energy, Volume 130, March 1998, Thomas Telford, London.

Herbert D.M. (1996) "Overtopping of Seawalls: a Comparison between Prototype and Physical Model Data" Report TR 22, HR Wallingford.

Kimura K, Fujiike T, Kamikubo K. Abe R & Ishimoto K (2000) "Damage to vehicles on a coastal highway by wave action" Proc. Conf. Coastal Structures '99, Santander, June 1999, publn. A.A. Balkema, Rotterdam.

Meer, J.W. van der, Tonjes P. & de Waal J.P (1998) "A code for dike height design and examination" Proc. ICE Conf. Coastlines, Structures & Breakwaters, T.Telford, London.

Owen, M.W. (1980), "Design of seawalls allowing for overtopping", Report EX924, Hydraulics Research, Wallingford.

Owen, M.W. (1982), "Overtopping of Sea Defences", Proc. Intl. Conf. On Hydraulic Modelling of Civil Eng. Structures, Coventry, pp469-480, BHRA, Bedford.

Pearson, J., Bruce, T. & Allsop, N.W.H. (2001), "Prediction of wave overtopping at steep seawalls – variabilities and uncertainties", Proc "Waves '01", San Francisco (ASCE).

Pearson, J., Bruce, T. & Allsop, N.W.H. (2002), "Violent wave overtopping – measurements at large and small scale", Proc. 28th Int. Conf. Coastal Eng. (ASCE) Cardiff.

Pullen T.A. Allsop, N.W.H. Bruce, T. & Geeraerts, J. (2003) "Violent wave overtopping: CLASH Field Measurements at Samphire Hoe" Proc. Coastal Structures 2003, ASCE.

Richardson, S. Pullen, T. & Clarke, S. (2002) "Jet Velocities of Overtopping Waves On Sloping Structures: Measurements and Computation" Paper 347 at ICCE 2002 Cardiff, July 2002 , publn ASCE, New York.

Rouck de J., Allsop N.W.H., Franco L. & van der Meer J.W. (2002) "Wave overtopping at coastal structures: development of a database towards up-graded prediction methods" Proc 28th Int. Conf. Coastal Engineering (ASCE), Cardiff, pp 2140-2152.

Waal, J.P. de Tonjes, P. & van der Meer, J.W. (1996), "Overtopping of sea defences" Proc 25th Int. Conf. Coastal Eng. (ASCE), pp2216-2229, Orlando, publn ASCE, New York.

Improving coastal flood forecasting services of the Environment Agency

R.Khatibi, Environment Agency, UK
B. Gouldby, South Pacific Applied Geoscience Commission, Fiji
P. Sayers, Engineering Systems Group, HR Wallingford, UK
J. McArthur, Coastal and Rivers, Posford Haskoning, UK
I. Roberts, Atkins Consultants, UK
A. Grime, Weetwood Services, UK
A. Akhondi-asl, Environment Agency, UK

Abstract

Preventive flood works are susceptible to overtopping and therefore associated with a residual risk of flooding, during which prime objectives are to save life, minimise damage, reduce disruption to communication lines, mitigate trauma and restore normality. Residual flood risks are managed by provision of information through flood forecasting and warning (FF&W) services for the population at risk of flooding. Lead-time plays a crucial role in the effective delivery of this service. Currently the lead-time is a minimum of 2-hours in England and 1-hour in Wales and the Environment Agency (the Agency) for England and Wales has the lead to provide FF&W services where cost-effective and technically feasible. This paper outlines a review of the Agency's capability on coastal flood forecasting (CFF) and the potentials for improving the lead-time of warning messages. It reports that existing CFF capabilities have normally been based on judgement and empirical approaches, signifying that these services have largely been reactive. An ongoing R&D project is integrating the modelling capability developed by coastal scientists with the wealth of experience gained by the fluvial forecasting community. A potential outcome is a new capability to improve lead-time by detecting the impending incidents, not from the water level at the coastlines alone but, with an insight into the whole processes of the sea-state. This paper outlines best practice in terms of the following elements:

- Modularisation of the service into separate processes with well defined interfaces.
- Seeking generic definitions for accuracy, timeliness and reliability.
- Categorisation of modelling techniques and physical systems.
- A risk-based and defensible approach to selecting a modelling solution.
- Integrating FF&W services to other flood management measures.

Key words floods, coasts, forecasting, realtime, lead-time, accuracy, timeliness, reliability, categorisation, best practice, 1st-3rd generation

Introduction

Floods are natural phenomena but pose risks to man normally mitigated by flood protection works. The key issue is then to manage the risks of their failure during operations including emergency times. However, flood protection works are "finite measures" due to being designed using prescribed annual probability values. Thus, risks to operation and management of flood protection works are always intertwined with residual risks to man by overtopping, even if these works are fully operational. Residual risks arise because experience and science confirm the emerging consensus that the total elimination of flooding risks by finite measures is impossible. As protection works are physical/structural in their nature, the mitigation of residual risks is through information promoting the goal of restoring normality through a provision of Flood Forecasting and Warning (FF&W) services against impending incidents or those incidents that are already rolling out. This paper outlines these services but with a particular focus on Coastal Flood Forecasting (CFF).

The professional culture of modelling has been embraced in flood forecasting by using the information contained in realtime data to manage residual risks of flooding to save life, reduce damage to properties, minimise disruptions to communication lines, to mitigate trauma and to restore normality. Flood forecasting is one process within a sequence of systemically arranged processes of detection, flood forecasting, warning, dissemination and response, where systemic is defined as the system reduced into subsystems interconnected at clearly defined interfaces. This model is shown in Figure 1. Without flood forecasting, the whole service is reduced to a reactive response but with accurate, timely and reliable flood forecasting a proactive capability emerges to respond to impending flood incidents. The main focus of this paper is on strengthening the science base of CFF but at the same time **(i)** flood forecasting is presented within the context of the five processes of flood warning services; **(ii)** the paper promotes the integration of managing risks and residual risks posed by floods.

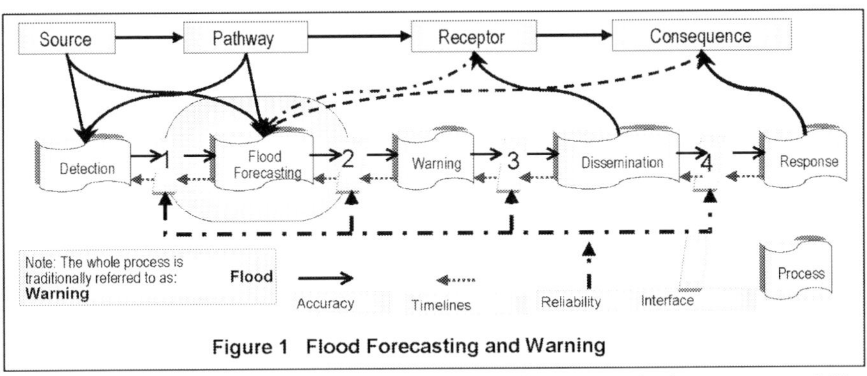

Figure 1 Flood Forecasting and Warning

The paper reports the recommendations of an R&D project in developing a new capability to improve lead-time by detecting impending incidents, not from the water level at the coasts alone but, by understanding the whole processes of the sea-state. This new capability aims to develop best practice in CFF. Best practice in modelling/forecasting is not a single powerful technique but a framework interconnecting a range of building blocks, as discussed by Khatibi *et al* (2003*b* and 2003*c*). The novelty of best practice as reported in this paper is in consolidating the growing experience in computational modelling of shelf seas and coastal areas, and through using categorisation as a tool to unravel generic patterns by deliberately disregarding minor individual differences among the various techniques. The effective delivery of FF&W services also depends on relevant organisational arrangements and a range of other factors including the following elements:

- Modularisation of the service into a number of separate processes with well defined interfaces.
- Seeking generic definitions for accuracy, timeliness and reliability.
- Categorisation of modelling techniques and the physical system.
- A risk-based and defensible approach to selecting a modelling solution.

As modelling skills have diversified, the use of the term model has become rather confusing. This paper aims to avoid the confusion with a consistent usage of the following qualifiers: modelling techniques, to refer to individual set of equations; modelling approach to refer a category of modelling techniques; model datasets to refer to the actual data used to run models; and modelling software tool. The paper argues that traditional model selection suffers from *ad hoc* procedures and a culture similar to civil engineering design is yet to be transferred to model selection. The paper reports that a decision tree is being developed to selecting "modelling options" and thereby select a modelling solution with minimum risks.

Review of coastal flood forecasting in the UK

Developmental background

The FF&W service is evolving from *ad hoc* responses based on duty of care to a science-based practice in many countries. The 1928 floods killed 18 people in London and led to the provision of Alert Messages by the forerunner to the Met Office. The devastating flood of January 1953 took the lives of over 300 people in East Anglia. This prompted the Waverley Report (1953), which recommended setting up the Storm Tide Warning Service (STWS). The STWS was funded by the predecessor of the Department for Environment, Food and Rural Affairs (Defra) and delivered by the Met Office. Since then there have been many developments including the emergence of modelling and real-time forecasting capabilities. However, the main landmarks of coastal flood forecasting in England and Wales have been the establishment of organisational arrangements with the delivery of the service based on mostly empirical techniques. The Agency has the lead in the delivery of FF&W services and the forecasting of coastal floods is largely based on the Storm Tide Forecasting Service (STFS) provided by the Met Office (STWS was renamed to STFS in 1998).

Overview of the organisational arrangements

The Agency operates a flood warning service across much of England and Wales under Section 166 of the Water Resources Act, 1991. Following a Ministerial Directive in September 1996, the Agency took over the responsibility from local police forces and since then has assumed the lead role in disseminating flood warnings to people at risk of flooding.

Flood warning teams in the Agency work with professional partners and with its own emergency response teams responsible for operating strategic flood defences. These organisations form a network whose aim is to mitigate the effects of flooding by co-ordinating efforts. Although the organisational arrangement for FF&W within the Agency is likely to be reorganised, its current arrangement since 2000, has been as follows:

- The 26 Area Offices are responsible for creating readily intelligible information messages referred to as 'staged warning codes' designed to trigger clear response actions, Figure 2. They are also responsible for the delivery and dissemination of these warnings through a wide range of communication channels.
- The eight Regional Offices plus Thames Barrier collect and process forecasting data and transform them into flood forecasting results, although one region is yet to conform in 2005. These results are provided as advice to 26 Warning Teams, as described above.

- The Agency recognises the strategic importance of **(i)** the development of best practice in FF&W through co-ordinating R&D projects, and **(ii)** other activities including marketing, communication and business planning. These activities were fulfilled by National Flood Warning Centre in 2000-3 but recently, introducing the staged warning codes in 2000, as shown in Figure 2. Now the development of best practice is responsibility of the policy, process and other integrated teams of the Agency.

Figure 2 Staged Warning Codes

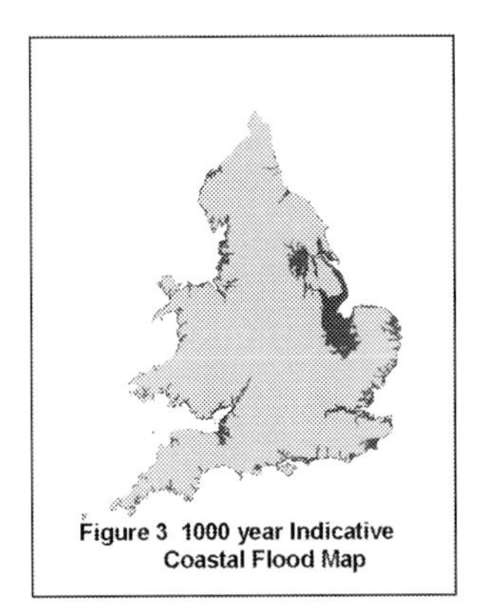
Figure 3 1000 year Indicative Coastal Flood Map

The Agency works in partnership with a wide range of organisations to deliver the FF&W services. While the Agency is responsible for the operations and the delivery of FF&W services, it also works with the Defra, as the owner for the policy of FF&W. In particular, the partnership with the Met Office is very important for producing flood forecasting data including surge and wind data. Proudman Oceanographic Laboratory (POL) also plays a role in developing and refining modelling capabilities, e.g. the POLCOMS model. The most important stakeholder is the population at risk of flooding. Other stakeholders include the emergency services, the police, fire and life rescue services, local authorities, flood wardens and local flood associations.

The extent of coastal flooding in England and Wales

The scale of coastal flooding in England and Wales is potentially extensive, as depicted in Figure 3 for annual probability of 0.1%. With the continuing threat of climate change and associated sea level rise, a better understanding of these risks are collectively important drivers for the development of best practice in the CFF.

The survey of past studies

A series of initiatives is underway to develop best practice as a need-driven approach for improving the effectiveness of the service. These initiatives often have two components **(i)** an incremental contribution in the form of generic solutions towards the development of best practice, and **(ii)** specific solutions meeting the needs of a particular project. A summary of their outcomes is presented in Table 1.

Survey of existing practices

The data used in the Agency may be considered as a key to gain an insight into the CFF in the Agency. Questionnaires were distributed in the Agency identifying more than 10 generic types of data to survey their usage. The quality of data was also assessed in terms of practitioners view (very good, good, fairly good, poor, and very poor). The data types are presented in Table 2 and the survey results are presented in Figure 4.

Table 1 Outline of Past Studies on Coastal Flood Forecasting and Warning

Project Description
National Tidal Flood Forecasting Project (1998) – included waves and surges **The sources of information:** • Information from the sea defence survey 1990 • Data from the EA flood forecasting units; and EA assessment of flood risks and forecast performance data **Factors contributed to the study** • Of the 209 identified Flood Warning Areas, the risk factors were "at risk" properties, at risk open-coasts, inadequate defences for floods with annual probability of 0.5% • Floods may arise due to tides, surge residuals, winds and waves • Existing flood forecasting capabilities were constrained in many different ways, as the full potentials of coastal modelling capabilities were unavailable in the CFF **The Action Plans included short- and medium-term proposals** with the following proposals: • **Specific proposals** – improvements to specific models, e.g. the Severn Estuary model, etc • **Generic proposals** – Renaming STWS to STFS; provisions of a national alert system & a 365 day forecasting service; the development of an integrated tide gauge network; confidence factors to surge forecasts; a new data management system; treat accuracy and reliability of surge models; impacts of meteorological errors on flood forecasts; onshore wave forecasts and wave overtopping models **Progress on the Action Plans:** A review in 1999 concluded that most of the actions were complete but a few are outstanding and the project on developing best practice on the CFF is one such.
Review of extreme levels in Estuaries R&D Projects (1999-2002) **Terms of Reference**: A review of current methods; characterisation of estuaries and development of guidelines to recommend rigorous but practicable methods for realtime forecasting of extreme water levels in estuaries. **Outcomes**: The project abandoned sensibly the pursuit of identifying rigorous techniques and embarked on selecting modelling solutions through categorising physical systems, categorising modelling techniques and matching them in a decision box. **Strength and Weakness:** The selection of modelling solutions through a decision box promoted a novelty in modelling practices (the strength) but the definitions of accuracy, timeliness and reliability were not tackled.
The Development of Generic Modelling Specification This is driven internally, encouraging defensible modelling practices through formalising the modelling procedures to promote minimum standards and auditable practices. The procedure is based on breaking down the practice into five stages of (**i**) developing preliminary models through using guidelines, (**ii**) value and risk management on the preliminary model, Khatibi *et al* (2002) (**iii**), calibration and validation, Khatibi *et al* (2003*a*), (**iv**) test controlling, and (**v**) application. For more details see Khatibi (2002).
Fluvial Flood Forecasting This project was similar to the "Review of Extreme levels in Estuaries" above but on fluvial systems and put the preliminary steps towards adopting the definitions of accuracy, timeliness and reliability as promoted by the Agency.
National Flood Forecasting Project (2003- ongoing) This project is developing an "open shell" as an effective step towards a need-driven modelling capability based on "open architecture", where a set of governing equations may be transformed into a module (computational module) attachable onto the shell without the involvement of the owner of the shell. Therefore, it will be very easy to assemble computational modules as needed but not as imposed by software producers.
POL Fine Grid Surge Model Evaluation Report, see Flather *et al* (2001) The report highlighted the need for nesting of fine-grid models (200-300m) into the large scale ones (12km).
The Agency's Vision Statement, see (EA 2002) The vision of the Agency includes targets (**i**) to ensure residents in flood risk areas take effective action; (**ii**) to ensure residents in flood risk areas improve the coverage of flood warning services. These are only feasible through the development of best practice, as discussed in this paper.
The ICE Learning to live with rivers, see ICE (2001) The report recommended a risk-based approach.

The survey of current CFF capabilities in the Agency revealed that most Regions use similar data, either from the same or similar sources. The CFF capability is based mainly on (**i**) the surge forecasts provided by the STFS (**ii**) astronomical predictions for a location, (**iii**) additional information such as overtopping utilising lookup tables for winds/waves/level-to-level correlation. Methods can vary in complexity within the Regions depending on the physical processes that lead to coastal flooding on an Area by Area basis and the location of Flood Risk Areas. Thus, the data required to perform flood forecasting depend largely on the physical processes that cause flooding in a risk area and as such each Region will have some variety in terms of data needs. Generally the three types of data and their use by Regions are:

- Water level information (measured and predicted using level-to-level correlation techniques) – Used frequently by all Regions
- Wind data (measured and predicted) – Used less frequently and only by some Regions, but important as used to upgrade warnings
- Waves (measured or predicted, inshore and offshore) – Used only by a couple of the Regions for specific stretches of coast

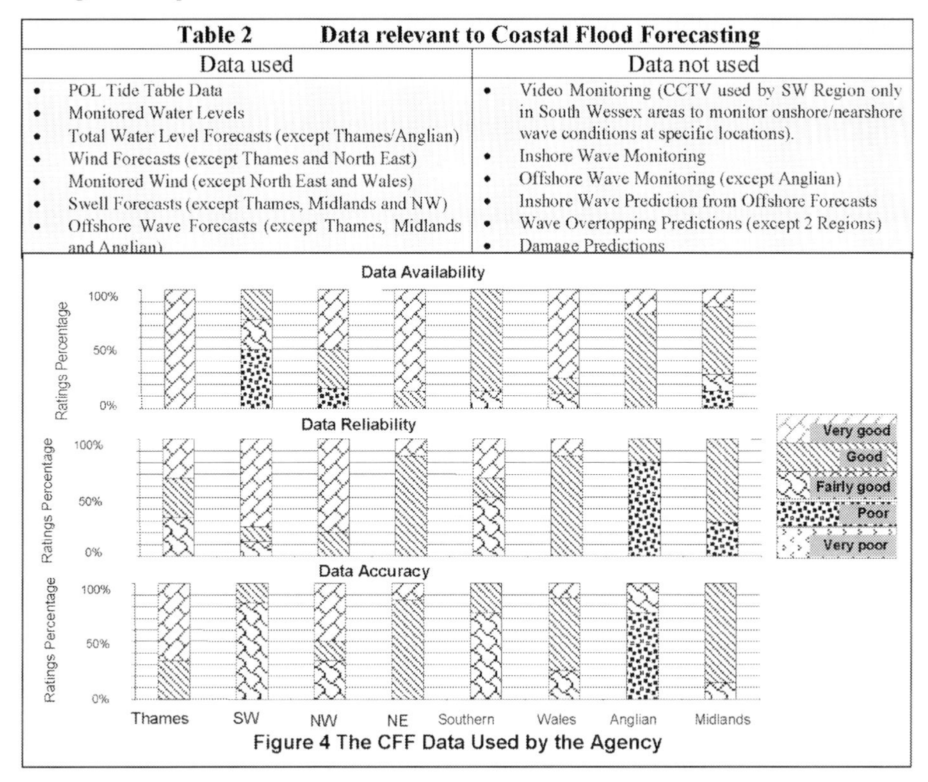

Table 2 Data relevant to Coastal Flood Forecasting

Data used	Data not used
• POL Tide Table Data • Monitored Water Levels • Total Water Level Forecasts (except Thames/Anglian) • Wind Forecasts (except Thames and North East) • Monitored Wind (except North East and Wales) • Swell Forecasts (except Thames, Midlands and NW) • Offshore Wave Forecasts (except Thames, Midlands and Anglian)	• Video Monitoring (CCTV used by SW Region only in South Wessex areas to monitor onshore/nearshore wave conditions at specific locations). • Inshore Wave Monitoring • Offshore Wave Monitoring (except Anglian) • Inshore Wave Prediction from Offshore Forecasts • Wave Overtopping Predictions (except 2 Regions) • Damage Predictions

Figure 4 The CFF Data Used by the Agency

There are indications for a need to improve timesteps and grid-size resolutions of the data provided by STFS. At present, the 12km spatial resolution dictates the prediction limits for the POLCOMS model and surge model developed by POL, run daily as in the STFS, providing 48-hour forecasts, McArthur (2001). The response to the questionnaires suggests practitioners' expectations for further improvements in the STFS include:

- Developments of more reference ports along the coast for surge prediction
- Data provision to include more information e.g. descriptive forecast
- Information on accuracy of surge forecasts (confidence bands) – combined need for data sharing between STFS and EA
- Continuous provision of data throughout tidal cycle (not just peaks)
- Secondary Alert (**i**) to provide information for closure of coastal defences and (**ii**) a check on the accuracy of surge models
- The need for formalising tide table data or astronomical predictions
- Joint STFS/POL/EA approach to developing *site specific* surge models for areas that are difficult to forecast e.g. the Wash in Anglian Region

Justification for further improvements

Assessing the likely risk-based flood forecasting requirements for at-risk areas is a requirement that will impact on the selection of CFF modelling solutions. In a range of cases the use of simpler methods such as lookup tables can be sufficient to provide accurate, reliable and timely forecasts. The concern in most of these cases is the basis on which these tables have been derived. Most Regions surveyed in this study suggested that these tables had been derived in the past either through modelling studies of some sort of historic events, and any shortfalls that occurred were adjusted by shifting trigger levels. This undocumented and rather convoluted development, though generally sufficient for specific Areas, relies heavily on a local knowledge and understanding of the particular vagaries of an Area. As the CFF moves towards centralisation within Regions, loss of this local knowledge could lead to inconsistent and inadequate prediction and further revision of flood forecasting levels.

The above capabilities and statements of requirements clearly demonstrate the need for a science-based framework to interconnect the various issues under a coherent and consistent methodology. The science of simulating the sea-state has made considerable advances. Over many years, oceanographic and coastal models have been developed and applied in a range of problem areas. For example, national meteorological organisations have developed models capable of predicting the development and propagation of waves over large ocean areas (many 1000's km). The primary function of these models was initially for forecasting wave conditions for mariners and marine operations. Until relatively recently, the coastal science and engineering communities have been primarily focussed on developing smaller scale regional models with the purpose of investigating how waves have influenced coastal areas in the past and may do so in the future, but not in a real time forecast sense. CFF capabilities require knowledge of both large scale and regional variations in wave and water level conditions in real time forecasting environments. The development of such a capability is overdue and can therefore be seen as the marrying of two historically different disciplines.

Drivers for developing Best Practice in the Agency

It is morally indefensible not to provide protection to the public against major hazards over which they have no control, Rowe *et al* (1995). Owing to this moral obligation, best practice is often opted for by default but many organisations would not suffice to default. The Agency has a range of drivers for the development of best practice and these include:

- National Flood Warning Strategy for England and Wales in 1997.
- The second phase of the National Flood Warning Dissemination Project (1997-2002) and the publication of a National Tidal Flood Forecasting Joint Action Plan in 1998.
- The Bye Report prompting the Easter Floods Action Plan of November 1998.
- National Flood Warning Service Strategy for England and Wales in 1999.
- Publication of "Reducing Flood Risk - A Framework for Change" in 2001, (EA 2001).
- Agency's science plan and the perception of the need for a corporate modelling strategy
- The Agency Vision Statement and the Making it Happen (MiH) documents (EA 2002)

In parallel to these are the following economic and socio-political drivers: (*a*) potential flood damage is increasing, in real terms, to goods stored in domestic, retail and industrial properties in flood risk zones. (*b*) Damage may be significantly reduced with adequate warning. (*c*) Disruption to traffic and public services can be reduced in times of flood if warnings are given in time to set up diversions. (*d*) There is a greater need to give warnings in flood risk zones where mitigation works have been carried out. (*e*) Evidence of recent incidents confirms that the public, industrialists, public services etc now expect substantially improved flood warning services.

Architecture of Best Practice for coastal flood forecasting

A brief overview of legacy modelling capabilities indicates two broad trends: (**i**) modelling techniques are proliferating and diversifying and (**ii**) each or a selection of these modelling techniques are often transformed into modelling software tools. In the prevailing modelling culture, modellers suffice to selecting one software tool. Khatibi (2002) argues that:

- Legacy practices in selecting a modelling technique are *ad hoc*, as currently there are numerous modelling techniques available to describe the dynamic-state of water in a diverse range of physical systems but the applicability of particular modelling techniques to particular physical systems is often overlooked.
- In the absence of a formalised model selection procedure, the choices may stem from personal preferences, historic reasons and familiarity with the choice, skill-base of the organisation or cost minimisation.
- In a risk-aware society, it is increasingly difficult to make defensible decisions on such *ad hoc* model selection procedures and therefore the time is right to move on.

It is a serious modelling problem that the practice is aided by intuitive knowledge of the modellers without any common ground between different modellers or between different model developers/users. Thus, existence of many *ad hoc* elements in modelling is tantamount to an important gap yet to be filled. The solution presented in this paper is to be recommended to the Agency, so the solution should be treated in tentative terms at this stage.

The oceanographic and coastal modelling communities have responded to many challenges including the mathematical treatment of a diversity of hydraulic processes, diversity of bathymetric features, a wide range of numerical schemes to solve the governing equations and managing massive amounts of data. The emerging consensus on modelling is that on the one hand models are valuable tools but on the other hand perfect models simply do not exist. Khatibi *et al* (2003*c*) outline a number of frameworks for the development of best practice in modelling and these include the following: (**i**) generic performance measures facilitating feedback mechanisms, (**ii**) the lifecycle of model, (**iii**) formalising the lifecycle of information management, and (**iv**) other frameworks such as handling data. This paper outlines performance measures and presents model selection for the CFF in a greater detail but the other frameworks are outside the scope of this paper.

Performance measures

Although flood forecasting and warning has traditionally been broken down into a number of processes, e.g. Haimes *et al* (1989), understanding of the subject is evolving. The performance criteria, depicted in Figure 1, are defined in terms of accuracy, timeliness and reliability. Currently, there is a perceived need to move away from *ad hoc* to a systemic arrangement of the processes of detection, flood forecasting, warning, dissemination and response processes to create transparency by inserting the following interfaces:

- Interface 1 between Detection – Flood Forecasting,
- Interface 2 between Flood Forecasting – Warning (decision-making),
- Interface 3 between Warning – Dissemination, and
- Interface 4 between Dissemination – Response.

Accuracy is taken to represent information flows from the technical to social dimensions, expressing the technical performance of the system at Interfaces 1–2–3–4. This may be explained through the example of the accuracy of flood forecasting with the Interfaces 1 and 2. The detection process will provide the various realtime input values to the flood forecasting process and the flood forecasting process will provide input to the warning process. There is little quantitative or qualitative knowledge on the way accuracy is cascaded from Interface 1 through Interfaces 2 to 4 and this explains partially the difficulties involved in improving the confidence in forecast results. However, first steps are being taken in a study commissioned recently to investigate the performance measures. Timeliness represents the information flow from the social to the technical dimensions and is defined in terms of the expected requirements of the population at risk of flooding at Interfaces 4-3-2-1. This is based on the recognition that the population at risk of flooding is heterogeneous and naturally the requirements of each category of the at-risk population varies statistically. Arguably, the performance measure of timeliness should account for the statistical distribution of the time required for effective mitigating actions. The timeliness requirement at Interface 4 has to allow for the processing time within the Dissemination, Warning, Flood Forecasting and detection processes at their respective interfaces. There are various definitions of reliability but this is likely to be defined in terms of accuracy and timeliness at the interfaces.

Model selection

This paper presents a model selection methodology, referred to as **substantive modelling** – commonly referred to as "horses for courses". This is defined as interconnecting a selected modelling solution to the complexity of the physical system through **(i)** categorising modelling techniques, **(ii)** categorising physical systems and **(iii)** matching their respective categories for the various performance criteria. As depicted in Figure 5, it is widely recognised by the coastal modelling community that complexity of the sea-state varies along the following broad categories of offshore, nearshore, overtopping and inundation zones.

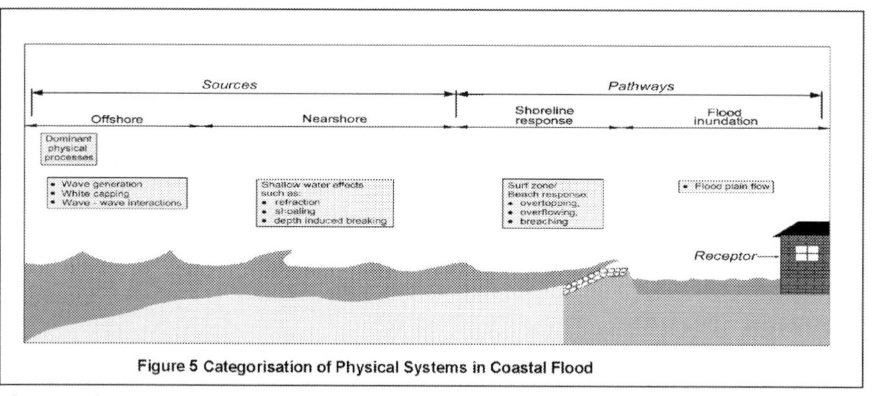

Figure 5 Categorisation of Physical Systems in Coastal Flood

Each zone is modelled using a wide range of modelling techniques, divided into five categories (each may be referred to as a modelling approach). These categories of modelling techniques are outlined in Figure 6. The specific features of each category is not covered here, as they are readily understood among the coastal modelling community. A guideline is under development to offer a formalised science-based approach for selecting modelling solutions taking into account the following:

- The needs for effective timely responses of particular at-risk areas;
- Identification of potential lead-times associated with each category of modelling techniques;
- The cost-effectiveness of a particular category of modelling solutions to meet the needs;
- Data availability; and
- Risks of not selecting more costly modelling techniques.

The various elements of the guideline are not detailed here but the architecture of the model selection procedure is depicted in Figure 7 and outlined below:

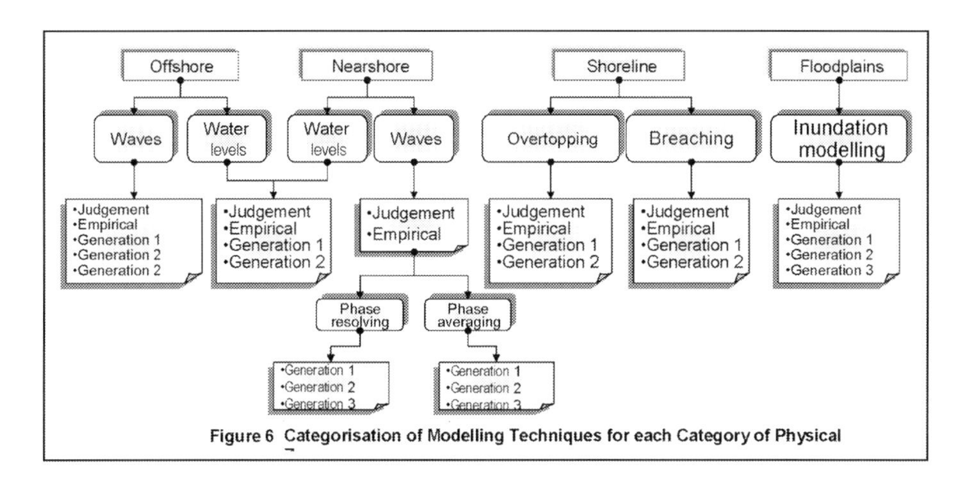

Figure 6 Categorisation of Modelling Techniques for each Category of Physical

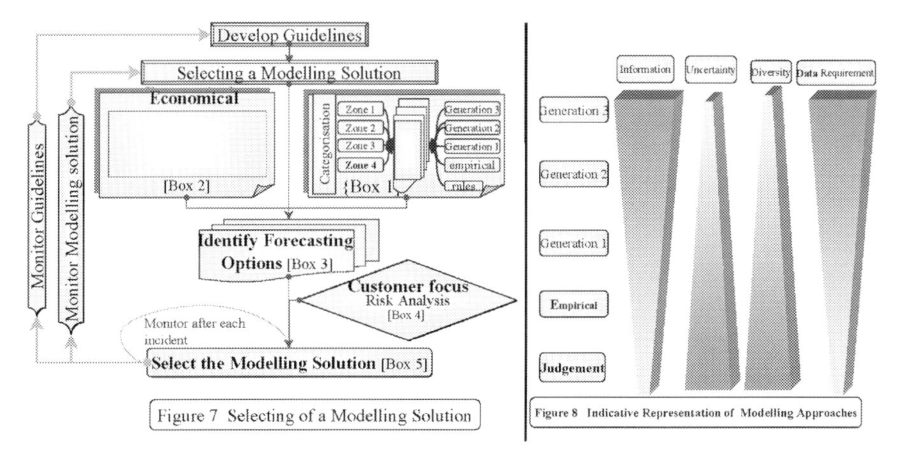

Figure 7 Selecting of a Modelling Solution

Figure 8 Indicative Representation of Modelling Approaches

- **Box 1** depicts substantive modelling in the form of matching each zone to a particular category of modelling techniques. This acts as a pointer towards improvements, should there be any need for improvements.
- **Box 2** depicts a procedure for estimating the benefits of a modelling solution.
- **Box 3** depicts a required procedure for selecting modelling options, as it is likely that there will be more than one category filling the requirements.
- **Box 4** signifies that the selected modelling solution needs to be scrutinised in the light of a formal decision-making procedure.
- **Box 5:** promotes three cycles of iterations: a short-term iteration in terms of measuring the performance of the modelling solution after each incident; a medium-term iteration measuring the effectiveness of the selected modelling solution; and a long-term iteration measuring the effectiveness of the guideline.

Overview

The CFF in an operational environment in the Agency follows the same legacy as in any modelling environment that modelling techniques are used on the basis of "one-for-all" but often irrespective of the complexity of the physical system (zones in the CFF). A range of alternatives is available to model each zone. Often modelling techniques are similar in their primary capability but differ in representing the physical processes and/or mathematical complexities. Recent initiatives in model selection have been formalising their science-base. This paper contributes to improving the situation through using categorisation as a tool. Categorisation unravels generic patterns by deliberately disregarding minor individual differences and is a particularly effective tool when the various methods/tools/products/ outputs diversify. Carried out in a meaningful manner, categorisation can relieve the burden of memorising the purpose and function of various models and assist in the selection of the most appropriate approach.

The method of categorisation, presented in this study, explains the complexity inherent in the modelling techniques developed over the years by the oceanographic and coastal modelling communities. As depicted in Figure 8, the proposed categorisation explains the information contents, uncertainty, data requirements and diversity associated with each category. It is normally the case that models with higher resolutions can offer a greater information content than those based on judgement or empirical rules, as these take into account more physical processes. Likewise, their data requirements are higher with high-resolution models than low-resolution ones but diversity of the techniques is often larger in low-resolution techniques than high-resolution one, with the exception of judgement. High-resolution techniques often have similar numerical formulations but differ in their implementations.

Discussion

A number of flood management measures has emerged since the Industrial Revolution (mid 18th-20th centuries). These include **(i)** hard or preventive (or finite) measures, **(ii)** maintenance programs, **(iii)** development control and **(iv)** flood forecasting and warning services. In current practices, the FF&W service is a discrete measure and without a capability to be integrated into other flood defence measures, even though the perceived synergy with other measures is significant. Khatibi and Haywood (2002) discuss some of the gaps to be filled and another gap is the intertwinement of the appropriate risk management techniques. The study reported in this paper takes one step to this end, as discussed below.

Managing risks to operations addresses the full assets and their operations including emergency responses to ensure that the assets remain fully operational during high flows including flooding times. The Agency is transforming the various practices to risk-based approaches through the commonly adopted conceptual risk model of *Source-Pathway-Receptor-Consequence* (*S-P-R-C*). This model, originally presented by DETR (DETR 2000) and adopted by the Agency (EA 2000); represents systems and processes that lead to a particular consequence. For an operational risk to arise there must be a hazard that consists of a *source* or initiating event; a *receptor* (person or property); and a *pathway* that links the *receptor* to the *source*. In the context of coastal flooding these terms have been identified in Figure 1. This is used as an incident-management model to mitigate risks to operational systems, pollution, flooding etc, which is intertwined with the systemic model of managing residual flooding risks through the five FF&W processes, depicted in Figure 1.

The equations governing the sea-state are often deterministic but this conventional wisdom has come of age and is not credible anymore. The state-of-the-art in modelling fosters uncertainty movement to create an information-rich modelling environment through ensemble modelling. While this project advocates the transfer of knowledge developed by the oceanographic and coastal modelling communities to the CFF, it is equally important to advocate the transfer of knowledge accrued by the treatment of uncertainty. Equally, it is often feasible to obtain streams of realtime data at an array of selected gauging sites. Realtime data streams can be incorporated during the model runs to reduce discrepancies between the forecasted results and corresponding gauged values. This is referred to as updating in flood forecasting and assimilation in meteorological forecasting. It is very important that these capabilities are also transferred to the CFF.

Forecasting breaches must be considered as an integral component of the CFF. However, the science-base of forecasting breaches is often rudimentary or expensive in CPU time and the development of CFF capabilities can offer an impetus to improving breach forecasts. In the Netherlands breach is being undertaken using 2D/3D models for specific examples. However, this is often numerically intensive and impractical for large areas.

This paper is not only advocating the transfer of knowledge from one field of modelling to another, but it aims for a cross-fertilisation of ideas. For the benefit of the CFF, the study recognises that each zone is associated with a diversity of modelling techniques and recommends their consolidation through categorisation. This has an implication on developing software tools that rather than the current practice of developing a monolithic software tool embodying a cascade of modelling techniques, modular facilities can be developed for each of these categories. This paper promotes the open architecture philosophy in software development, where each of these categories of modelling techniques can be developed as a computational module and attached into the system, see Khatibi *et al* (2001) and Khatibi *et al* (2003*d*). This gives the modeller freedom to experiment with wide choices and truly make use of the guideline developed here. The benefits of the integrated CFF capabilities are significant and include the emergence of proactive capabilities on coastal FF&W and improved reliabilities. The synergies between FF&W and other flood management measures can then be explored helping sustainable development.

Conclusion

Selecting a modelling solution is an emerging culture and only one attribute in the development of best practice in flood forecasting. This paper outlines a review of current coastal flood forecasting practices in the Agency and identifies the need for the transfer of knowledge, as developed by the coastal science and engineering communities. It is argued that there is an opportunity here for cross-fertilisation of cultures and these included:

- The transfer of knowledge developed by the coastal science and engineering communities
- The elements of best practice in flood forecasting as fostered by the Agency
- Using categorisation as a tool and developing need-driven guidelines, which match categories of modelling techniques to categories of physical systems, referred to the substantive model selection method. The guidelines in addition incorporate cost-effectiveness of the categories of modelling technique and identify risks in not selecting the costly high-resolution models.
- Other attributes of best practice are also discussed in the paper.
- This paper presented a tentative account of model selection guidelines under development by a consortium of consultants for the Agency.

When the future impacts of this study are envisaged, it appears likely that the Agency may develop cascading models for the inundation, overtopping and nearshore zones nested with the offshore models and meteorological models run by the Met Office. With a better network of coastal gauges, the culture of updating can also be established in the CFF. With the development of open architecture, the various modelling techniques can be transformed into working tools aiding the modellers to create information-rich professional environments. All these can help the reliability of the forecasts and the development of a need-driven CFF.

Acknowledgement
The Best Practice in Coastal Forecasting project is funded as part of the joint Defra/ Environment Agency Research and Development programme. The project involves a consortium of consultants (HR Wallingford, Posford Haskoning and Atkins) and led by HR Wallingford. The solutions presented in this paper are tentative at this stage and the opinions shared by the authors do not necessarily reflect the policies of their respective organisations.

References
Bates, P, De Roo, A.P.J. (2000), "A simple raster-based model for flood inundation simulation", Journal of Hydrology, Vol.236, 54-77.

Bye, P. and Horner M., (1998), *The Bye Report, Easter 1998 Floods*, Final Assessment by the Independent Review Team.

DETR (2000), DETR, Environment Agency, Institute for Environmental Health, *Guidelines for environmental risk assessment and management*, ISBN 0-11-753551-6.

EA (2000), Environment Agency Report No. FD 2302/TR1, HR Wallingford Report SR 587, HR Wallingford, *Risk, performance and uncertainty in flood and coastal defence - A review.*

EA (2001), "Reducing Flood Risk - A Framework for Change".

EA (2002), "The Agency Vision Statement" and the following "Making it Happen".

Flather R., Williams J., Blackman D., & Carlin L. 2001. *Fine grid surge model evaluation*, POL Internal Document No 141.

Haimes Y. Y., Li, d., And stakiv, E. Z., (1989), "Selection of optimal flood warning threshold", Proc. 4th conference on "risk-based decision making in water resources", ed. Haimes and stakhis, ASCE.

ICE, (2001), *Learning to Live with Rivers*, Institution of Civil engineers.

Khatibi, R., Haywood, J., Akhondi-Asl, A., Whitlow, C., Wade, P. and Harrison, T., (2001), "Open architecture in flood forecasting systems", In Falconer, R. A. and Blain, W. R. River Basin Management, WIT Press, Southampton.

Khatibi R.H. (2002), "Formulation of modelling and systems strategy - End users initiatives on improving flood forecasting", a paper presented at the 2nd MITCH workshop in Barcelona, available at: http://www.mitch-ec.net/news-event.htm.

Khatibi, R. and Haywood, J., (2002), "The role of flood forecasting and warning on sustainability of flood defence", Special Edition of Municipal Engineering of the Proc. Inst. Civ. Engrs., Vol. 151, Issue 4, pp 313-320.

Khatibi, R. H., Moore, R. J., Booij, M. J., Cadman, D. & Boyce, G. (2002), "Parsimonious river flow modelling", the Proceedings of the iEMSs conference in Lugano; Ed. Rizzoli, A.E., and Jakeman A.J.; (2002); pp 399-405.

Khatibi, R., Jackson, D., Pender, G., Whitlow, C., Werner M., and Harrison, T., (2003*a*), "A Critical review of calibration", submitted to the 30th IAHR Conference 2003.

Khatibi, R., Haggett, C., Cook, L., Haywood, J., Jackson, D., and Kerman, M., (2003*b*), "Towards best practice in flood warning", to be printed in the *Proc. of River Basin Management, Cardiff*, Proc. of the conference on River basin Management.

Khatibi, R., Jackson, D., Cadman, D., Harrison, T., Price, D., and Haggett, C., (2003*c*), "Defining best practice in Flood forecasting", a paper presented at the first EFFS International Conference in Rotterdam – the paper is submitted for publications in the proceedings of HESS.

Khatibi, R., Jackson, D., Curtin, J., Whitlow, C., Verwey, A., Samuels, P., (2003*d*), "Vision statement on open architecture for hydraulic modelling software tools", **in press** to be published by the J. of Hydroinformatics

McArthur, J., (2001), "Comparison of shelf seas model and surge model water level predictions", Met Office, Ocean Applications Internal Report No. 38.

Pinardi N. & Woods J., (2002) *Ocean Forecasting - Conceptual Basis and Applications* Springer Verlag,

Rowe N.A., O'Connor, T., and Wroclawski A.C. (1995); "Integration and co-ordination of services"; Conference transactions of IMechE on *Emergency Planning and Management*; pp. 67-74; published by Mechanical Engineering Publications Limited for the Institute of Mechanical Engineering, London.

Flood risk assessment for Shoreline Management Planning

R. J. Dawson, Department of Civil Engineering, University of Bristol, UK
Dr. J. W. Hall, Department of Civil Engineering, University of Bristol, UK
P. B. Sayers, HR Wallingford Ltd, Wallingford, UK
Dr. P. D. Bates, Department of Geographical Sciences, University of Bristol, UK

Abstract
Modelling and decision-support tools are becoming available that will make it possible to conduct quantified risk-based appraisal of options for shoreline management. These methodologies form part of a hierarchy of risk-based methods that also include a national-scale methodology for flood risk-assessment and detailed methods for appraisal of individual coast protection and flood defence projects. This paper focuses on assessment of coastal flood risk as part of the Shoreline Management Planning process. A methodology is proposed for analysis of the performance and interaction of systems of defences that protect low-lying coastal areas from flooding. The methodology includes statistical analysis of waves and water levels, simplified reliability analysis of flood defence systems, rapid hydrodynamic modelling of inundation of coastal floodplains and evaluation of the economic and social impacts of flooding. The method is demonstrated in a flood risk analysis at Towyn Bay on the North Wales coast.

Introduction
Considerable progress has been made in recent years in the development of methodologies for risk assessment and risk-based management of the coast. Risk assessment provides a rational basis for the development of coastal flood management policy, allocation of resources and monitoring the performance of coastal management activities of local, regional and national scales in a transparent and auditable manner. Developments in risk assessment have included methods for assessment of coastal flood risks (Meadowcroft *et al.,* 1996, Reeve, 1998), as well as methods for evaluating the risks of coastal erosion and soft cliff retreat (Hall *et al.*, 2000, Lee *et al.*, 2001, Hall *et al.*, 2002, Lee and Clark, 2002). However, the focus of these methods has been the appraisal and design of individual schemes on the coast. Because of the considerable data and computational requirements, until recently it has not been possible to apply these methods on the broad scale required for SMPs.

Because at the time there were no readily accessible methodologies for strategic and regional-scale risk analysis on the coast, the first round of Shoreline Management Plans (SMPs) were not explicitly risk-based and involved rather varied levels of technical analysis (Newcastle University *et al.*, 2000). The recent acquisition of datasets, including FutureCoast (Burgess *et al.*, 2002), Lidar and InterMap digital elevation models, the Environment Agency's National Flood and Coastal Defence Database, and AddressPoint and Focus datasets of domestic and commercial properties mean that some of the data required to conduct coastal risk assessments at the scale of coastal cells is now becoming available. This means that future

Coastal Management 2003, Thomas Telford, London, 2003.

generations of SMPs can be supported by quantified, albeit approximate, analysis of the risks associated with shoreline management options. This is attractive in that it brings appraisal of options at SMP level into line with appraisal for strategy plans and projects (DEFRA, 2000), meaning that there should in the future be fewer discrepancies between the management options that are identified in the SMP and the strategies and projects that are subsequently implemented.

Risk-based appraisal at the scale of SMPs is by no means straightforward. The behaviour under extreme conditions of defences against coastal erosion and flooding is still only partially understood. In addition, at a strategic and planning level *systems* of defence have to be considered in risk analysis rather than just individual cross-sections. A given coastal floodplain may be protected by a number of different defences, whose collective integrity needs to be analysed. Moreover, sediments on beaches and in coastal waters provide a mechanism of interaction between different defence sections. The behaviour of beaches is a key determinant of the reliability of coastal defences because beaches have a critical role in modifying incident waves and because toe scour is a critical failure mechanism. Strategic and planning appraisal of coastal options must therefore in some way take account of these interactions.

An objective of SMPs is to identify management options that are sustainable in the long term in that coastal management should as far as possible work with rather than against natural processes (DEFRA, 2001). To achieve this objective of long-term sustainability it is necessary to understand the long term morphological processes at work on the coast. This was an objective of the FutureCoast project. Yet SMPs need to move even beyond the broad evidence-based assessment of current processes in FutureCoast to quantify the impacts of a range of proposed changes on the coast. Broad-scale modelling of coastal change over decadal time-scales is a significant challenge, but significant progress has been made in this direction in recent years (Walkden and Hall, 2003, Ohl *et al.*, 2003). Assessment of morphological change for SMPs need not be detailed or precise and indeed naïve up-scaling of detailed sediment transport models that work on time-steps of minutes of hours may not yield credible results (de Vriend, 1991a, 1991b). The principles of sediment continuity upon which sediment budget models are based provide a more robust basis for morphological appraisal at SMP level, but these need to be supplemented by process-based understanding of coastal morphology. Even sediment budgets may be problematic to establish where there are significant off-shore or estuarial sources or sinks.

The objective we are seeking to address is the development of robust and accessible risk-based methodologies that will make quantified regional-scale appraisal of SMP options a routine activity in future. The computer-based decision-support tools we are seeking to develop will make quantified analysis of the risks of coastal flooding and erosion and the broad-scale morphological interactions that determine long-term change more accessible to planners and other decision-makers. We do not under-estimate the difficulty of this task and the essential requirement for site-specific expertise. However, recent progress in decision-support tools for Catchment Flood Management Plans (HR Wallingford *et al.*, 2002, Ramsbottom *et al.*, 2002) has illustrated the benefits that can be achieved by partially automating options appraisal and by encoding expert knowledge in decision support tools. The regional-scale coastal modelling tools under development are:

- **risk-based**, in that they consider systems of defences, their probability of failure and the consequences of failure;
- **scenario-based**, in order to challenge assumptions of the status quo and encourage imaginative thinking about sustainable coastal management;

- **model-based**, in that they promote the use of standard coastal dynamics and risk modelling tools to enable quantified analysis of coastal management options.

The focus of this paper is on appraisal of coastal flood risk, using newly developed methods for reliability analysis of sea defences. The methods provide a snap-shot of flood risk at present or in a future scenario of defence geometry and condition. Broad-scale modelling of coastal morphology, which is necessary to establish beach levels in the future, is not addressed in this paper, though, as has been stressed above, quantified (albeit approximate) morphological prediction is an essential aspect of analysis of coastal flood risk in the long term.

Tiered flood risk assessment

The amount of resources put into data acquisition and analysis committed to a risk assessment should reflect the importance of the decision(s) that are being informed. Coastal management take place at a number of levels, ranging from national policy decisions to planning decisions in a river catchment or coastal cell down to individual scheme design and day to day operational decisions. Strategies are identified to adhere to policies and intervention options are selected in order to enact strategies.

A tiered risk assessment methodology, building on the work of Meadowcroft *et al.* (1996), that supports decision-making at multiple levels has been devised. Each tier provides a progressively more thorough and accurate assessment of flood risk appropriate to the decision being taken (Table 1). Insights into the uncertainty associated with higher levels of analysis can be obtained by comparing with more detailed analysis. The *High Level* analysis of flood risk is performed on a national scale and can therefore only use the limited data that is available for the entirety of England and Wales. Whilst topographical and land usage datasets exist at quite a high resolution, information on the standard of protection provided by defences is frequently very uncertain. The high level method is used to provide a national flood risk assessment (HR Wallingford *et al.*, 2003) to support resource allocation within central government.

The *Intermediate Level*, which is the focus of this paper, is more analytical and incorporates additional information on loading, floodplain topography and defence structure. This level is used to assess the flood risk associated with coastal cells and can therefore be used to support decisions for SMPs. The *Detailed Level* will combine the most accurate data available with continuous simulation models of loadings and system response to calculate flood risk. For example, reliability methods developed in the Netherlands have been applied to analysis of the flood defences at Caldicot on the Severn Estuary (Buijs *et al.*, 2003), demonstrating one approach to detailed analysis. A further feature of the detailed analysis is the use of real or synthetic time series data to drive coastal and flooding models. This computationally expensive approach is attractive because it can capture variability in antecedent conditions in a more satisfactory manner than the event-based approach, which is adopted at the high and intermediate levels. Johnson and Hall (2003) have demonstrated this type of approach by application to assessment of the risk of breaching the shingle barrier beach at Pevensey Bay on the South coast of England. Their approach involves simulating very large numbers of synthetic time series of 25 years wave and water level data and is now being implemented on a parallel computer. It is attractive in that it enables the detailed appraisal of alternative strategies for beach management, including responsive strategies that react to observed evolution in beach behaviour.

Table 1 Hierarchy of risk assessment levels, the decisions they inform, the data needed and methodologies used to implement them

Level	Decisions to inform	Data sources	Methodologies
High	National assessment of economic risk, risk to life or environmental risk Prioritisation of expenditure Regional planning Flood warning planning	Defence type Condition grade Standard of Protection Indicative flood plain maps Socio-economic data Land use mapping	Generic probabilities of defence failure based on condition assessment and crest freeboard Assumed dependency between defence sections Empirical methods to determine likely flood extent
Intermediate	*Above plus:* Flood defence strategy planning Regulation of development Maintenance management Planning of flood warning	*Above plus:* Defence crest level and other dimensions where available Joint probability load distributions Flood plain topography Detailed socio-economic data	Probabilities of defence failure from reliability analysis Systems reliability analysis using joint loading conditions Modelling of limited number of inundation scenarios
Detailed	*Above plus:* Scheme appraisal and optimisation	*Above plus:* All parameters required describing defence strength Synthetic time series of loading conditions	Simulation-based reliability analysis of system Simulation modelling of inundation

The three tiers of flood risk assessment share common elements. Each level requires:

- a description of loadings (water levels and wave heights),
- an estimation of defence failure probabilities,
- an inundation model, and,
- an estimation of consequences (people and property in the flooded area).

The methods used to achieve these aims vary according to the required detail of analysis. The high level methodology is now complete (Hall *et al.,* 2003a) and has been applied nationally for England and Wales (HR Wallingford *et al.,* 2003). This paper introduces the intermediate level methodology, which is suitable for assessment of flood risk at the scale of a coastal cell and for appraisal of the broad coastal management options that are considered in SMPs.

Flood risk assessment for Shoreline Management Planning

A flood risk assessment requires estimation of likelihood and consequence of flooding. An economic flood risk assessment provides a measure of Average Annual Damage (AAD). However, purely economic analysis has been criticised for neglecting human distress and the social and environmental impacts of flooding (Bye and Horner, 1998, ICE, 2001). Flood risk assessment is therefore moving towards a more multi-attribute approach (HM Treasury, 2003).

A risk assessment methodology that aims to support an SMP should therefore be able to:

(1) quantify the risk to people and the developed, historic and natural environment,
(2) facilitate identification of the preferred policies for managing these risks in the future,
(3) quantify the consequences of the policies to people and the developed, historic and natural environment,
(4) enable monitoring of these risks to ensure the effectiveness of these policies,
(5) quantify uncertainties associated with the risk assessment, and,

(6) provide a commentary on the sources and sensitivity of these uncertainties.

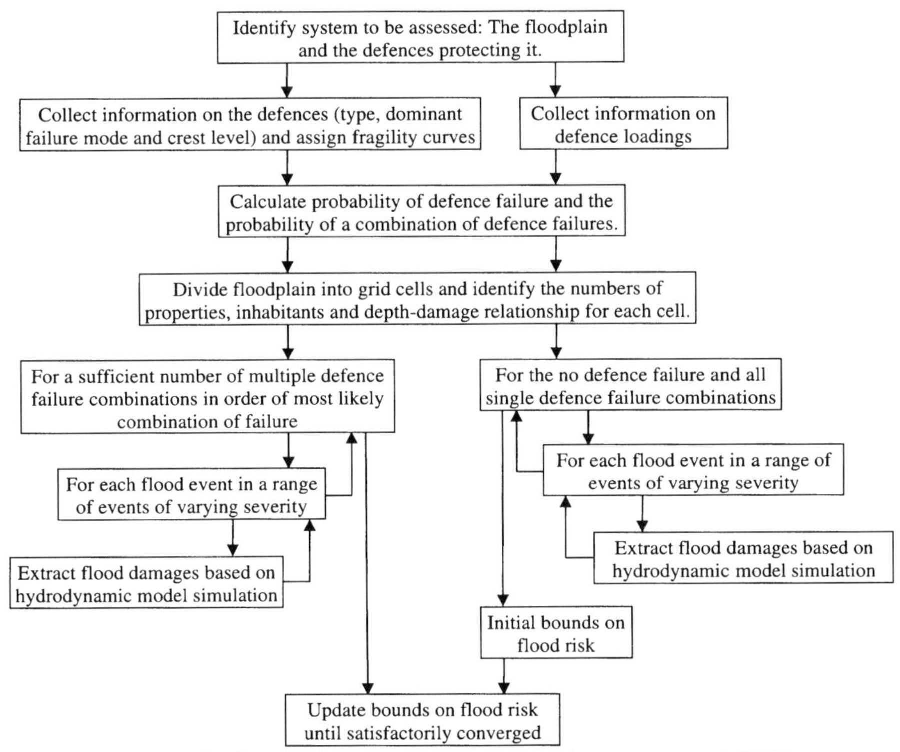

Figure 1 Overview of a flood risk assessment methodology to support SMPs.

The approach to flood risk assessment, described in more detail below, satisfies all these requirements and is based only on data that should be realistically available during preparation of an SMP. Defence strength is described probabilistically using fragility curves (Dawson and Hall, 2002a, 2002b). Wave heights and water levels are estimated using joint probability methods. The extent and depth of flooding resulting from this failure is estimated using an inundation model. Consequences can be calculated based on the extent and depth of flooding from the inundation model. An initial estimate of flood risk, in the form of an upper and lower bound, is made by modelling only a limited number of defence failures, as more combinations of defence failure are considered these bounds converge. The key stages of the methodology are described in more detail below and an overview is provided in Figure 1.

Loadings

Joint-probability distributions of wave height and water level were obtained using standard statistical approaches (HR Wallingford and Lancaster University, 2002). Overtopping volumes were calculated using guidance in the Overtopping Manual (HR Wallingford, 1999).

Defence fragility

Defence behaviour is described using fragility curves (Figure 2). The fragility of a structure is the probability of its failure conditional on its loading. This may be either water level, wave height or a combination of both. A fragility curve is a plot of these conditional failure probabilities over a complete range of possible loadings. A reliability-based method that maximises the use of all available information to generate fragility curves, such as that described by Dawson and Hall (2002a, 2002b) are only appropriate at a detailed level of risk assessment. At the intermediate level, except where better information is already available, a likely dominant failure mechanism is established from a visual inspection of the defence by experts. Fragility curves are assigned based on established criteria and limited analysis of limit state functions.

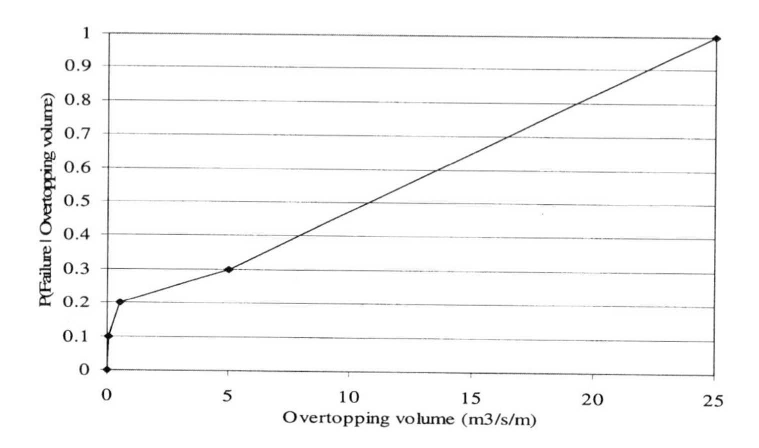

Figure 2 Fragility curves that establish the relationship between overtopping volume and conditional failure probability for an embankment and a revetment (HR Wallingford, 2003).

The failure probability of defence *i*, $P(D_i)$, can be established by integrating the fragility curve over the loading distribution. If the probability of failure is dependent on only one loading parameter, it is given by:

$$P(D_i) = \int_0^\infty p(l)P(D_i \mid l)dl \tag{1}$$

where $p(l)$ is the loading distribution and $P(D_i|l)$ is the fragility.

Hydrodynamic modelling

Hydrodynamic modelling has been performed using the LISFLOOD-FP software (Bates and De Roo, 2000). LISFLOOD-FP was adopted for the purposes of the flood risk assessment methodology because of its ability to realise flood simulations an order of magnitude quicker than finite element based models, yet provides a similar level of performance (Horrit and Bates, 2001). However, it is important to note that the risk assessment methodology is not limited to any specific hydrodynamic model.

In LISFLOOD-FP channel flow is handled using a one-dimensional St. Venants equation described in terms momentum and continuity (Knight and Shiono, 1996). Floodplain flows are similarly described in terms of continuity and momentum equations, discretized over a grid of square cells (Figure 3), which allows the model to represent 2-D dynamic flow fields on the floodplain. We assume that the flow between two cells is simply a function of the free surface height difference between those cells:

$$\frac{dh^{i,j}}{dt} = \frac{Q_x^{i-1,j} - Q_x^{i,j} + Q_y^{i,j-1} - Q_y^{i,j}}{\Delta x \Delta y} \tag{2}$$

$$Q_x^{i,j} = \frac{h_{flow}^{5/3}}{n}\left(\frac{h^{i-1,j} - h^{i,j}}{\Delta x}\right)^{1/2} \Delta y \tag{3}$$

where $h^{i,j}$ is the water free surface height at the node (i,j), Δx and Δy are the cell dimensions, n is the effective grid scale Manning's friction coefficient for the floodplain, and Q_x and Q_y describe the volumetric flow rates between floodplain cells. Q_y is defined analogously to Equation 3.32. The flow depth, h_{flow}, represents the depth through which water can flow between two cells, and is defined as the difference between the highest water free surface in the two cells and the highest bed elevation. Q_y is defined analogously to Equation 5 which is also used to describe flow between the channel and the floodplain. Flow over defences is modelled using a weir equation (British Standards Institute, 1981).

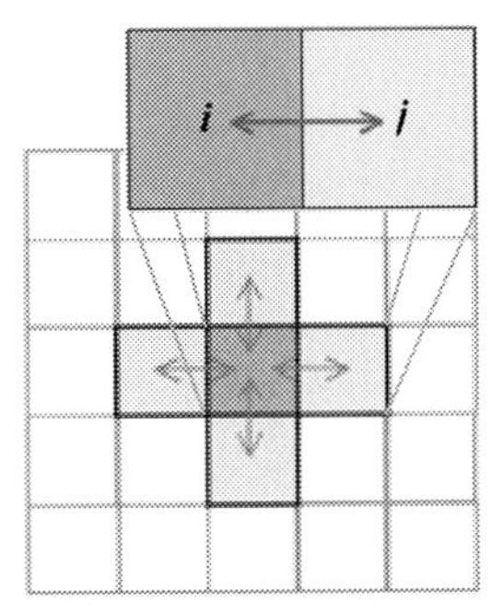

Figure 3 Representation of flow between floodplain cells

LISFLOOD-FP has been adapted to take time-dependent overtopping or breach discharge at a sea defence in order to estimate the extent and depth of coastal floodplain inundation.

Consequences

A national database of the locations of residential and non-residential properties is combined with damage data is taken from the recently published Multi-Coloured Manual (Penning Rowsell *et al*., 2003). This allows depth-damage relationships to be established for each raster cell. The initial implementation of the methodology quantifies only economic risks to property. Additional risks can be considered by using an appropriate measure of consequence.

Systems failure analysis

Defences can be in either a failed or non-failed state. For a defence system made up of n defences, there are 2^n possible system states. Each system state is a possible combination of defence failure or non-failure, henceforth referred to as a failure scenario. These combinations range, in the extremes, from all defences remaining intact to all defences breaching. The probability of a typical scenario, $P(S_i)$, in which defences D_1 to D_r fail and D_{r+1} to D_n do not fail is:

$$P(S_i) = P(D_1) \cap \ldots \cap P(D_r) \cap P\left(\overline{D_{r+1}}\right) \cap \ldots \cap P\left(\overline{D_n}\right) \tag{4}$$

The large number of failure scenarios means for any realistic system it will not be possible to run a hydrodynamic model for all of these scenarios. Therefore, bounds on flood risk have to be established. As more failure scenarios are modelled, these bounds converge. To obtain the most accurate estimate of risk given the available computational resources an algorithm has been established that optimises the convergence of these bounds.

An initial estimate of flood risk is obtained from modelling scenarios in which no defences fail and only individual defences fail (first order scenarios). The risk R_i associated with each scenario, S_i, is:

$$R_i = P(S_i).C_i \tag{5}$$

Summing the risk associated with the scenarios that have been modelled will produce a lower bound on flood risk, R_L, as the contribution from a large number of scenarios has not been considered.

$$R_L = \sum_{i=0}^{r} P(S_i).C_i \tag{6}$$

where r is the number of scenarios that have been analysed (n+1 initially). To estimate an upper bound on flood risk, a worst case flood extent is identified. This requires establishing a worst case flood depth in each raster cell. The maximum flood depth of the scenarios that have been analysed thus far is identified. This is used to estimate the maximum flood damage, C_{max}. The upper bound on flood risk is therefore:

$$R_U = R_L + C_{\max}\left[1 - \sum_{i=0}^{r} P(S_i)\right] \tag{7}$$

As r is increased through analysis of more defence failure scenarios, the lower bound increases whilst the upper bound decreases. The rate of convergence of the bounds is optimised by analysing the most likely combinations of defence failure first (*ie*. the highest probability, $P(S_i)$, of the defence failure combinations that have not yet been simulated). This is repeated until the bounds on flood risk have converged sufficiently.

Case study

The case study site selected is Towyn in North Wales (Figure 4). The town is built on large areas of coastal lowland that were reclaimed during the 18th century. The area modelled corresponds to section 3 of coastal sub-cell 11a. Towyn was inundated in February 1990 when 467m of seawall was breached by a 1 in 500 year event when a 1.3m storm surge coincided with high tide and 4.5m high waves. A lack of natural protection meant that the seawall, which had been targeted for maintenance in the near future, felt the full force of the waves. The nature of the topography resulted in the flood reaching as far as 2km inland with a maximum depth of 1.8m. Although there were no direct fatalities, 5000 people were evacuated from nearly 3000 properties. Immersion of agricultural areas resulted in damage to crops. The total cost of the flood was estimated as being in excess of £50million.

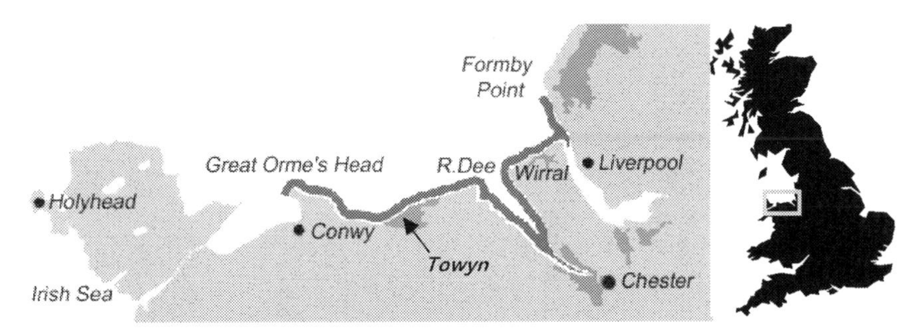

Figure 4 A map showing Towyn, its location within the UK and the extent of coastal sub-cell 11a (low-lying areas such as Towyn are a darker shade)

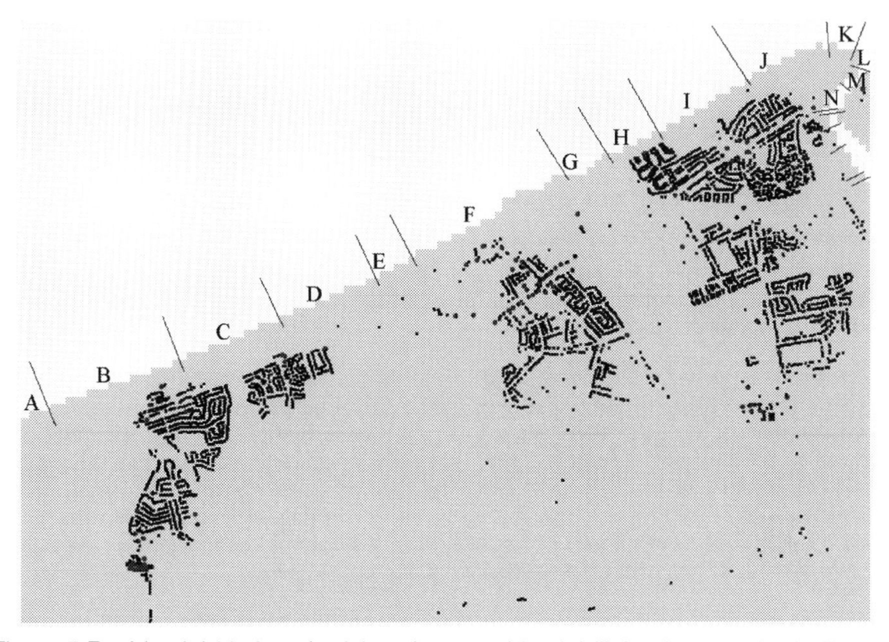

Figure 5 Residential (darker shade) and non-residential (lighter) properties in Towyn and the defence system that protects them.

Towyn is situated on the estuary of the river Clwyd. A previous study (HR Wallingford, 1985) indicated that water levels in the estuary are controlled by astronomical tides alone. A more recent study (HR Wallingford, 2003) reports that these estuarial defences are not prone to overtopping even during extreme events and are ignored for the purposes of this case study. The remaining defence system comprises of 14 coastal defences (marked A to N in Figure 5 - defence F was breached during the 1990 floods). These are currently all protected by a shingle beach that is recharged in places. The defences vary in type from sea walls to dunes with crest heights ranging from 7m to over 9m.

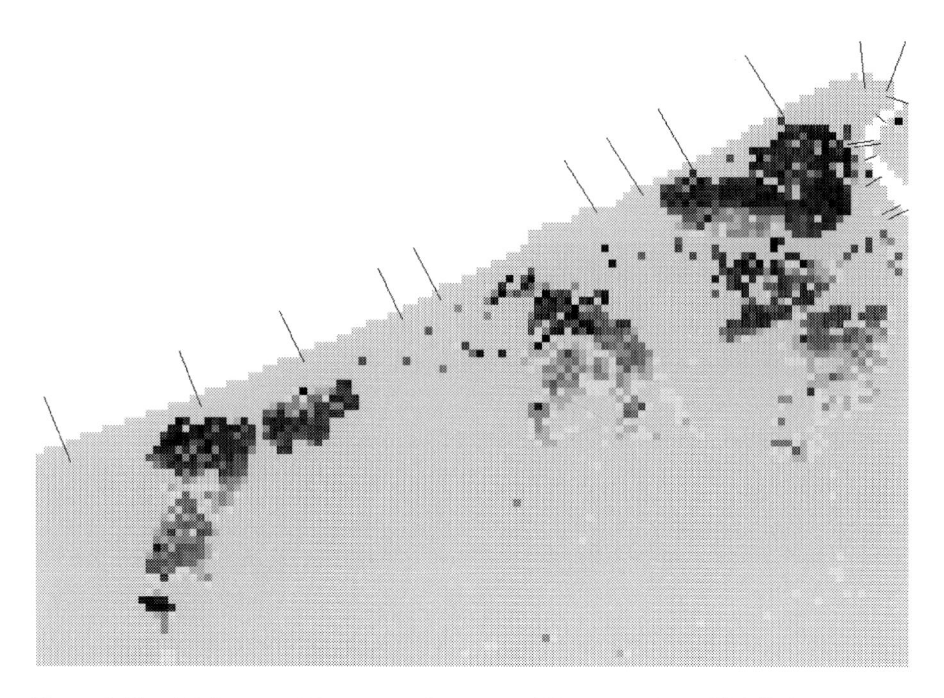

Figure 6 Flood risk map of Towyn plotted on a 50m raster grid. Darker shades indicate a higher economic flood risk.

Recent flooding provided a means of validating the hydrodynamic model for the site. A recent study undertaken by HR Wallingford (2003) provided information on crest levels, wave and water level distributions, overtopping volumes and defence fragility. The flood risk assessment methodology as described above has been implemented using the case study data. The hydrodynamic model LISFLOOD-FP (Bates and De Roo, 2000) was used to implement the methodology. For the purposes of the study, the floodplain was discretised into 50×50m raster cells. The economic flood risk for each raster grid cell was calculated and is shown in Figure 6. Only the economic risk to property is shown.

Assuming that property is written off at 3m depth of flooding, then the maximum flood damage calculated for Towyn is £450million (approximately 5000 properties). After analysis of the non-failure and single defence failure scenarios, the risk was calculated as being between £20.3million and £105.3million. This represents analysis of 15 scenarios and just over 80% of probability of all the scenarios. The total flood risk, calculated as an average annual damage, lies between £25.8million and £66.8million. This represents the bounds on flood risk after running 21 failure scenarios. This accounted for just over 90% of the total probability contributions. The additional analysis of just six failure scenarios substantially reduced the uncertainty associated with the estimate of economic flood risk. To reduce the uncertainty between the upper and lower bound to below £5million 58 more failure scenarios would need to be analysed. Analysis of a further 124 scenarios would reduce the uncertainty to £1million. Clearly the decision-maker needs to balance the benefits of closer bounds with the time and resources required to perform this analysis.

Other queries of the performance of the system can be extracted from this analysis. The contribution made by individual defences to the flood risk can be measured. This does not necessarily correspond to the defence most prone to failure, but also reflects the value of the system that the defence protects.

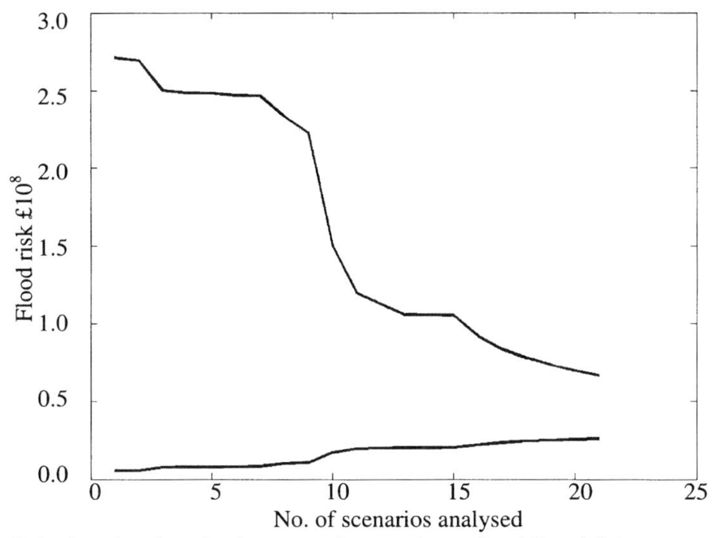

Figure 7 A plot showing the lower and upper bounds of flood risk converging as more flood defence failure scenarios are analysed. Optimisation starts after scenario 15 when the non-failure and single defence failure scenarios have been analysed.

Discussion

The results from the analysis clearly compliment the needs of the shoreline management planner outlined previously. This initial case study has been on a relatively small scale and considering only a limited number of factors in order to demonstrate the applicability of the methodology. This section discusses the results, the usefulness of the methodology for strategic planners and future developments that will provide increasing amounts of risk-based analytical support for SMPs.

The flood risk assessment can be used to support a wide variety of decisions a number of which are listed in Table 1. The effect on flood risk of a number of natural or human changes to the flood system are shown in Figure 8. Alteration of model parameters allows the change in flood risk to be quantified. For example, maintaining or replacing defences will result in an increase in defence fragility and consequently a decrease in flood risk. The methodology described in this paper is sufficiently straightforward to enable risk-based analysis of a range of coastal management options.

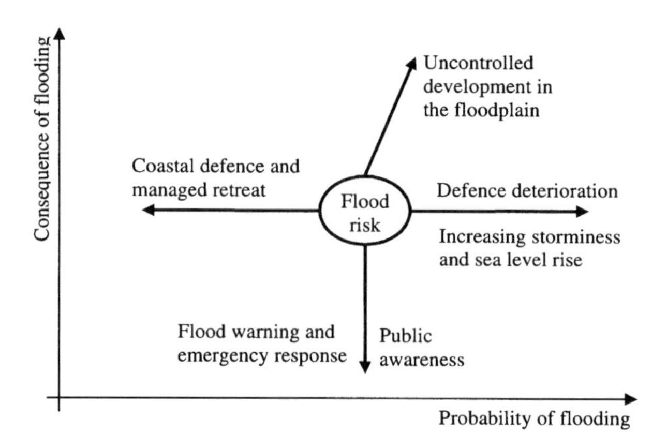

Figure 8 The effect of human and natural intervention on flood risk (Hall *et al.,* 2003b)

Whilst in the case study only economic flood risk has been quantified, the same methodology can be used to estimate social and health risks from flooding. This can be achieved by multiplying an appropriate indicator by the probability of inundation. Population at risk can be estimated from census data. The Social Flood Vulnerability Index (SFVI) Tapsell *et al.* (2001) can be used to identify communities that are vulnerable to flooding. Social vulnerability is ranked from "very low" to "very high" and is a function of the number of lone parents, elderly population, long term sick, non-homeowners, unemployed, non-car owners and population density. The risk of social impact is obtained as a product of the probability of flooding to a given depth and the SFVI, providing a comparative measure for use in policy analysis. Similarly, using appropriate indicators, environmental risks can be quantified.

Displaying the results in a GIS format provide a visual overview of the distribution of risk. Targeting resources at areas of higher risk provides a sound basis for identifying preferred policies. Conversely, it may be possible to re-allocate resources from areas of low risk.

Uncertainty in flood risk assessment can be considerable. The methodology described previously quantifies uncertainty in an easily accessible manner by providing interval bounds on all uncertain quantities. The decision-maker can identify the major sources of uncertainty and gauge the sensitivity of the risk assessment to the uncertain parameters. Quantification of uncertainty also allows the justification of data collection strategies. The cost of collecting new or more accurate data can be weighed against the reduction in uncertainty gained.

It is clear that this methodology can provide a useful snapshot of flood risk and uncertainty. However, SMPs need to consider the change in flood risk over a number of years. Prediction of future flood risk introduces additional factors that need to be accounted for. Some factors, such as maintenance or replacement of defences are within human control and can be modelled by altering the initial value of input variables. The influence of natural phenomena such as climate change can be modelled by altering the probability distribution of sea level. Climate change predictions range greatly and the increase in uncertainty over a long period of time may be considerable. However, it is important to consider this uncertainty as it is a reflection of our understanding of the system and its future behaviour.

As discussed in the introduction to this paper, broad-scale morphological modelling should complement the coastal flood risk analysis described in this paper. Coastal erosion threatens over 100,000 properties with a capital value at nearly £8billion (Halcrow *et al.*, 2001). Erosion of cliffs and beaches often supplies critical sediments down drift, however, this process can be interrupted and drastically altered by the construction of coastal defences or erosion protection schemes. Loss of beach material can significantly increase the loading on the defences, thereby increasing the risk of flooding. Clearly, flood risk and morphological changes are interlinked and this interdependency needs to be considered over the duration of the SMP. Future integration of flood risk assessment and erosion assessment models will provide an important advance for supporting long term strategic planning.

Flood risk is not just a function of long term morphological and infrastructure interactions, but also includes socio-economic interactions which can have a significant effect. These changes are influenced by political and societal behaviour. Previously these have been considered by modelling a number of 'storylines' that predict changes under a set of pre-defined rules (Holman *et al.*, 2002). For example, one storyline might involve improving public education and reducing floodplain development, whereas a contrasting storyline will have uncontrolled development. The relative likelihood of these storylines is controlled by society, however, using evidence from risk-based models the benefits of improving management of floodplains can be quantified and demonstrated to policy makers.

Conclusions

There is considerable scope for improvement of the tools used to support flood risk assessment in SMPs. A risk assessment methodology appropriate for supporting Shoreline Management Plans has been described. It includes statistical analysis of waves and water levels, reliability analysis of flood defence systems, hydrodynamic modelling of inundation of coastal floodplains and evaluation of the economic and social impacts of flooding. The reliability methodology involves analysis of an increasing number of defence failure scenarios. As more scenarios of defence failure are analysed, the bounds on the flood risk estimate converge. A case study in the town of Towyn has demonstrated the applicability of the methodology. This approach to risk assessment provides a powerful tool for flood risk managers and satisfies the needs of the SMP as a practical method for evaluating the impacts of alternative coastal management options in quantified, albeit approximate, terms.

A significant advancement in effective management of coastal risks is achievable by taking a more explicitly systems based approach. This system includes the construction, operation and maintenance of flood defences; flood forecasting and warning; management of drainage systems and development control (Hall *et al.*, 2003b). This is a complex task, but the systems-based approach to assessing the risk of flooding from defence failure described above provides a useful platform on which to build. This paper has provided an insight into one aspect of an integrated risk management framework. Future stages of development have been identified in order to provide a more holistic and analytical approach to risk assessment for shoreline management planning.

Acknowledgements

The research described in this paper formed part of a project entitled "RASP: Risk assessment for flood and coastal defence systems for strategic planning", funded by the Environment Agency within the joint DEFRA/EA Flood and Coastal Defence R&D programme. Dr. Hall's research is supported by a Royal Academy of Engineering post-doctoral research fellowship. The authors would like to thank Dr. Mohamed Hassan at HR Wallingford for providing the date used for the Towyn case study.

References

BATES, P. D. and DE ROO, A.P.J. (2000), A simple raster-based model for flood inundation simulation, *Journal of Hydrology*, 236: 54-77.

BRITISH STANDARDS INSTITUTION (1981), BS3680: Methods of measurement of liquid flow in open channels, weirs and flumes, BSI, London.

BUIJS, F.A., VAN GELDER, P.H.A.J.M., VRIJLING, J.K., VROUWENVELDER, A.C.W.M., HALL, J.W. and SAYERS, P.B. (2003), Application of Dutch reliability methods for flood defences in the UK, in *ESREL '03: Proceedings of the European Safety and Reliability Conference*, Maastricht, Netherlands, June 15 - 18, 2003, in press.

BURGESS, K., BALSON, P., DYER, K. ORFORD, J. and TOWNEND, I. (2002), FutureCoast – The integration of knowledge to assess future coastal evolution at a national scale, in *Coastal Engineering: Proc. 28th Int. Conf.*, Cardiff.

BYE, P. and HORNER, M. (1998), *Report by the independent review team to the board of the Environment Agency*, Volumes 1 and 2, Environment Agency, Bristol.

DAWSON, R. J. and HALL, J. W. (2002a), Improved condition characterisation of coastal defences, *Proceedings of ICE Conference on Coastlines, Structures and Breakwaters*, pp 123-134, Thomas Telford, London.

DAWSON, R. J. and HALL, J. W. (2002b), Probabilistic condition characterisation of coastal structures using imprecise information, in *Coastal Engineering: Proc. 28th Int. Conf.*, Cardiff UK, July 8-12, 2002, in press.

DEFRA (formerly MAFF) (2000), *Flood and Coastal Defence Project Appraisal Guidance Volume 3: Economic Appraisal*, DEFRA, London.

DEFRA (formerly MAFF) (2001), *Flood and Coastal Defence Project Appraisal Guidance Volume 2: Strategic Planning and Appraisal*, DEFRA, London.

HALCROW, HR WALLINGFORD, JOHN CHATTERTON ASSOCIATES (2001), *National Appraisal of Assets at Risk from flooding and coastal erosion including the effects of climate change*, DEFRA, London.

HALL, J.W., LEE, E.M. and MEADOWCROFT, I.C. (2000), Risk-based benefit assessment of coastal cliff recession. *Water and Maritime Eng.* 142:127-139.

HALL, J.W., MEADOWCROFT, I.C., LEE, E.M. and VAN GELDER, P.H.A.J.M. (2002), Stochastic simulation of episodic soft coastal cliff recession. *Coastal Engineering*, 46(3):159-174.

HALL, J. W., DAWSON, R. J., SAYERS, P., ROSU, C., CHATTERTON, J. and DEAKIN, R. (2003a), A methodology for national-scale flood risk assessment, *J. Water and Maritime Engineering*, accepted.

HALL, J. W., MEADOWCROFT, I. C., SAYERS, P. B. and BRAMLEY, M. E. (2003b), Integrated flood risk management in England and Wales, *Natural Hazards Review*, in press.

HM TREASURY (2003), *The Green Book: Appraisal and evaluation in central government*, The Stationery Office, London.

HOLMAN, I.P., LOVELAND,P.J., NICHOLLS, R.J., SHACKLEY, S., BERRY, P.M., ROUNSEVELL, M.D.A, AUDSLEY, E., HARRISON, P.A. and WOOD, R. (2002), *REGIS: Regional Climate Change Impact Response Studies in East Anglia and North West England*, DEFRA, London.

HORRITT, M. S. and BATES, P. D. (2001), Predicting floodplain inundation: raster-based modelling versus the finite-element approach, *Hydrological Processes*, 15: 825-842.

HR WALLINGFORD (1985), *Conwy estuary crossing field data collected by Hydraulics Research*, Report EX 1251.

HR WALLINGFORD (1999), *Overtopping of Seawalls: Design and Assessment Manual*, R&D Technical Report W178.

HR WALLINGFORD (2003), *Conwy Tidal Flood Risk Assessment, Stage 1 – Interim Report*, Report EX 4667, HR Wallingford.
HR WALLINGFORD and LANCASTER UNIVERSITY (2000), *The Joint Probability of Waves and Water Levels: JOIN-SEA, A rigorous but practical approach*, Report SR 537.
HR WALLINGFORD, FHRC, CEH WALLINGFORD, EALES, R. and CRANFIELD UNIVERSITY (2002), Catchment Flood Management Plans: Development of a Modelling and Decision-Support Framework, Report EX4495, HR Wallingford. http://www.mdsf.co.uk/.
HR WALLINGFORD, UNIVERSITY OF BRISTOL, HALCROW and JOHN CHATTERTON ASSOCIATES (2003), *National Flood Risk Assessment 2002*, Report EX 4722, HR Wallingford.
ICE (2001), *Final report of the ICE's presidential commission to review the technical aspects of flood risk management in England and Wales: Learning to live with rivers*, Institution of Civil Engineers, London.
KNIGHT, D. W. and SHIONO, K. (1996), *River channel hydraulics and floodplain hydraulics*, in Floodplain Processes, M. Anderson, D. Walling, P. Bates (eds), Wiley, London.
LEE, E.M., HALL, J.W. and MEADOWCROFT, I.C. (2001), Coastal cliff recession: The use of probabilistic prediction methods. *Geomorphology*, 40:253-269.
LEE, E.M. and CLARK, A.R., *Investigation and Management of Soft Rock Cliffs*. Thomas Telford, London, 2002.
MEADOWCROFT, I.C., REEVE, D.E., ALLSOP, N.W.H., DIMENT, R.P. and CROSS, J. (1996), Development of new risk assessment procedures for coastal structures, *Proc. ICE Advances in Coastal Structures and Breakwaters '95*, Thomas Telford, London.
NEWCASTLE UNIVERSITY, PORTSMOUTH UNIVERSITY, SHORELINE MANAGEMENT PARTNERSHIP and ISLE OF WIGHT COUNCIL (2000), *A review of shoreline management plans 1996-1999: Final Report*, DEFRA, London.
JOHNSON, E.A. and HALL, J.W. (2002), Simulation-based optimisation of a beach nourishment concession, in *Coastal Engineering: Proc. 28th Int. Conf.*, Cardiff UK, July 8-12, 2002, in press.
OHL, C. O. G., FREW, P. D., SAYERS, P. B., LAWTON, G., FARROW, P. A. J., WALKDEN, M. J. A. and HALL, J. W. (2003), North Norfolk - a regional approach to coastal erosion management and sustainability in practice, *in this volume*.
PENNING-ROWSELL, E.C., JOHNSON, C., TUNSTALL, S.M., TAPSELL, S.M., MORRIS, J., CHATTERTON, J.B., COKER, A. and GREEN, C. (2003), *The Benefits of Flood and Coastal Defence: Techniques and Data for 2003*. Middlesex University Flood Hazard Research Centre.
RAMSBOTTOM, D., VON LANY, P. and PENNING-ROWSELL, E. C. (2002), The Modelling and Decision Support Framework for Catchment Flood Management Plans, in *Proc. 37th DEFRA Conf. River and Coastal Engineering*, Keele University.
REEVE, D. E. (1998), Coastal flood risk assessment, *J. Waterway, Port, Coastal and Ocean Engineering*, 124(5):219-228.
TAPSELL, S. M., PENNING-ROWSELL, E. C., TUNSTALL, S. M. and WILSON, T. L. (2002), Vulnerability to flooding: health and social dimensions, *Phil. Trans., R. Soc. Lond. A360*: 1511-1525.
VRIEND, H.J. DE (1991a), Mathematical modelling of large-scale coastal behaviour, part 1: physical processes. *Journal of Hydraulic Research*, 29(6): 727-740.
VRIEND, H.J. DE (1991b), Mathematical modelling of large-scale coastal behaviour, part 2: predictive models. *Journal of Hydraulic Research*, 29(6): 741-753.
WALKDEN, M.J.A and HALL, J.W. (2002), A model of soft cliff and platform erosion, in *Coastal Engineering 2002, Proceedings of the 28th International Conference,* Cardiff UK, July 8-12, 2002, in press.

Evaluation of coastal process impacts arising from nearshore aggregate dredging for beach recharge – Shingles Banks, Christchurch Bay

Bradbury AP, Channel Coastal Observatory, New Forest District Council, UK
Colenutt AJ, Channel Coastal Observatory, New Forest District Council, UK
Cross J, Channel Coastal Observatory, Southampton University, UK
Eastick C Channel Coastal Observatory, New Forest District Council, UK
Hume D New Forest District Council, UK

Abstract

Although monitoring is a requirement of all aggregate production licences (within the UK), this generally relates to examination of physical or biological impacts within, or immediately adjacent to, the dredging area. Relatively few licences require detailed monitoring of the coastal zone, as aggregate dredging-areas are usually in relatively deep water (>20m) and some distance from the shoreline.

Post-dredging responses of the Shingles Banks dredging area; shoreline; and nearshore bathymetry of Christchurch Bay and Western Isle of Wight, are examined, in relation to a nearshore shallow water aggregate dredging programme for beach recharge of Hurst Spit. Data from an intensive 6-year programme of field monitoring and analysis has been compared with pre-dredging coastal process impact assessments, based largely on numerical modelling. Surveys have been integrated within best practice local and regional schemes of shoreline management and monitoring, to identify patterns of erosion and accretion. Changes have also been monitored at control sites, outside of the potential influence of the dredging area.

Conditions in the shelter of the offshore banks have been shown to be less severe than numerical modelling methods have previously suggested. Conversely, the offshore wave climate, to seawards of the dredging area, has been more severe than long-term statistical modelling methods have predicted; this suggests increased storminess over the analysis period. Shoreline responses demonstrate that beaches on the shoreline of the Isle of Wight have not been affected by dredging activity. Beaches have remained stable, despite increased storm activity. Offshore bathymetric surveys have demonstrated considerable movements and continued evolution of a dynamic offshore bank system, in response to wave and current action; this has remained in context relative to historical patterns of movement

Some of the risks, uncertainties and conflicts, associated with nearshore aggregate dredging for beach recharge purposes, are highlighted. The benefits of intensive aggregate-licence monitoring requirements, and co-operation between local authorities are demonstrated, providing a robust analytical approach to coastal process impact assessment of nearshore dredging.

Coastal Management 2003, Thomas Telford, London, 2003.

Introduction

Aggregate dredging within the coastal zone may create a range of conflicts, uncertainties and opportunities. Demand for high quality aggregates from offshore sources is high, especially for beach recharge. There is considerable concern however, particularly from Local Authorities, that the dredging will have adverse impacts on the coastal zone.

Coastal process impact studies are required to inform aggregate production licence applications within the U.K. Numerical models are usually used as pre-dredging impact-evaluation tools, yet the results and robustness of these studies are often questioned by the coastal community at the time of application (Simons and Hollingham, 2001). When significant changes occur at the shoreline following dredging activity, it is difficult to prove or disprove that the impacts have arisen as a direct consequence of the dredging. Intensive pre- and post- dredging monitoring of: wave climate; dredging area responses; nearshore bathymetry and shoreline changes can provide the necessary tools to assess the validity of coastal process impact studies. In many instances post-dredging monitoring programmes are of limited value however, since lengthy pre-dredging control data sets are infrequently available for comparison.

This paper examines results from a monitoring programme, in compliance with licence conditions for the Shingles Banks dredging area (Figure 1). The results are examined in context with predictions derived from pre-dredging coastal process impact studies. Trends are examined by reference to control data sets, derived in conjunction with Local Authorities' best practice coastal zone management programmes. The significance of high quality baseline data, validity of modelling methods and appropriateness of measurement techniques are examined.

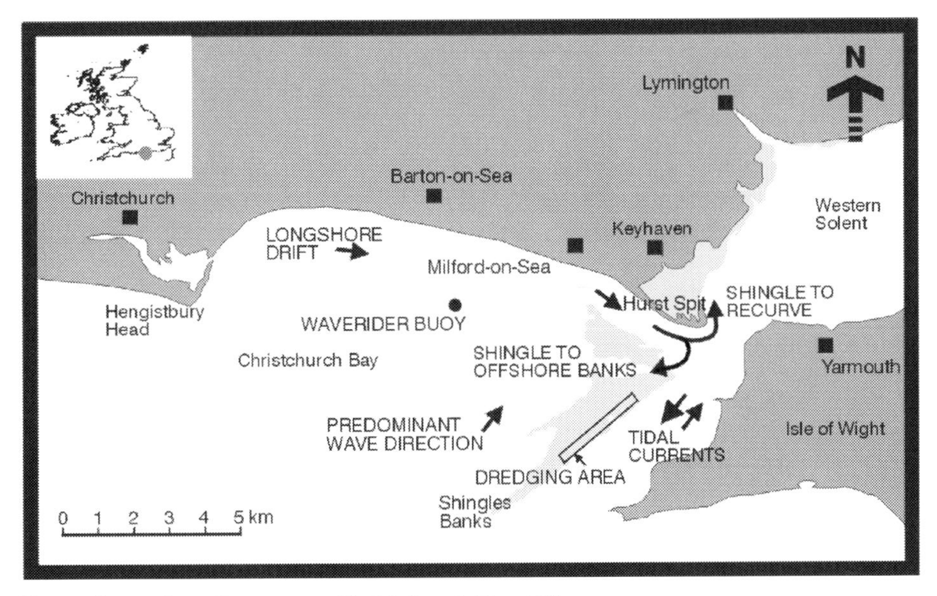

Figure 1 Location map – Christchurch Bay UK

Pre-dredging coastal process impact assessments

The conventional UK approach to pre-dredging coastal impact assessment was adopted for these investigations (Simons and Hollingham, 2001). Procedures included (a) assessments of potential changes in nearshore wave conditions, arising from the bathymetric changes caused by dredging, (b) the consequent potential impacts on the shoreline and (c) assessments of the impacts of dredging on sediment transport paths and seabed mobility.

Dynamics of the study area – historical evolution

The Shingles Banks system comprises mobile sand and shingle and is the major offshore sediment sink in Christchurch Bay. The total volume of the Shingles Banks system was estimated at approximately 42 million m^3, during prospecting investigations (Velegrakis, and Collins, 1991, 1992). A small surface emergent area of mobile sediment is exposed regularly; this represents a volume of approximately 15,000m^3. The local perception is that this surface emergent component is the most significant element of the Shingles Banks; in reality this area represents a tiny fraction of the total mobile bank volume (0.04%). A pre-dredging assessment of bathymetric changes in Christchurch Bay, based on historical chart analysis (1882-1988), identified that the Shingles Banks system was highly dynamic (Bradbury, 1992). Net growth of more than 3 million m^3 was measured for the analysis period, although large-scale spatial and temporal variations in patterns of erosion and accretion occurred, and the accuracy of surveys is uncertain. Field studies of seabed mobility, including sidescan sonar, have concluded that the system is highly mobile and that sediment is derived from the beaches of Christchurch Bay (Velegrakis, 1994). Trends were extrapolated forward, assuming that natural evolution would result in continued growth. The proposals for dredging from the Shingles Banks, for use as a beach recharge source, was essentially a recycling solution.

Wave modelling

The basis for prediction of the impacts of pre- and post- dredging bathymetry scenarios on nearshore wave climate is numerical wave transformation modelling. Transformation of offshore wave conditions to a series of nearshore locations (Figure 2) identifies comparative changes in wave direction, period and nearshore wave height, arising from hypothetical modifications to the bathymetry. Comparisons of a range of dredging scenarios allowed an examination of the impacts for the removal of up-to 500,000m^3 of material (Wimpey, 1993). A factor of safety is included within these assessments, since only 300,000m^3 was to be dredged. A long-term synthetic offshore wave climate (Hydraulics Research, 1989a,b), extrapolated to determine a range of extreme conditions, has provided input conditions for wave transformation modelling (Bradbury, 1998). Extreme wave conditions are used, since these identify the greatest changes to nearshore wave conditions arising from modified bathymetry. If transformations arising from extreme conditions show no changes for alternative scenarios, then less severe events are highly unlikely to impact on near-shore wave climate.

Criteria for an acceptable impact at the coastline is such that any change in nearshore wave climate, predicted as a consequence of dredging, is considered of significance; in such an instance the licence will be refused. A nil effect is assessed by reference to the limitations of the modelling methodology, and scatter of data within the limits of model resolution (Simons and Hollingham, 2001). A threshold change of less than 3% is typically stated to identify limits of the accuracy and scatter of the modelling methodology, although this may be greater in situations where non-linear processes, such as bed friction and wave breaking, are significant. The complexity of the bathymetry across the Shingles Banks limits the precision of numerical modelling at this site. Modelled wave transformations, determined for pre- and post- dredging bathymetry scenarios, suggested that the dredging programme would have no significant impact on nearshore wave conditions. Many consultees view the modelling process with some degree of scepticism; this is perhaps not surprising since the wave transformation processes and modelling techniques are complex. The outputs of the numerical models and the real coastal impacts of post dredging are rarely tested, however. Transparent validation of the modelling techniques would provide greater confidence in the coastal process impact assessment procedures.

Pre-dredging shoreline evolution

Historical records of coastal evolution provide the basis for long-term predictions of coastal change, coupled with regional wave climate studies and modelling. Local practice within Christchurch Bay provided a solid monitoring baseline, dating back to 1987, against which future changes could be assessed. Analysis of Christchurch Bay beach-profile data identified declining beach volumes prior to dredging, hence the need for recharge at Hurst Spit. No such control data sets were available for the Western Isle of Wight, however.

Dredging licence monitoring programme overview

Intensive post-dredging monitoring was undertaken along the shorelines of both Christchurch Bay and the Western Isle of Wight, between 1996-2003; this programme followed a pre-dredging baseline survey, which took place in 1996. Dredging was conducted over a six-month period commencing in August 1996 (Bradbury and Kidd, 1998). Quarterly beach surveys and annual nearshore bathymetric surveys have been analysed together with detailed bathymetric surveys of the dredging area (quarterly) and Shingles Banks system (biannual). Measurements of morphodynamic responses have been supplemented with continuous tide and wave recording, further numerical modelling and biological sampling within the dredging area.

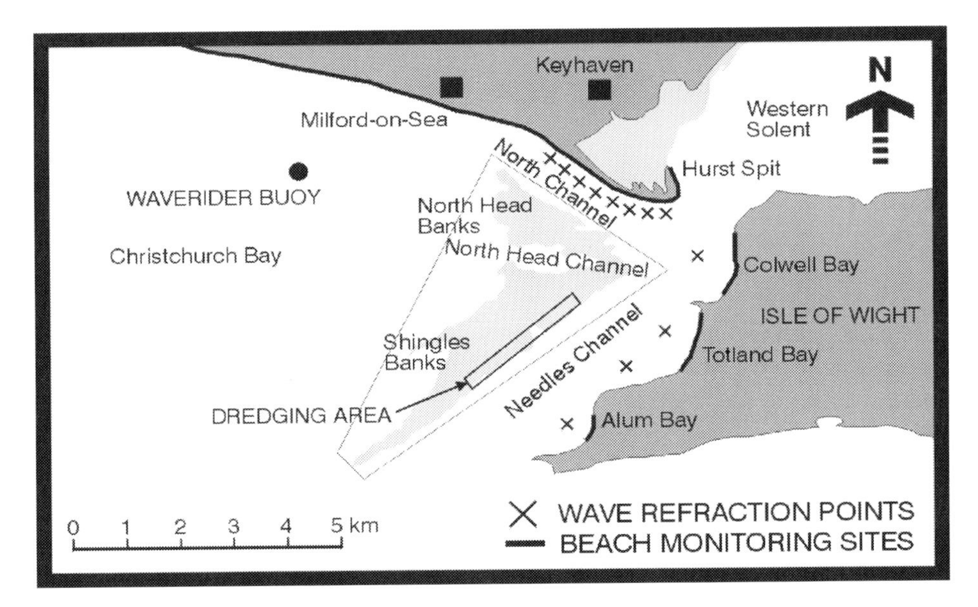

Figure 2 Layout of monitoring and modelling programme

Wave climate

A long-term waverider buoy deployment was sited in approximately 10-12m water depth, west of the dredging area (Figure 2); this is outside of the possible influence of the Shingles Banks dredging area, as the predominant wave direction is from the southwest. As such, the waverider is located at a local control site. Wave measurement has not previously formed part of a post-dredging coastal-impact monitoring programme. Data from the wave buoy site has been compared with synthetic offshore wave data that has subsequently been transformed to the wave buoy site, by numerical modelling. A comparison of a one-month sample of measured and synthetic data (Figure 3) indicates a strong correlation between measured and modelled conditions (Figure 4); this provides confidence in (a) the offshore synthetic wave data (b) the numerical wave transformation process and (c) the wave buoy measurements.

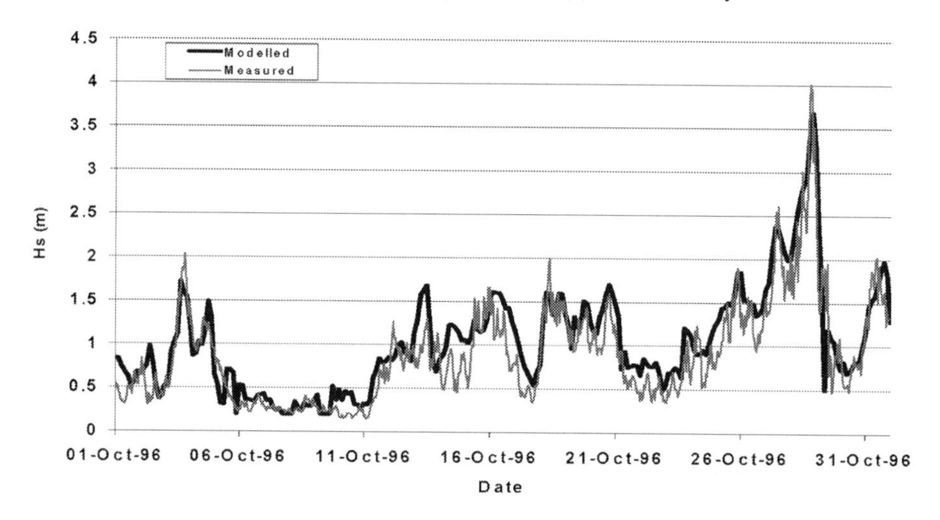

Figure 3 Comparison between modelled and measured wave heights at Milford-on-sea Waverider site

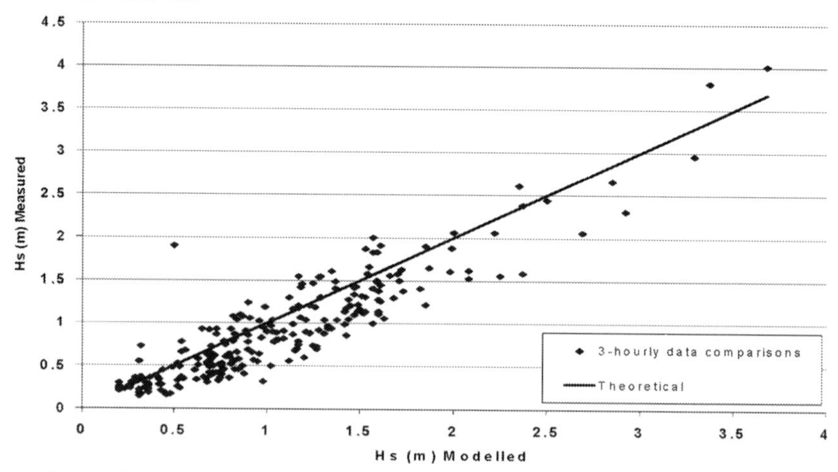

Figure 4 Correlation between modelled and measured wave heights at Milford-on-sea Waverider site

Pre-dredging wave climate studies were completed in 1989 on the basis of 16 years synthetic wave data (Hydraulics Research, 1989b). Extrapolation of a 3-parameter Weibull distribution provides extreme conditions; these were transformed to the waverider buoy site. Subsequently, the 1:100 year return period event (3.32m) calculated for this site (Hydraulics Research, 1989a), has been exceeded on 16 occasions (Figure 5); this demonstrates that either (a) the time series used for determination of extremes was too short, (b) the frequency of extreme events has increased locally during the past 13 years or (c) both. The increased severity of nearshore wave conditions implies that a higher degree of erosion might be expected under these circumstances, than during the pre-dredging control baseline period.

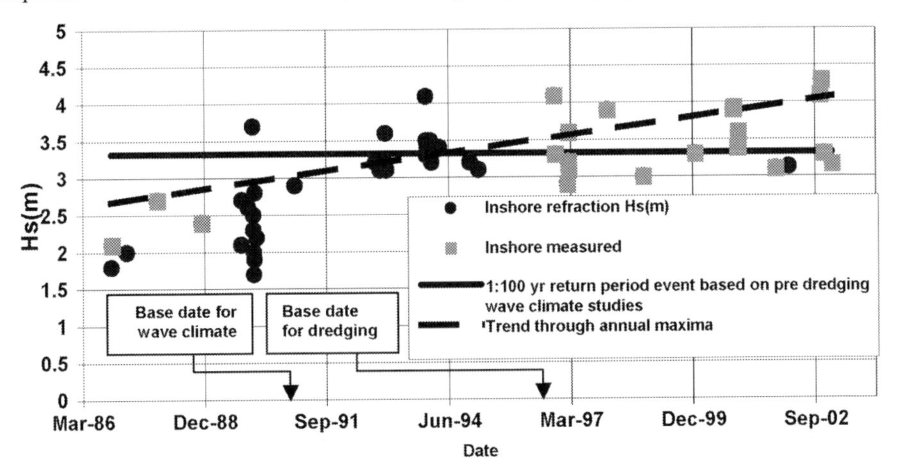

Figure 5 Temporal distribution of nearshore storm events resulting from storms with an offshore significant wave height >5m 1986-2003

A short-term pre-dredging deployment of the buoy, in the shelter of the Shingles Banks, adjacent to Hurst Spit demonstrated that the same numerical model provides a significant overestimate of nearshore wave heights at this site (Figure 6).

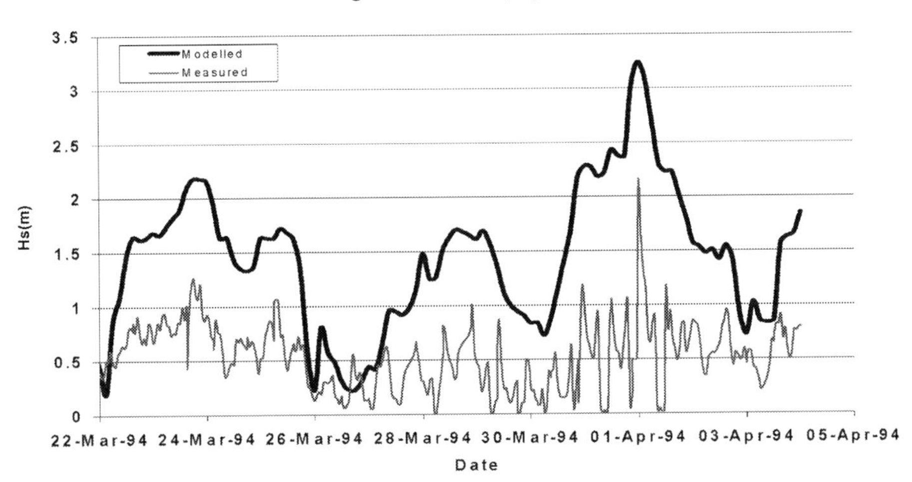

Figure 6 Comparison between modelled and measured wave heights at North Channel Waverider site

The poor correlation at the North Channel location is explained by the presence of non-linear wave transformation processes that are not well represented within the model. The complexity of the bathymetry and shallow water conditions across the Shingles Banks are such that bed friction and wave breaking are significant controls on wave height. A clear tidal signature is evident within the wave time- series, at the buoy site. Despite the limitations of the modelling methodology for this complex location, the modelling process provides conservative results.

Shoreline response

The mixed shingle-sand beaches of Hurst Spit and Milford-on-Sea (Figure 1) have been monitored since 1987, as part of a long-term beach management survey programme, thereby providing temporal controls on pre-dredging trends. Control sites at Milford–on-Sea are on broad unmanaged beaches that are outside of the influence of both the Shingles Banks and coastal structures. The control sites generally demonstrated a declining beach volume prior to dredging. Hurst Spit is a mixed shingle-sand barrier beach, whilst the fringing beaches of the Western Isle of Wight are backed by cliffs (Colwell and Alum Bays) and by a nearly vertical seawall with a recurve (Totland Bay). The shorelines of both Hurst Spit and the Isle of Wight are protected by the offshore Shingles Banks system. Regrettably no high quality pre-dredging beach performance data was available for the Isle of Wight shoreline. Beach cross-section changes were monitored relative to an initial master profile baseline survey, conducted immediately prior to the dredging programme (1996). The shoreline response data is compared with (a) Pre- and post-dredging beach trends; (b) Changes in the nearshore wave climate arising from dredging activity and (c) Changes in the patterns of nearshore wave climate adjacent to, but outside of the influence of, the dredging area.

Milford-on-Sea control beach site

Beach profile trends indicate that the control beach at Milford-on-Sea has continued to erode following dredging (Figure 7). There is no discernable increase in the rate of loss of beach material, despite the apparent increase in severity of wave conditions since 1990. Major departures from the trend line are indicative of storm events.

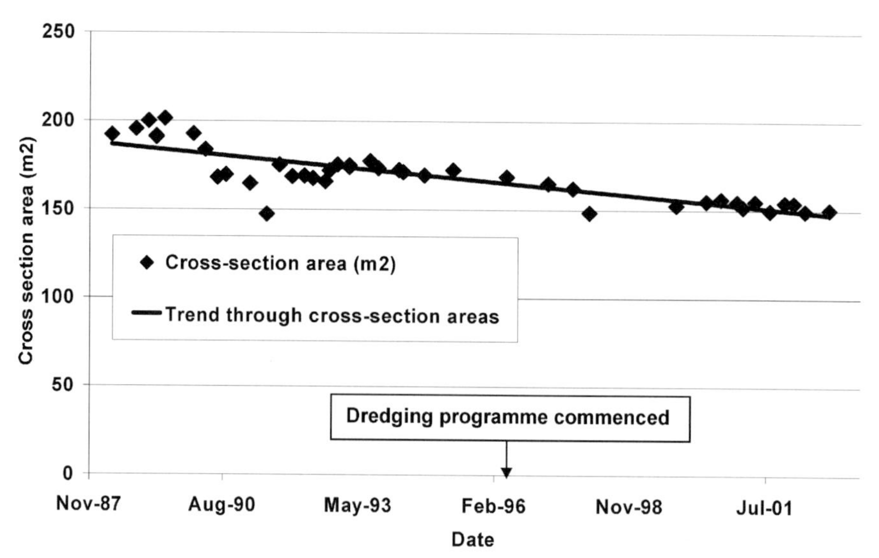

Figure 7 Beach profile trend analysis at Milford-on-Sea control site

Hurst Spit recharge site

Hurst Spit has been subject to major beach recharge and recycling, as part of a 50-year beach management plan, since 1996. Estimates of annual losses from the system have been made on the basis of (a) Pre-recharge numerical modelling (b) Physical modelling and (c) Long term field measurements. Comparisons between design stage projections of post-recharge losses, and measured responses are shown (Figure 8); these suggest that despite the increased severity of wave climate at the site, beach losses have been less than originally projected.

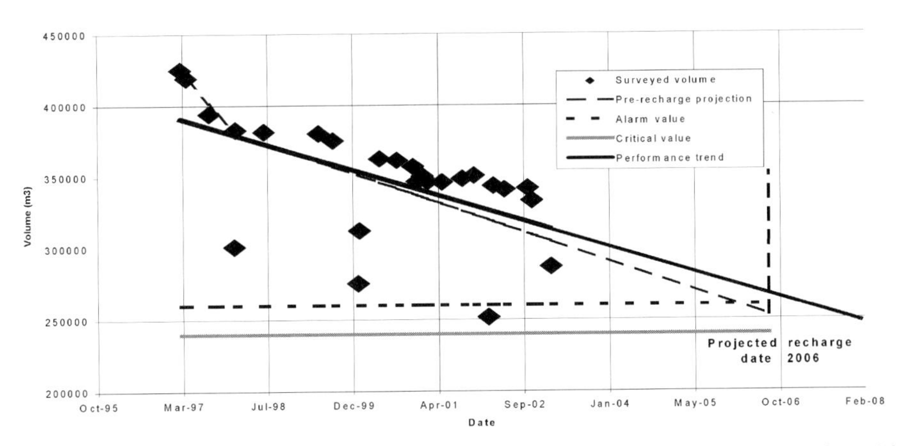

Figure 8 Comparison of surveyed and projected post recharge volumetric changes at Hurst Spit (above 0mODN)

Although extensive field data (since 1987) provided a pre-dredging historical baseline, the beach recharge activity has interrupted the natural processes and altered the system such that it is no longer in equilibrium. Beach responses are likely to be very different following recharge, whilst the beach adjusts to the new conditions. It is not reasonable to use this site for examination of either the influence of dredging, or increased storminess, due to the overriding impacts arising from beach management intervention.

Isle of Wight

Spatial analysis of 30 beach profiles indicates that accretion is the predominant post-dredging trend. Only three profiles measured exhibited a net downward trend. A longshore trend is evident from Alum Bay, through Totland Bay and Colwell Bay. Erosion is evident in the south (Alum Bay), whilst beaches to the north-east are either stable or accreting; however, there is net accretion throughout the system. Evidence of sediment transport to the north-east is demonstrated by patterns of change within groyne compartments of Totland Bay. Although the frequency of severe offshore wave conditions has increased, the beaches do not show any signs of a reduction in volume or performance. Overall the system appears remarkably stable.

As expected, seasonal variations in beach volumes are evident at all sites: higher volumes are evident in the summer months and lower volumes in the winter. Although the overall morphodynamic trends are similar, seasonal beach responses within Totland Bay are much more volatile than within Colwell Bay. Typical trend analyses for the two beaches are shown in Figures 9 and 10. Beaches of Totland Bay had a generally low initial beach cross-sectional area. Significant beach structure interaction is evident at this site; the shape of the post storm profiles and rapid changes in profile evidences this. Application of parametric profile models suggests that a dynamic equilibrium profile is unable to form anywhere within Totland Bay, even for frequently occurring conditions. It is suggested that the rigid seawall structures, in combination with the low beach volume, have restricted natural evolution of the beach.

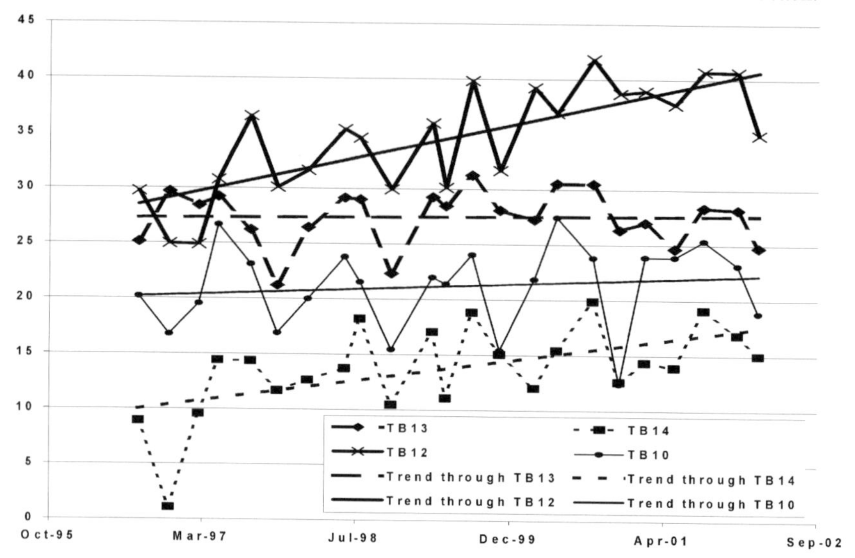

Figure 9 Example post-dredging beach profile trends in Totland Bay

By contrast, the initial beach cross section at Colwell Bay (Figure 10) is generally suitably large to enable a dynamic equilibrium profile to form under most conditions, without beach-cliff or beach-structure interaction.

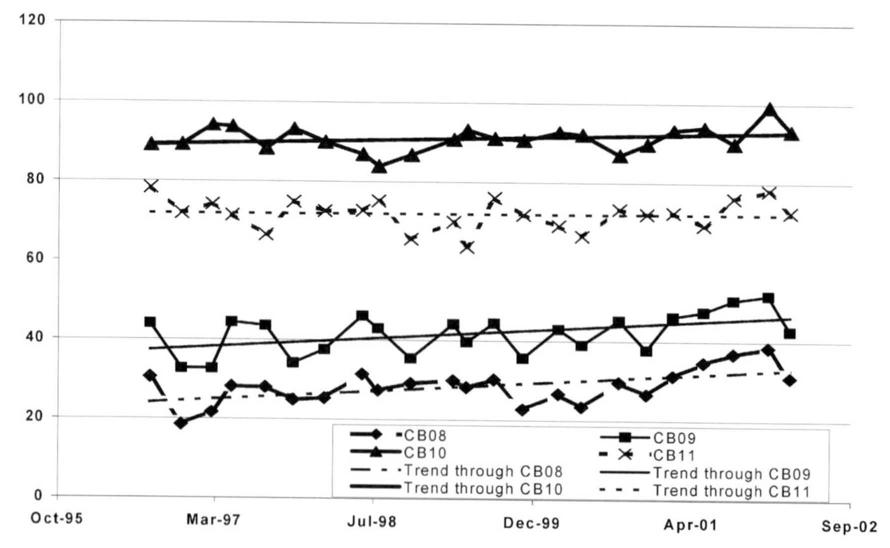

Figure 10 Example post-dredging beach profile trends in Colwell Bay

The small beach at Alum Bay has experienced a loss of beach volume. Interaction of the beach with the soft cliff geology and groundwater driven cliff failures complicates the patterns at this site.

Concerns that dredging would have an adverse impact on the beaches of the Isle of Wight cannot be supported by the beach profile evidence. In fact, the beaches exhibit remarkable stability for the wave climate conditions. Regrettably, no high quality pre-dredging beach performance data was available for comparison with the post-dredging programme, for any Isle of Wight beaches. On this basis, any post-dredging trends observed during the monitoring programme, (either coastal erosion or accretion) cannot be linked with earlier trends, or directly with dredging activity.

Offshore morphological changes

The Shingles Banks system consists of two distinctive areas, separated by the narrow and shallow North Head Channel: the main body of the Shingles Banks, and the smaller North Head Bank. The system is bounded to the east by the Needles Channel, with depths ranging from 15 to 60mCD, and is separated from Hurst Spit to the north by North Channel (Figure 2). Pre-dredging chart analysis demonstrated that the total volume of the system had generally grown since 1882. Subsequent bathymetric surveys (1996-2002) show that rapid changes have continued to occur.

The main body of the Shingles Banks exhibits an asymmetric morphology; it comprises a relatively gently sloping and stable western flank, highly mobile ephemeral shoal areas, evolving channels in the central section, and a steeply sloping eastern bank that faces towards the Needles Channel (Velegrakis, 1994). Analysis of historic charts and the recent bathymetric surveys have enabled evolution of specific morphological features of the Shingles Banks system to be analysed in detail.

Shingles Banks

Historic bathymetric charts and recent bathymetric surveys indicate that the central section of the Shingles Banks system is very mobile; the shoals and adjacent channel features are subject to large and rapid changes due to the local wave climate, tidal currents, and storm events. The location of drying shoals on the Shingles Bank system (see C Figure 11) has remained consistent, although their area varies. In particular, the shoals have accreted during storm events; they become surface emergent over the full spring tidal cycle on occasions.

A submerged-bar feature extending in a south-westerly direction (see D Figure 11) is clearly present on the western flank of the Shingles Banks system from 1882 to 1995. The feature was most prominent between 1945-1965, but had decreased in size and length by 1995. The bar feature was not evident in 1997, when an adjacent channel appears to have been filled with sediment. Disappearance of the feature coincided with dredging of the Shingles Banks. Surveys in 2001 and 2002 indicate redevelopment of the channel and bar with a similar form and extent to that measured in 1995. The short-term, localised morphological changes to this feature suggest that it is ephemeral and highly sensitive to hydrodynamic changes.

North Head Bank

The North Head Bank region (see A Figure 11) has declined in area and increased in depth since 1882. The 1882 Admiralty survey indicates that the North Head Bank was comprised of two areas separated by a channel that were joined along their western flank. The 1972 survey indicates that the Bank had evolved into a distinct single area, which had reduced in size.

Further reductions in area had occurred to the North Head Bank by 1997. Post-dredging monitoring has identified a continued reduction in size of the North Head Bank. This long-term evolutionary trend is indicative of continual adaptation to wave climate, tidal currents and sediment supply. Its location is distant from the dredging area and dredging processes are unlikely to have impacted upon the evolutionary trends.

North Head Channel

The North Head Channel (see B Figure 11), demarked by the 5mCD contour, was a pronounced feature in 1882; this Channel gradually extended and lengthened eastwards, as indicated in the 1988 survey, although the North Head Bank remained connected to the Shingles Banks at the eastern end of the channel. The North Head Bank was completely separated from the Shingles Bank system, as a result of eastward extension of the North Head Channel, between 1988 and 1995. The 1995 survey indicates a narrow passage separating the North Head and Shingles Banks systems at its eastern end (see B Figure 11). The channel continued to widen and deepen until July 2002.

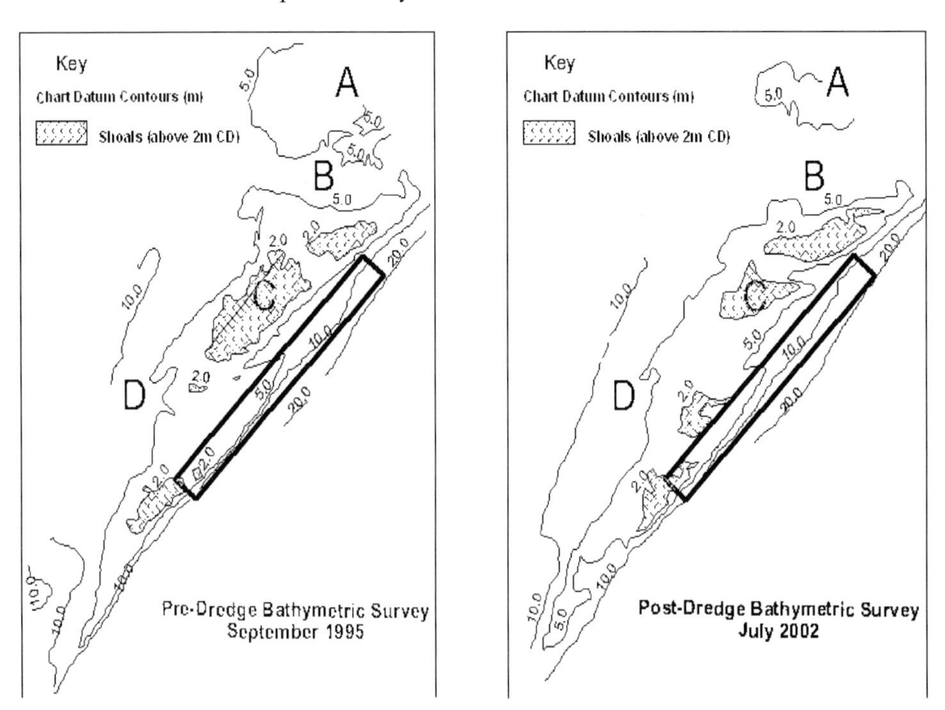

Figure 11 Evolution of the Shingles Banks system following aggregate dredging

Post-dredging volumetric changes

Beach recharge material was dredged from the eastern flank of the Shingles Banks, (August 1996 to January 1997). Dredging operations were monitored continuously by the onboard Crown Estates' Electronic Monitoring System (EMS), which indicated that most of the dredging activity occurred in depths between 5 and 15mCD. A total of 300,000m^3 sediment was dredged.

Unlike most aggregate dredging areas, the Shingles Banks are not a relic deposit. Storm events, particularly those from the southwest, rework and re-deposit the marine gravels within the system. Patterns of deposition are dependant on the occurrence, direction and, duration of storm events. Bathymetric surveys provide a series of volumetric snapshots of a highly mobile and dynamic system. The measured differences in volume between surveys need to be analysed in conjunction with regional wave climate data and tidal current patterns, therefore. Even with this data, interpretation of changes is extremely difficult. The potential accuracy of bathymetric survey techniques, typically +/- 0.15m vertical and +/-1m horizontal, is a significant issue in this environment. As the survey area is large (13 hectares), measured changes in volume of more than 2 million m^3 may lie within the expected scatter of survey error. Bathymetric survey line spacing for the dredging area is at 50m, whereas the line spacing for the surrounding Shingles Banks area (see Figure 2) is at 200m. The line spacing for the two areas provides sufficient information to describe plan shape changes of morphological features. The closer line spacing within the dredge area is more appropriate for volumetric calculations. Volumetric analyses of post-dredging bathymetric surveys (Colenutt, 1997-2001) indicates that the dredging area, and the Shingles Bank system as a whole have experienced substantial gains and losses of material (Figures 12). Wave conditions measured over the last 10 years, at a control site outside of the influence of the Shingles Bank system, suggest an increase in the frequency, intensity and duration of storm and wave activity.

The combination of intense periods of wave activity, strong tidal currents and the mobility of the coarse-grained sediment causes rapid and significant deposition and erosion of material within the dredged area. Patterns observed appear consistent with sediment transport pathways suggested by Velegrakis (1994). It is suggested that deposition occurs within the dredging area during periods of intense wave activity; this results as the shoals of the bank system roll north-eastwards. Ebb dominant tidal currents winnow out fine material and transport this to the south-west, resulting in losses from the dredging area, during calm periods.

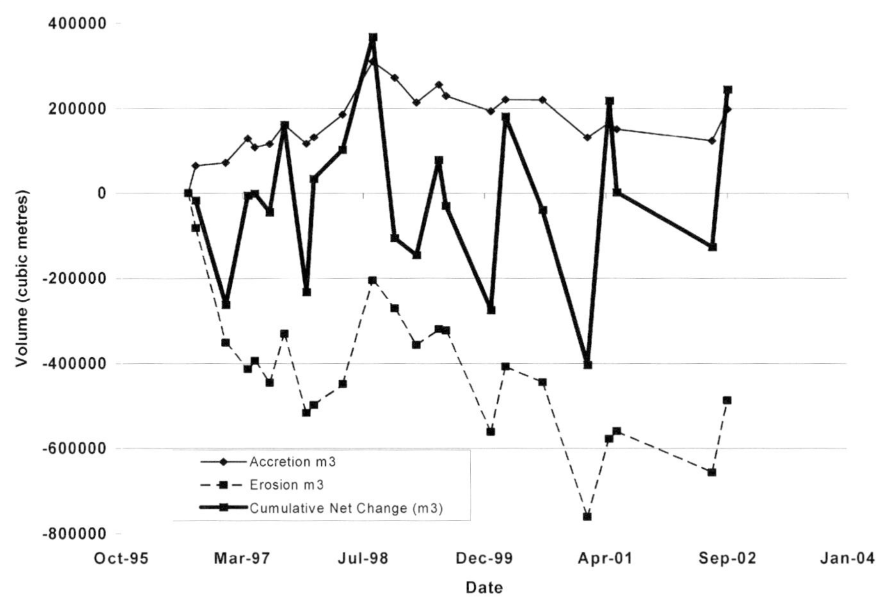

Figure 12 Volumetric Changes for Shingles Banks Dredging Area 406

Analysis of the post-dredging bathymetric surveys of the dredging area, plus a 300m wide bounding box of the surrounding area suggests that substantial fluctuations of accretion and erosion occurred within this area (Figure 12). Survey accuracy may also factor in the volumetric variations. Volumetric changes indicate both deposition and erosion, with no clear trend, although there is some evidence of deposition following stormy periods. The dredging area is part of a highly dynamic system and the mobile sediments respond rapidly to the prevailing hydrodynamic conditions, and storm events.

Post-dredging volumetric analysis of the whole Shingles Banks system demonstrates a volatile response of the system. Accumulation appears to be evident following more energetic periods. No obvious trend is evident however, with cumulative volumetric changes swinging from erosion to accretion between surveys. Trend analysis of the data sets shown in Figure 13 suggests that there is no statistically valid trend. A cumulative net loss of 1,625,000m^3 of material arising as a balance from 4,034,000m^3 of erosion and 5,660,000m^3 of accretion was evident on the occasion of the last survey. The net change is the equivalent of a 0.12m thickness over the entire Shingles Banks area; this quantity can be consumed by the expected error of a hydrographic survey. Although the measured volumes are subject to some considerable uncertainty, evolutionary trends are clearly evident within the system.

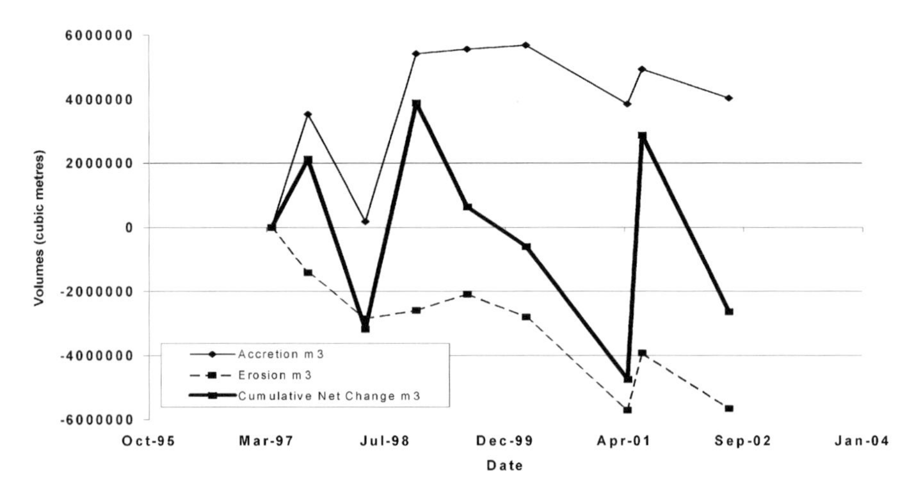

Figure 13 Volumetric Changes for Shingles Banks Whole Area

Benthic Surveys

A pre-dredging benthic survey was conducted in August 1996, and post-dredging benthic surveys conducted in March 1997 and September 1997 (Serpell-Stevens, 1996,1998). The two summer surveys are directly comparable, whilst the March 1997 survey shows significant seasonal mortality of the indigenous species. The frequency of individuals and diversity of species, measured on each survey, were not significantly different. Although the post-dredging survey indicated a significant reduction in both count and diversity of species, this may reflect the impacts of seasonal mortality as well as disturbance by dredging. A rapid recovery of species is evidenced by the September 1997 survey. The dredging operation appears to have had a short-term impact on the benthos of the dredging area, but recovery has been rapid.

Discussion

Despite the fact that local wave conditions have been significantly more severe, and storm conditions more frequent than anticipated at the environmental assessment stage, there is no evidence of a decline of the beaches on the shoreline of the Isle of Wight. The control site in Christchurch Bay has exhibited similar trends to those prior to dredging. Beach monitoring at Hurst Spit demonstrates a declining beach, in line with pre-recharge prediction. However, the influence of the beach recharge operation has been overwhelming at this site. Large-scale post-dredging evolution has continued within the offshore bank system although there is no evidence of changing patterns.

The Milford-on-Sea wave buoy data provides some confidence in the numerical modelling methodology used for coastal process assessments. Whilst its location is ideally suited to a validation exercise for the modelling, it cannot be used to prove whether wave conditions have become more severe in the shelter of the Shingles Banks, since post-dredging measurements were not made at these sites.

These observations collectively imply that the nearshore dredging area has not had an adverse impact on the shoreline. This cannot be shown conclusively however, because of the lack of suitable controls at either Hurst Spit or on the Isle of Wight. Significant improvements to the understanding of causal linkages can be made only when long-term pre-dredging control data is present at appropriate locations.

The coastal manager has a responsibility to conduct monitoring routinely, as part of best practice shoreline management, and to understand natural coastal evolution and changes arising from the introduction of coastal defences. This responsibility has been addressed within the south-east of England by the introduction of a strategic regional coastal monitoring programme. Only when comprehensive baseline programmes are in place at appropriate locations, can additional monitoring data relating to dredging activity be put to best use. Conclusive linkages with dredging activity cannot be made unless adequate temporal control baselines are in place. The responsibility for achieving this approach lies with both the operating authorities and with the organisations responsible for dredging.

Conclusions

- Beaches on the Western Isle of Wight have been stable or accreted since dredging on the Shingles Banks.
- Large-scale morphodynamic changes have continued within the Shingles Banks system. Patterns of change are consistent with those observed since 1882.
- The benefits of the integration of best practice long-term coastal monitoring, arising from coastal management initiatives and dredging licences, are demonstrated.
- The occurrence of severe nearshore wave conditions has increased within Christchurch Bay. Regular exceedence of the predicted 1:100 year wave height condition has occurred.
- Regular reviews of wave climate statistics are required, to identify temporal trends in extreme conditions, particularly during periods of perceived climate change.
- Comparisons of modelled and measured wave data have indicated a strong correlation for the Milford on sea wave buoy site, but a poor correlation with the North Channel site.

Dedication

This paper is dedicated to the memory of Darren Hume, a revered colleague of the authors, who died tragically in summer 2002.

References

Bradbury, A.P., and Kidd, R., (1998). Hurst Spit stabilisation scheme - design of beach recharge. *Proceedings 33rd MAFF Conference of River and Coastal Engineers.*

Bradbury, A.P., (1998). Response of shingle barrier beaches to extreme hydrodynamic conditions. Unpublished Ph.D. Thesis, University of Southampton.

Bradbury, A. P. (1992). Hurst Spit Stabilisation Scheme: Assessment of Bathymetric Changes in Christchurch Bay. Report CR0192. Unpublished Report.

Colenutt, A. J. (1997, 1998, 1999, 2000, 2001). Hurst Spit Stabilisation Scheme: Dredging Area 406: Annual Reports. Unpublished Reports

Hydraulics Research., (1989a) Wind-wave data collection and analysis for Milford-on-Sea, *HR Report EX 1979. 20pp.*

Hydraulics Research., (1989b)., Christchurch Bay - Offshore Wave Climate and Extremes. *HR Report EX1934. 9pp.*

Velegrakis, A.F., (1994). Aspects of the morphology and sedimentology of a transgressional embayment system: Poole and Christchurch Bay, Southern England. Unpublished Phd Thesis, Department of Oceanography, University of Southampton.

Velegrakis, A.F., and Collins, M.B., (1992). Marine Aggregate Evaluation of Shingles Bank, Christchurch Bay. Unpublished Technical Report, SUDO/TEC/92/14C 33pp.

Velegrakis, A.F., and Collins M.B., (1991). Coarse-grained Sediment Deposits in Eastern Christchurch Bay: A Preliminary Report. Unpublished Technical Report, SUDO/TECI9l/19C. Department of Oceanography, Southampton University. l5pp.

Serpell-Stevens, A. (1996 and 1998). Pre-dredge, Post-dredge and 2nd Post-dredge Benthic Grab Survey (Shingles Bank). Unpublished Reports.

Simons, R. & Hollingham, S. (2001). Marine Aggregate Dredging: A Review of Current Procedures for Assessing Coastal Proceses and Impact at The Coastline. Technical Report HYD10401 by Civil and Environmental Department UCL. Report for SCOPAC and BMAPA.

Wimpey Environmental (1993). Hurst Spit Stabilisation Scheme: Environmental Impact Assessment. Report TRMZ5198.

Thought that is silenced is always rebellious – An argument for argument in consultation

J.G.L.Guthrie Principal Engineer, Posford Haskoning, Peterborough, UK
A.Battison Engineer, Posford Haskoning, Peterborough, UK
R.Cottle Senior Environmental Scientist, Posford Haskoning, Peterborough, UK
R.Hopewell Flooding and Coast Protection Co-ordinator, North Ayrshire Council, Irvine, UK.

The need for good consultation

Nowhere is the idea of 1:20:200 more relevant than in coastal engineering. The idea that if the effort put in to delivering the correct concept (design) is at a scale of 1, the scale of cost in implementing the ideas may be 20 but the potential long term influence on the coast, its users and interests are at a scale of 200. We have responsibility in the first stage to ensure that the concept for management of any area of the coast is correct and delivers what is needed.

The ultimate client for any coastal engineer or coastal manager is not the Coast Protection Authorities or the Environment Agency, nor even government, but the communities and broader society that these organisations serve. Coastal managers have expertise in engineering, geomorpholgy, ecology, coastal processes, economics or tourism and recreation. We have the ability to predict and to engineer solutions, even the ability, sometimes, to understand what we are doing, but the real experts at living with the coast, live and work out there on the coast. They have knowledge of what happens, but possibly not the understanding of why, they understand their interests and what might impact on those interests but not necessarily the understanding of the broader consequences of action or inaction in coastal management. More often than not, however, they have the ability to understand, if issues are explained, and to contribute productively to sustainable and inclusive management.

As a generality, consultation fails to properly engage consultees. It is considered acceptable to confirm decisions on the basis that consultation has taken place; regardless of the quality, quantity or clarity of responses. The procedure has often become more important than the result; a hurdle to be crossed, a waste of money and a delay to implementing the preferred option. It is considered acceptable to accept the response "This organisation has no interest in your study"; infrequently do we question this and where there is felt to be important issues is there the response in turn "Yes you have, for these reasons, and your input is vital". More often, it is a sigh of relief that one less point of view has to be built into an already complex process of analysis; one less burden on an already stretched budget of initial design and approval.

Typically, there is a poor return of relevant information due to a lack of understanding either of what the consultee is being asked or a lack understanding of what they are actually saying.

We have a tendency either to knock the fight out of consultees or to back them into disgruntled corners; simply by not consulting at a stage when there is opportunity for it to

Coastal Management 2003, Thomas Telford, London, 2003.

make a difference. At worst, there is a perception that views are being ignored. The attitude has to change from one where consultation is a burden of gaining approval of what is already determined to one where consultation is considered an integral part of the design process; a situation where we want to consult, where we consider there is real benefit to be gained. We want to hear about the issues and interests of people, of the environment, of commerce, of communities, because we want to develop policy, strategies or schemes that really deliver long term benefit. If we cannot deliver for economic, sustainability or legislative reasons, the consultees have a right to know and a right to know the reason. Transparency in decision making comes not from following set procedures but through engaging those on whose behalf we are working.

Good consultation is a two way process, aiming at merging knowledge of the expert with those involved with life on the coast, supported by a good background of information and understanding. In this, consultation may have to be a continuing, adaptive and iterative process.

The responsibility is on us to demonstrate the value of real consultation, to take the problems to consultees, to inform and to listen.

Development of consultation

Consultation with respect to coastal defence schemes has improved over the last decade but, while certainly broader, does not necessarily obtaining the detail of the many of the local discussion of fifty years ago. In many cases we appear to have sacrificed detail in our attempt to provide a broader consensus and in focussing on the organisations rather than individuals. To take forward a change in attitude it is helpful to consider how this progression towards better consultation has developed. This progression may be described under the following headings.

I will listen but stop me if you can!

Such an approach relied on two stages of consultation.

- Gathering information.
  Initial contact is made with all who may have an interest in the area of study, informing consultees of the need for the study and requesting information with respect to each individual's interests.
- Confirmation.
  Formal consultation requesting confirmation and comment that a preferred approach was acceptable; or at least was not illegal.

The approach, although in decline, is still widely used, modified slightly, nominally to include, as part of the initial stage, a description of generic options; at policy level: Do Nothing, Hold the line, etc. and, at a scheme level, to provide a list of all possible forms of defence known to mankind. Is it any wonder that responses would be littered with irrelevant information such as:

- groynes cause downdrift erosion (when what was really requested was a discussion of how ecological integrity might be affected by such erosion);
- beach recharge is an acceptable form of soft engineering (when really what was required was a discussion of the important use of a beach for tourism which might be detrimentally affected by regular recharge).

Such responses (actually received during consultation), hopefully, add little to the knowledge of most coastal engineers but may actually help set hostility to options in the minds of those consulted before ever understanding the issues; especially when the most appropriate solution is, for the sake of argument, groynes.

Because, within an SMP, options have already been prescribed, this modified approach frequently became the mode of consultation for the first round of SMPs. An assumption was made that consultees would understand. That, from the clear, transparent (and standard) description of each policy option, such as:

Hold the Line; by maintaining or changing the standard of protection.

and from the summarised impacts that such a policy might imply, all consultees would fully understand the implications of an option on their interest and livelihood. Argument focussed not on the objectives upon which the SMP was developed, the intent or implementational consequences of the policy, but rather fatuously on the words describing the policy. This led to modification of policy descriptions such as Managed Realignment instead of Retreat or the curious Selectively Hold the Line, without really addressing the substance.

Inform and defend!

This development in consultation added a new second stage, whereby the standard project appraisal is used in consultation to demonstrate the range of options considered before arriving at a preferred option. Although, importantly, helping to identify the reasons behind the selection of a final solution and, to a degree explaining the decisions made for rejecting other options, this approach still lacks an involvement of consultees in actually selecting and modifying alternatives.

The process is typically undertaken at a time when significant effort has already been carried out in appraisal, where objectives are fully established and where, alongside consultation, the process for funding approval is advancing. There is a massive inertia at this stage to see through the preferred option.

With opportunity to change course closed down, with little opportunity for consultees to question basic assumptions or to offer alternatives, consultees are frequently put in the position of criticism of what is on offer rather than being allowed the opportunity for introducing new ideas.

The benefits, however, of this now, more widespread approach is that a better understanding of what might be feasible can be generated. Real awareness could be created and a genuine attitude of partnership could result. Like good comedy, good consultation relies on timing.

Estuary management

Management within estuaries highlights and heightens awareness of all the similar issues relating to the open coast. Estuaries tend to have a concentration of interests: those of ecology, recreation, commerce and community aspirations. There is often strengthened interaction between these interests and in the physical behaviour of the estuary. The consequence of management, or attempts to address site specific issues, can be wide reaching both in impact and in location. Nowhere, is it more critical that the interests and knowledge of those involved with an estuary be incorporated within management. This point is clearly understood and is being developed within many estuary situations around UK. Examples of this show through the Firths initiative in Scotland, research into estuary management, through programmes such as Emphasys (Whitehouse 2000) and the subsequent Estuaries Research Update Project (Brew 2002).

On the Humber an innovative approach was taken to the analysis of how the estuary might respond in the future and the consequences this might have on management of interests (Ponttee and Townend 1999). In this two parallel strands of analysis were undertaken; one examining the function of the estuary the other focussing on resources. The complex system of analysis took an estuary wide view of how management practice should be undertaken to address the long-term development and how management of individual units within the estuary could best meet shorter term objectives based on economics and initial consultation.

This approach highlighted the areas of conflict in management practice, allowing much better informed consultation to follow. Essentially, the understanding and atmosphere of discussion engendered by this approach, involving and informing consultees before decisions were made has established the framework of trust from which the Humber Estuary management structure has been allowed to develop. It is important to appreciate, however, that it has taken years to arrive at the situation existing today; a heavy but very worthwhile commitment to consultation.

This closer involvement of consultees was also a key feature of the Suffolk Estuarine Strategies (SES). Element of SES are discussed below, highlighting some of the problems but also advantages that were experienced.

Suffolk Estuarine Strategies

Three estuaries were considered as part of this study: the Deben, the Alde/Ore and Blyth. The purpose of the study was to develop a strategy for sustainable flood defence which would, working within the constraints of the Environment Agency's powers and funding, take due account of the use and interests of the estuary. The study has previously be described in detail (Pettitt et al 1999).

An initial study of the physical behaviour of the estuaries was undertaken by ABP Research and Consultancy Ltd. and this work formed the baseline line assessment from which to develop the strategies. The strategies (Posford Duvivier in association with HR Wallingford) focussed principally on the line of the defence being maintained, analysing initially how various defence lengths and the exclusion or inclusion of defended areas within the estuary system influenced the ability of the estuary to respond to growing pressures due to on-going change and potential sea level rise. An essential element of this work was understanding the current patterns of use and interests within the estuaries.

A traditional first stage approach was taken to consultation. This entailed a comprehensive search for information, extending over a six month period. The extent of those consulted, although well researched in association with the Environment Agency and English Nature, was increased by the simple technique of requesting those first contacted to identify others with interests. The interests were collated on a simple ACCESS database and, as the process of assessing how the defence structure influenced the physical behaviour of the estuaries, were analysed to identify potential benefits or conflicts that might arise from different management scenarios. The response showed, quite strongly, that in the majority of cases people felt an appropriate balance of use existed; why change?

Following this first stage of study a second round of consultation was undertaken. At this stage, no formal appraisal had been undertaken as to specific options or management scenarios. The aim of this second stage consultation was specifically:

- to inform consultees as to the nature of the estuaries in which they had an interest,
- to highlight where stress within each system might, as a result of natural change, threaten those interests,
- to discuss the constraints associated with defence management, but also,
- to discuss the opportunity for change.
- To include consultees and interest groups in the process.

Meetings were arranged, with the assistance of the local authorities, the Alde/Ore association and the parish or community councils. Additional meetings were held at the request of local societies and organisations, at their request, throughout the area.

In general these meetings were well received. However, two general attitudes had also to be countered, demonstrating the expectation prevalent among many consultees at that time:

- "This meeting is a waste of time as you are not telling me what you plan to do."
- "We are being softened up and you are planning to abandon defence in the estuary."

The second could relatively easily be addressed through discussion. The first, generated from an expectation of experience of past consultation, was far harder. It demonstrates the need, even before direct consultation, to restore confidence in consultees that they have an important role to play. These problems are not unique (Brooke 1999) but if these responses had been anticipated, information and presentations could have been adapted more effectively.

Key issues were discussed and argument focussed around:

- An understanding that estuaries do change; regardless of, as well as because of defence management.
- A better understanding of legislation relevant to the environment, engendering often a more sympathetic attitude to environmental issues.
- The financial and economic justification for defence schemes and the issues of compensation.

In many areas, such discussion went beyond the scope of the study to resolve, but allowed specific issues to be raised in the strategy reports to be taken up at a broader forum. In other cases, the understanding of the situation, the possible threats to livelihood or to interests, prompted consultees to be far more detailed in their responses. This provided additional information and the correcting of misconceptions both immediately, as a response to these initial meetings, and through the course of developing the study. Although considerably more time consuming, the information as to the interests of those who might be affected was far more detailed than would otherwise have been obtained. Importantly, this information, shaping the decisions within the study, came from those with a stake in getting it right, building confidence that all issues were being considered correctly.

Subsequent consultation was tough. Consultees were better informed and were willing to argue issues on the basis of what they had already been told. The additional consultation, far from bringing on fatigue, encouraged involvement throughout the process and developed a transparency and consensus to the final conclusions of the study.

Two further rounds of consultation were undertaken formally, although in practice consultation, in terms of direct access to the study team, was continuous. Further meetings, some 26 in total, were arranged with parish and community councils and finally as full public meetings organised by the local authorities. A formal consultation period on the draft strategies was undertaken with reports available to all.

The estuary strategies have provided a platform from which to examine specific schemes, in context of a broader sustainable management of the each estuary as a whole. Beyond this, the strategies have identified important local issues, issues important to individuals, which can now be explored in detail, from a common baseline of information and understanding.

The open coast

Although estuaries frequently have a concentration of issues in a highly interactive manner, the need for better communication and consultation is apparent on the open coast. Two examples are discussed below; first quite briefly highlighting the opportunity taken by the community of Borth, West Wales, to influence the strategy for defence along their frontage and secondly, in more detail the development of the scheme at Largs, North Ayrshire.

Borth Visioning

Over the last two years, a coastal defence strategy has been under development for the village of Borth (Posford Haskoning 2001). It has always been recognised the life in Borth is closely linked with its coastline and the manner in which the defence of the village from flooding and erosion is provided. Over the years, leading up to the strategy study, this has been reflected in the various presentations and discussions that have taken place over coastal issues. In this regard, the community of Borth is well informed, taking a keen interest in how the coast has developed and the history and management of defences.

Neil Caldwell Associates were engaged by Borth Community Council, with the assistance of Cyngor Sir Ceredigion, to help it actively involve members of the local community in a 'visioning exercise' designed to clarify the ways in which future developments at Borth might be pursued with the maximum support of its residents. In particular, the exercise was prompted by the development of the coastal strategy, recognising the opportunity to influence design at a critical stage. This process went beyond but built upon the more traditional consultation being undertaken for the strategy.

The form of an evening meeting was carefully structured and facilitated to maximise the opportunity for those attending to contribute their ideas about the way that the engineering works could respond positively to the community's views and priorities. Initial group sessions were held to discuss and identify aspiration and specific ideas for the improvement of the village. These were grouped by general attitude and perception of how the village might be developed to advantage and by issues people felt might influence or be influenced by the development of coastal defences. Finally, all participating, some 107 people, were allowed to vote on those issues which they felt were of greatest priority.

Although many of the issues related to maters beyond the scope of management of the coast, several important issues were identified in relation to the strategy. These were presented in the form of what people would like to see the strategy achieve, not surprisingly the top priority was an appropriate defence of the village, and also what concerns people had and what they wanted to ensure the strategy avoided.

Among the first category, it was hoped that the defences would help impose some zoning of activities; water sports, especially water skiing, being a key issue, if only be creating marker point along the coast. Other opportunities included improved surf conditions and establishing better harbour facilities.

Of those issues of concern were the impact on adjacent frontages, damage to ecology interests, visual intrusion and spoiling the view and ensuring continued safe family use of Borth's wonderful beaches.

The visioning results were reported and included within the development of the coastal strategy. Subsequent, consultation has been undertaken to discuss further these issues. As a direct result one of the enhancements the Coast Protection Authority, Ceredigion is examining are the opportunities for of improving surfing facilities.

The exercise shows, among others, two points:

- that the initiative for consultation in coastal engineering need not necessarily come from those undertaking studies into coastal defence. The awareness created by the efforts of the Council over many years, leading up to the strategy development, has paid dividends in encouraging involvement at a critical time when outline design is still being formulated.
- the importance of getting behind the thinking of some of the more generic statements made in response to typical forms of consultation, arriving at the issues that are of real concern to communities on the coast.

This second point is developed in considering the consultation for the Flood Prevention Scheme at Largs.

Largs Flood Prevention Scheme

Largs is situated on the west coast of Scotland in the outer reaches of the Firth of Clyde. The Largs sea front is considered to be an important amenity and recreational asset, important to the Town, with significant social and environmental value. The views from the promenade, over to Great Cumbrea are as important as the promenade itself.

Largs Seafront

Overtopping of the Fort Street seawall in Largs often occurs, with minor flooding of the sea front area being a regular result. On more severe events the whole town centre can be flooded causing substantial damage to both commercial and residential buildings. An example of this occurred in 1999, when major flooding affected in excess of 100 properties, causing access problems to and through the town centre and resulting in significant economic damage. Posford Haskoning was therefore engaged by North Ayrshire Council to investigate possible solutions to these flooding problems.

As part of the initial feasibility study consultation was undertaken by post and through meetings to identify environmental considerations. The Feasibility Study phase of the project included the assessment of the sea defences at Largs in North Ayrshire, and the development of a defence strategy for the central section of the town's seafront. The investigation involved a desk study of the coastal processes and flood routing and mechanisms, investigation of the likelihood of overtopping and the economic damages that would arise due to this, and the development of preliminary designs for the appraisal processes. However, it was always appreciated that the visual aspects of any scheme would be a key issue; both the retention of the character and attractiveness of the promenade and the views from the promenade. It was considered necessary to undertake full consultation at a local level to ensure these concerns would be properly covered before finalising a preferred approach to flood prevention.

The approach adopted, therefore, was to address, initially, the engineering issues; examining what was practically feasible before taking options to public consultation with the specific intent of raising aspects of potential schemes and engendering discussion.

Consultation, beyond the initial exercise of information gathering, was undertaken in three stages.

Public Display

A display, held in a building adjacent to the promenade, was extended over a period of four full days. Notices were displayed in local shops and in the press. Over 300 people attended during its course. Formal presentations were made at regular intervals as new visitors arrived to explain in full detail the history of flooding, the causes for the flooding and the possible defence options.

In addition, and as importantly, all visitors were encouraged to discuss the issues from their perspective, with either the consultant or a member of North Ayrshire Council staff. Views and attendance were recorded for further analysis.

It became apparent during the course of the exercise that there was a lack of representation from local businesses and this was addressed by visiting shops and offices in the area to promote involvement (proactively dragging consultees in off the street).

Two options were presented in detail aimed at drawing out comment relating and contrasting two different approaches to what had been felt were the key issues emerging from the feasibility study; these being the amenity and visual aspects of any scheme. Briefly described these were:

Option 6 - including a low flood wall set back from the main defence over the southern section of the frontage. In the northern section a new, higher flood wall replacing the existing seawall.

Option 7 - including a new rock revetment fronting a reclaimed area of land to the northern section of the frontage where the seawall requires complete replacement. A low flood wall would also be constructed, set back from the main defence, over the full length of the frontage.

Option 6 and 7 are shown in Figures 1 and 2 respectively. Artist's impressions were provided to assist in understanding how either option might change the appearance of the frontage. In both cases, the options intended to run a retired flood wall through the centre of the memorial green to the south of the frontage, making use of the slight undulation of the green, associated with landscaping, to reduce visual impact.

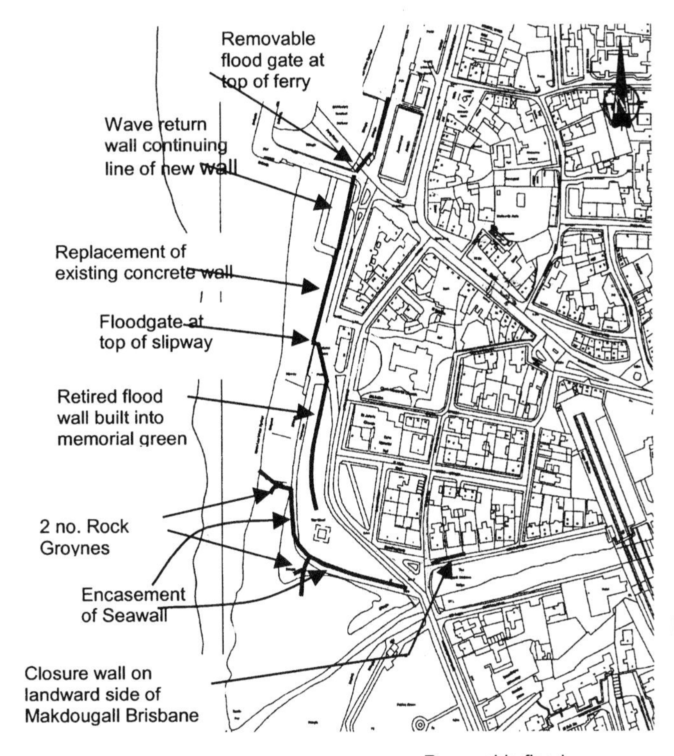

Figure 1. Option 6

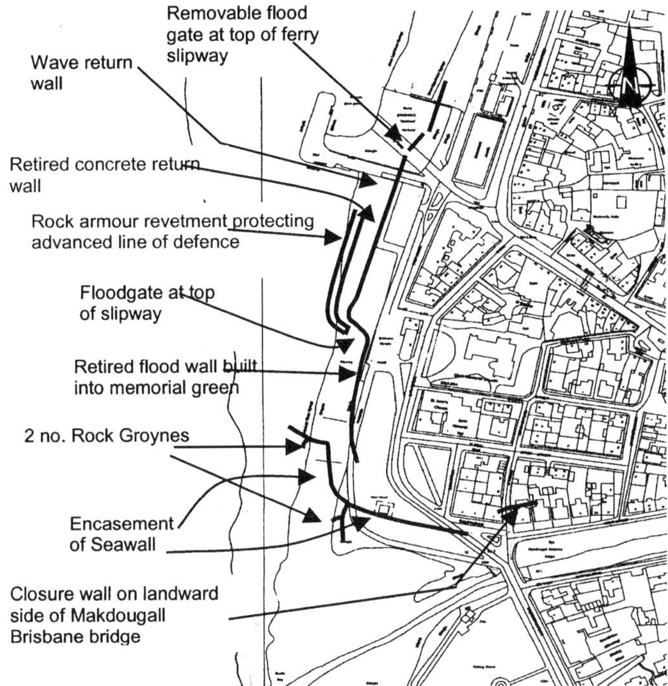

Figure 2. Option 7

Significantly, the artist's impression of option 7 showed an existing private car park to the north of the area. This was picked up by consultees as being a new car park on the reclaimed land and became one of the key features of the presentation, consultation and the subsequent responses on the questionnaires.

All those visiting the display were given both information leaflets and a questionnaire. The results from this questionnaire are discussed below.

Analysis of the questionnaire

When planning the consultation questionnaire the intention was to steer away from the 'tick box' approach, which was considered too rigid for this particular situation, as it does not encourage imagination or 'lateral thinking' from consultees. A 'tick box' questionnaire may be seen as guiding people towards answers that have already been decided upon, rather than allowing their own opinions to be expressed.

The questionnaire addressed the subjects as shown in Table 1:

Table 1. Questionnaire Headings

Question	Subject
a - d	Consultee details - name, address, telephone no., specific interests in the scheme.
E	Need for the scheme.
F	Specific concerns relating to the scheme in general, and how these might be resolved.
G	Comparison of options, including opinions on whether the options will have beneficial or detrimental impacts and the reasons for these opinions.
h-I	Issues identified in the feasibility study: importance of slipway and beach.
J	Key uses of the frontage that should be taken into consideration.
K	Possible enhancements to the frontage and existing problems that may be addressed as part of the scheme.
l - n	Comments on the consultation process

The questionnaire was simply constructed but aimed to elicit a discussion of issues. Issues could be raised in more than one place. All responses and opinions were considered in the analysis of the questionnaire responses, wherever the answers may have been recorded. In answer to many of the questions most of the consultees recorded more than one concern or opinion. All consultees recognised the need for improvements to the Largs sea defences, and agreed in principle with the need to develop a flood prevention scheme.

The specific concerns of the consultee's relating to the scheme were identified from answers to question (f) as well as from responses elsewhere in the questionnaire. These responses are summarised in Table 2 below.

Table 2. Summary of Concerns

Concerns

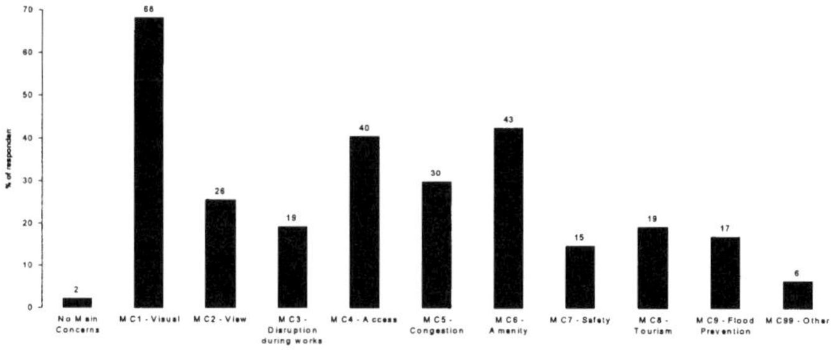

Concern	No. of responses	Percentage (%)
SC1 - Visual impact of scheme	32	68.1
SC2 - Effect on views	12	25.5
SC3 - Disruption during construction works	9	19.1
SC4 - Access to sea front (during construction and on completion)	19	40.4
SC5 - Traffic congestion (during construction and on completion)	14	29.8
SC6 - Effect on amenity value of sea front	20	42.6
SC7 - Safety (during construction and of completed scheme)	7	14.9
SC8 - Effect on tourism	9	19.1
SC9 - Effectiveness of flood protection provided	8	17.0
SC99 - Other	3	6.4
No Main Concerns	1	2.1

From this a detailed analysis was carried out, highlighting specific issues relating to the concerns. These were identified as:

- Additional Car Parking
- Memorial Green
- Old Fish Quay
- Interface between users of frontage

Further information was also obtained on the use of the Memorial Green. In particular, access to the War Memorial for Remembrance Services was mentioned as an important consideration. The Memorial Green was also considered an important amenity asset of the Largs sea front and consequently the schemes put forward were seen to have a detrimental impact on this asset (particularly in relation to family use of the open area).

Furthermore the issue of car parking had not been identified during the feasibility study, although North Ayrshire Council were well aware of the strength of opinion, both for and against additional car parking. This latter point was found to be significant in consultees response on options.

In comparing the responses of those favouring option 6 compared to option 7, opinion was split (49% for option 7, 21% for option 6) as shown in Figure 3.

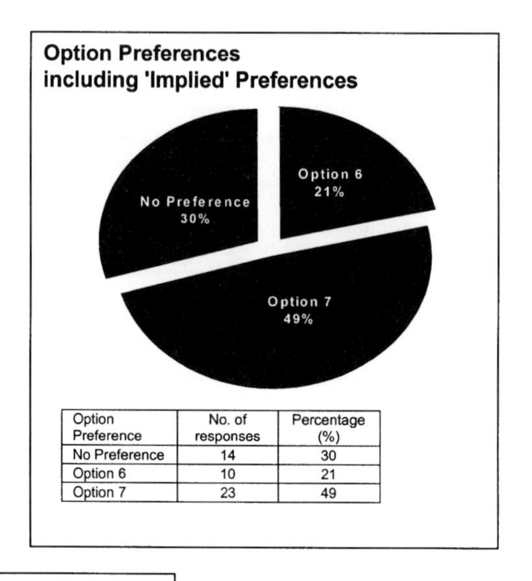

Option Preference	No. of responses	Percentage (%)
No Preference	14	30
Option 6	10	21
Option 7	23	49

Figure 3. Option Preferences

The responses indicated that the most important reason given for preference of option 6, which resulted in a narrower promenade, was that this would exclude any opportunity for additional car parking. Those in favour of option 7 identified the need for a wider promenade, reducing the impact of the floodwall. However, the car parking issue overshadowed other requirements of the scheme in the choice of options; such that a response might identify the need for a wider promenade to allow less conflict between users or to minimising the impact of defences and yet still opt for option 6 so as to exclude parking of cars.

The overwhelming reason for selecting neither option was the issues relating to the Memorial Green. The report on the consultation process set out a series of recommendations for developing a revised preferred option. This report was re-issued to all those who took part in the consultation for comment.

Public Meeting

Following the preparation of the consultation report, the outline design was substantially altered.

The defence in the area of the memorial green was brought forward to the landward side of the promenade, still allowing landscaping behind to reduce its visual impact. This minimised the impact on the green, maintaining it as a useful open area. Steps and seating were incorporated within the design to address issues raised by consultees.

The width of the promenade suggested in option 7 was reduced and vehicular access restricted so as to demonstrate that this area could not be used in the future for car parking.

These proposals were taken to a public meeting and were approved. The final scheme, which is now under construction was then submitted to formal consultation under the provisions of a Flood Prevention Scheme. Only one objection was raised and this in an effort to highlight the issues nationally, relating to standards of defence. This objection was withdrawn after further discussion.

At the commencement of the consultation, one consultee came armed with an umbrella, offering to sort out those intending to destroy the promenade. A mark of the benefit of good engagement and sensible argument was the fact that that person complimented the Council on its consultation process at the end of the public meeting, having also contributed herself to the eventual design.

Conclusions

The various examples of consultation given throughout the paper demonstrate that it is not enough to merely do one's duty. Good consultation has to be taken to all those who might have an interest. The process has to engage people, not just organisations, at a stage where their opinions and their knowledge can help develop appropriate answers; answers which address their concerns.

Development of strategies or schemes necessarily require an iterative process. This involves:

- An initial identification of interests.
- An analysis of where these interests may result in conflict either with other issues, the physical processes or legislation.
- Informing consultees of such conflict, raising their awareness of the problems and inviting them to become party to the solutions.
- Encouraging alternative options to be put forward and encouraging consultees to offer clear objectives from which imaginative options may be developed.
- Encouraging an ethos whereby consultees do feel that their views and interests are being listened to but can appreciated that not all can be incorporated into the final scheme
- Confirming that we, as coastal managers, have got it right.

Associated with this is the need to appreciate that consultation involves time and effort. This, however, is amply repaid in the short term by reducing the effort required in addressing formal objections or defending decisions at public enquiry, and in the longer term in meeting the functional requirements of those living on or involved with the coast.

Argument and even conflict are essential tools in consultation, opening the way to building consensus. The only consultation fatigue that is likely is that of the study team, not, if they are actively involved, amongst those whose views are being properly considered.

Consultation cannot always resolve all issues. Where conclusions are drawn that someone's livelihood is a stake, where there can be no offer of aid, where even further defence of their property will not be permitted, then there will be continued conflict. We do, however, have duty to provide a full and understandable explanation.

In many respects the highest compliment in consultation is reflected in a response from the work on the Suffolk Estuaries.

"I totally disagree, from a personal level, with the action you are proposing, but I understand and appreciate why such action will be taken; and you have given me knowledge to fight you all the way!"

References

Whitehouse.R., 2000. EMPHASYS Consortium, Modelling Estuary Morphology and Process, Study Report, MAFF FD1401, HRWallingford Report TR111. Dec 2000.

Brew D.S., 2002, Estuaries Research Programme Phase 1, Uptake Project. DEFRA FD2110, Posford Haskoning 2002.

Pontee N.I. and Townend I.H., The development of a framework for estuary shoreline management plans. ICE proceedings Coastal Management Conference, Sept. 1999, Thomas Telford.

Pettitt N., Cottle R.A, Guthrie J.G.L., Roberts W., Taking an integrated approach towards flood defence planning in estuaries. ICE proceedings Coastal Management Conference, Sept. 1999, Thomas Telford.

Brooke J., The Bristol Channel Marine Aggregates Resources and Constraints Research Project: keeping the consultation process clear of murky waters! ICE proceedings Coastal Management Conference, Sept. 1999, Thomas Telford.

Posford Haskoning 2001, Borth Coastal Study Project Report for Cyngor Sir Ceredigion.

The sustainable management of the Humber Estuary

Dr A.M.C. Edwards, Environment Agency, Hull, UK
P.J.S. Winn, Environment Agency, Leeds, UK

Introduction

The Humber Estuary with its ports and chemical industries plays a vital role in the UK's economy. It is also of outstanding value for wildlife conservation, particularly for water birds. Much industry, communications infrastructure and high-grade farmland is located on the tidal floodplain, where also live over 300 000 people - all are protected by tidal flood defences. Sea level is rising relative to the land and the rate of increase is expected to increase as a result of global warming.

This paper examines the sustainable management of the Humber so that economic and social aims are facilitated while at same time the environmental riches are enhanced. In particular it discusses cross-cutting matters such as partnership working, the integration of plans, protection of habitats and climate change in relation to water quality and the long-term planning for flood defence.

The Humber is one of the North Sea's principal estuaries. Its catchment is the largest in the British Isles, covering an area of approximately 24 472 km^2, which represents one fifth of the land area of England (figure 1); it has a population of nearly 11m. The estuary has the UK's largest complex of ports and one of the country's biggest clusters of the chemicals industry.

The Humber, which has a length of 62km, is formed by the confluence of the Trent and Yorkshire Ouse at Trent Falls. The tidal length of these rivers and their tributaries is 255km. The Humber is a dynamic estuary with a tidal range of up to 7m, and the channels and sandbanks that are continually shifting. It is renowned for its inter-tidal habitats and bird populations with the importance for nature conservation recognised by a number of national and international designations. The inter-tidal area has been greatly reduced since large-scale reclamation commenced in the seventeenth century; the drained floodplain now amounts to 90 000ha. There are developing fisheries and the area has a rich archaeological and historic heritage.

The management background

The overarching concept influencing much environmental policy and legislation is that of sustainable development. The Brundtland Report (World Commission on the Environment 1987) defines this as:

"The development that meets the needs of the present without compromising the ability of future generations to meet their own needs."

Sustainable development is comprised of non-inflationary economic growth, social cohesion through the access of all to employment and a high quality of life, and enhancement and maintenance of the environmental capital on which life depends. Sustainable development is, thus, concerned with the integration of human needs and the environment so that the quality

Coastal Management 2003, Thomas Telford, London, 2003.

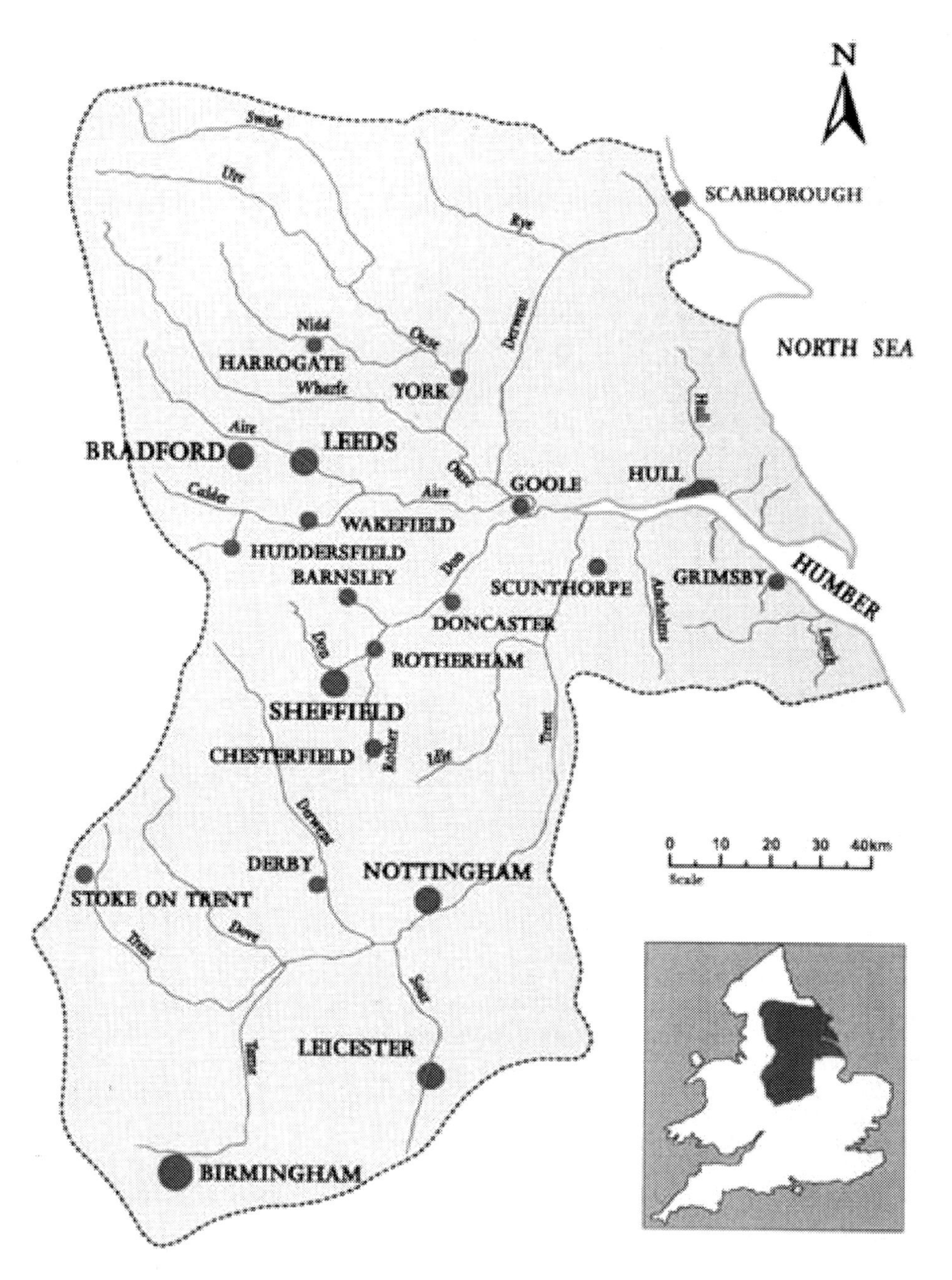

The Humber Catchment

Figure 1

of life is enriched. These concepts are encapsulated in the UK Strategy for Sustainable Development, "A better quality of life" (Department of the Environment Transport and the Regions, 1999) by the following aims:

- Social progress which recognises the needs of everyone;
- Effective protection of the environment;
- Prudent use of natural resources;
- Maintenance of high and stable levels of economic growth and employment.

Other principles of the UK's approach to sustainability are taking a long term perspective; integrated management; taking account of costs and benefits; making the polluter pay; following the precautionary principle; working in partnership – public sector, business and voluntary organisations; enabling people to participate in decision-making and be properly consulted on matters which affect them; basing decisions on the best available information and sound science; and being transparent, open and informative in all decision-making.

Sustainable development is not just concerned with balancing economic, social and environmental interests; in seeks wins for, or at least no adverse effects on, all three components – the triple bottom line. Compensatory activity may, thus, be needed to offset an unavoidable impact. An example relevant to the Humber is the creation of new inter-tidal habitat to replace that lost by engineering works.

The Vision

The Environment Agency's approach to its tasks is based on "A better quality life" and is set out in "The vision for our environment: making it happen" (2002a). This is translated into the following vision for the Humber:

- An unpolluted estuary albeit one which is naturally muddy;
- A healthy fishery with a sustainable run of migratory fish to and from the inland catchments;
- Thriving wildlife with an expanded inter-tidal habitat;
- Flood defences that provide an assured level of protection while maintaining the estuary's natural processes;
- Prosperous ports, industry and agriculture, which have minimised their environmental impacts and benefit from the diverse and healthy Humber, and the rivers and canals running to it;
- A landscape with outstanding archaeological and historic features that are enhanced by the Environment Agency's activities;
- A tourist economy that benefits from the Humber but is sensitive to its wildlife and historic riches;
- A vibrant community that understands, cares for and enjoys the Humber.

Regional and Sub-regional context

The Environment Agency's management aims and priorities are set out in its Humber Action Plan (1998) and the issues are further explored in the Humber State of the Environment Report (Environment Agency 1999). Since these publications were produced much work has been done on the functional plan for flood defence – the Humber Estuary Shoreline Management Plan (Environment Agency 2000a).

The Agency's plans and strategies are influenced by (and in turn exert influence on) the plans, strategies and management schemes of other organisations. Most of the tidal system is in the Yorkshire and Humber Region with the estuary itself being largely in the Humber sub-region based on four local unitary councils – East Riding of Yorkshire, Kingston upon Hull City, North Lincolnshire and North East Lincolnshire. A variety of regional and sub-regional plans are relevant for the Humber. They were largely produced for specific purposes but the aim is to gain the integration of such initiatives into a coherent whole. Working in partnerships is also vital, as there are many users of the estuary and a number of organisations with statutory powers contributing to its management.

The Government Office, the Yorkshire and Humber Assembly (comprising the 22 local authorities and a range of organisations representing, business, education, social welfare, the environment and the voluntary sector) and Yorkshire Forward (the Regional Development Agency) have developed "Advancing Together", three components of which are:

- Regional Planning Guidance (Government Office for Yorkshire and the Humber 2001), which provides the context for the land, use plans of the local planning authorities.
- Regional Economic Strategy (Yorkshire Forward 2003), which also has sub-regional action plans.
- Regional Sustainable Development Framework (Yorkshire and Humber Assembly 2001), which includes criteria and procedures to aid sustainability appraisal.

This over-arching framework aims to ensure that key strategic activities in the region and those of partners, including the Environment Agency, are co-ordinated. A very large number of organisations representing all facets of the region's community have the opportunity to participate through the Assembly's commissions and other partnerships. Sustainability is recognised as a key issue for economic and land use planning; sustainability appraisal was used in the development of the Regional Planning Guidance and Regional Economic Strategy.

Land Use

The whole of Goole and Hull are on the Humber's floodplain, as are appreciable areas of Cleethorpes, Grimsby, Immingham and other settlements. Likewise much industry and most of the strategic sites designated for large-scale industrial development are also in potential flood risk areas (and beside the Special Protection Area). Government policy on development and flood risk is set out in Policy Planning Guidance 25 (Department of the Environment Food and Rural Affairs 2001). The Agency has worked to gain policies to minimise new development in floodplains in the Yorkshire and Humber Regional Planning Guidance. These policies then influence the local and structure for the Humber sub-region. The four unitary local authorities are in the process of preparing strategic flood risk assessments to ensure that land use planning policies and land allocations reflect flood hazards.

Economic planning and regeneration

The Humber Forum is the public private partnership of the four unitary local councils and business, which leads the sub-regional elaboration of the Regional Economic Strategy. The Humber Economic Development Action Plan (Humber Forum 2003) identifies priority business clusters for chemicals and process industries, food, ports and logistics, and the digital and electronics sector. The Humber Forum also has progammes on Rural and Urban Renaissance, and the Fishing and Coastal Communities

The Humber Trade Zone (HTZ) (Yorkshire Forward 2001) is a component of the Humber Economic Development Action Plan. The aim of HTZ is to exploit the current and future potential of the Humber ports for the greater wealth of the UK whilst ensuring that the benefits are specifically captured in the Humber sub-region and wider Yorkshire and Humber region. The objectives are:

- To attract strategic inward investment;
- To promote increased use of the Humber ports by UK exporters and importers;
- To gain more "added value" in the sub region and region from goods going to or coming from the ports;
- To resolve transport blockages;
- To provide serviced sites;
- To improve labour skills.

Sustainability and environmental protection are central to the Economic Action Plan and HTZ. The Environment Agency works with Humber Forum and Yorkshire Forward to see that theses issues are addressed and to ensure that the Agency's local priorities reflect where practical those of the wider community.

It should be noted that major port developments are in progress or planned at Immingham and Hull. The Immingham Outer Harbour project will result in the loss of 22ha of inter-tidal mud and, will require new habitat to be created to compensate for the loss.

Conservation

The conservation value of the Humber influences all aspects of its environmental and economic management. The habitats of particular interest are coastal lagoons, Atlantic salt meadows, mudflats, sandbanks and the estuary itself, which has populations of the endangered river and sea lampreys. There is also a thriving grey seal colony at Donna Nook at the estuary mouth. A review is in progress of the wildlife designations - Humber Sites of Special Scientific Interest (SSSI), the possible extension of the Special Protection Areas for Birds (SPA) and the RAMSAR site, and the potential designation of the whole estuary east of Trent Falls as a Special Area for Conservation (SAC). The SAC and SPA are European designations that are incorporated into UK law by the Habitats Regulations 1994.

The Regulations require a review of all discharge consents, integrated pollution control authorisations, radioactive substances permits, waste management licences and abstraction licences that could have an effect on designated site. An appropriate assessment is needed before a new permit is granted, or flood defence or port works undertaken, if there could be an effect on the integrity of the site. Mitigation at the site must be providing for any unavoidable adverse impacts. Failing that and where there is an "over-riding public need", compensation may have to be provided elsewhere.

The Regulations require "relevant authorities" to produce a single management scheme covering day to day activities, rather than "plans and projects", for the Humber European Marine Site. There are 39 Humber Relevant Authorities – Environment Agency, English Nature, local authorities, internal drainage boards, sea fisheries committees, water companies, navigation and harbour authorities, and the Ministry of Defence (for the weapons range at the estuary's mouth). The preparation of the partnership's scheme is advised by a "package" of measures on the features of interest, conservation objectives and favourable status conditions (English Nature 2002). The management scheme should be published in 2004 and will consider the operations currently undertaken in relation to the conservation criteria and produce an action plan to reduce any impacts of activities that could compromise the site's conservation status. The Relevant Authorities work with the independent Humber Advisory Group, which has representatives from agriculture, community groups, conservation groups, fishing, water sports, wildfowling, industry and the recently formed Humber Industry Nature Conservation Association.

Water quality

The pollution of the Humber has reduced greatly in recent years. Water quality management is largely based on environmental quality standards laid down in European Directives and UK legislation plus the requirements for processes and emission standards of the Urban Waste Water Treatment Directive and Integrated Pollution Control (now being replaced by Pollution Prevention and Control). The objectives and approach is described more fully by Environment Agency (1999) and Edwards (2001).

The quality of the Humber is now generally good. In the past the rivers draining South and West Yorkshire and the Trent catchments were seriously polluted, and control over discharges to tidal waters was weak until the mid 1980s. The major sewage effluent discharges to the Humber now have two-stage treatment - Hull's sewage was discharged untreated until 2001. This clean up was largely driven by the European Directives - on 'Pollution from Dangerous Substances discharged to the Aquatic Environment of the Community', 'Titanium Dioxide', 'Urban Waste Water Treatment' and 'Bathing Waters' – and the introduction of Integrated Pollution Control. The Cleethorpes beach, the seaside resort within the outer Humber, now complies with the standards laid down for bathing waters. The estuary is largely in a "favourable condition" in respect to the protected habitats. The conditions of many consents and authorisations for the discharge of effluent are, however, now the subject of appropriate assessments for the Habitats Regulations as some species may benefit from higher quality standards. This task has to be completed by 31 March 2006.

The Humber has a muddy appearance because the mobile bottom sediments lead to a high concentration of sediment suspended in the water column. The turbidity maximum is located in the tidal Ouse, where suspended particulate matter can exceed 70,000 mgl^{-1} (Uncles and Stephens 1999). Depleted dissolved oxygen in the tidal rivers, particularly the Ouse, in summer at times of low freshwater flow is a long-standing problem. It the past much organic pollution was received from the inland catchments and direct discharges to the tideway. Dissolved oxygen levels have increased as a result of investment in effluent treatment plants (Figure 2) but have not fully recovered. This is partly a natural phenomenon resulting from the high suspended sediment concentration but a further reduction in effluent load is being sought.

The environmental quality standards for metal and organic pollutants are now generally met. In the past there was a large input of metals from industry on the banks of the Humber, including arsenic and cadmium from a former tin smelter west of Hull. The treatment of titanium dioxide wastes resulted in a rapid reduction of iron levels in seaweed and the disappearance of ochrous staining from sea walls. Metals are readily absorbed onto the fine-grained sediments and are either stored within the Humber's mudflat or pass into the North Sea.

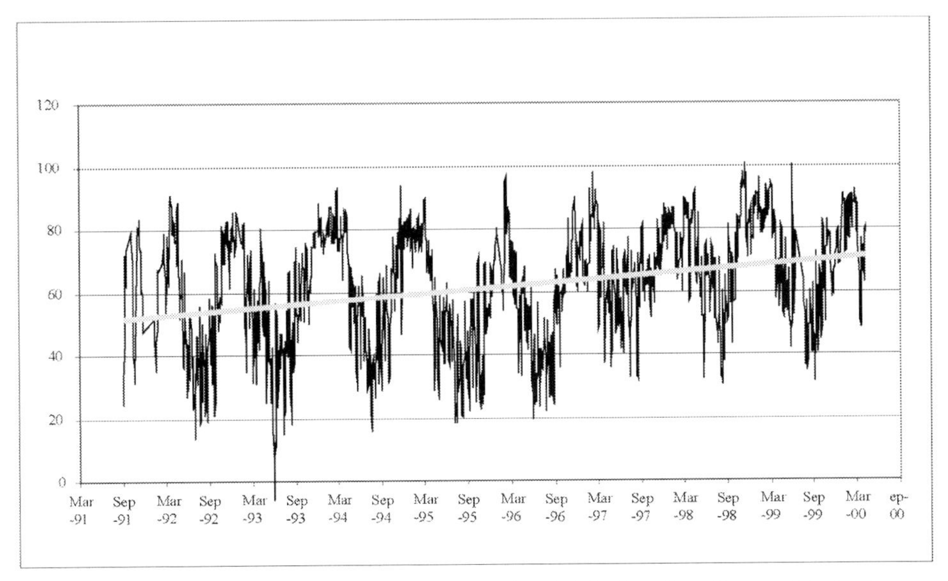

Figure 2 Trend in dissolved oxygen saturation (%) in the River Ouse at Blacktoft

The large flow of the Humber's and the concentration of people, industry and intensive agriculture in the catchment means that it is a significant input of metals, organic substances and nutrients into the North Sea (National Rivers Authority 1995). Cadmium, copper, zinc, lead, lindane and nitrogen and phosphorus compounds are of greatest significance in relation to the total land input to the North Sea, although by far the largest component comes from the continental rivers. The Humber is hypernutrified but there is no evidence of algal problems as plant grown is inhibited by low light penetration resulting from the high natural turbidity.

A number of the chemical companies have active waste minimisation programmes. Demonstration projects, including ones undertaken in the Humber catchment (Edwards and Johnson 1996), highlight the practicality of reducing the waste of water, materials and energy in the production process by modification of processes or products and by good 'housekeeping'. Frequently there is also a rapid payback to industry resulting from reduced material and effluent treatment costs. The minimisation of waste at source, rather than reliance on "end of pipe (or chimney) treatment" is a key concept for sustainable development as:

- It reduces the exploitation of raw materials, including water and energy;
- It reduces emissions and discharges;
- It reduces the risk of pollution incidents from process failure and other accidents as systematic procedures are adopted;
- It frequently saves businesses money.

The Agency with other partners are promoting widely the benefits of waste minimisation for business and the environment. Some firms need help to adopt these methods and the use of procedures such as environmental management schemes. Improving environmental performance is, thus, recognised as a benefit to the region's economy and there are projects to help industry, which are part funded by the Regional Economic Strategy
Systematic surveys of the benthic invertebrate biology of the intertidal and subtidal areas commenced in the mid 1970s. This monitoring shows that the biology of the Humber follows a typical estuarine pattern. The inner estuary with its mobile bed and large fluctuations in salinity over the tidal cycle is a hostile environment for macroinvertebrates. The abundance of taxa and individuals increases towards the mouth of the estuary and provides food for the birds and fish.

Fish stocks in the Humber have risen with the improvement in water quality. The outer Humber is a nursery ground for North Sea fish, particularly plaice. The commercial fishery within the estuary is of minor importance, although eels are caught and there is angling interest. There is a developing cockle fishery at the Humber's mouth. Over 80 species are recorded in recent years and salmonids are starting to return to the Trent and Yorkshire rivers. Lampreys are found in some abundance as they migrate into the North Yorkshire rivers. There remain other issues to be resolved to ensure a healthy and self-sustaining fish populations. These include removal of obstructions to the movement of fish into the tributaries and reducing their entrainment by some water intakes.

In 2001 The Humber ports and wharves handled 41.5m tonnes of bulk oil and chemicals. Approximately one fifth of the UK's crude oil is refined in two oil refineries in the South Bank industrial zone near Immingham. A major oil spillage would devastate habitats, birds and fish as well as despoiling the beaches and the coastline. Another partnership of Humber port interests, local authorities and environmental regulators has prepared contingency plans to minimise the risk of the occurrence of spillages and to reduce the impact of any such incident. There are regular exercises of the plans.

Flood Defence: Humber Estuary Shoreline Management Plan

Sea level is rising relative to the land as a result of global warming and the downward tilt of this part of the country. The average rate over this century is predicted to be 6mm per year, compared with 2mm per year over the 20th century. The Environment Agency is developing the Humber Estuary Shoreline Management Plan (HESMP) to counter the heightened flood risk and the associated increase in erosion and loss of mudflats. The work is funded by the Lincolnshire, Midlands and Yorkshire Flood Defence Committees and receives grant aid from the Department for the Environment Food and Rural Affairs.

The Objectives of the project are:

To develop a coherent and realistic plan for the estuary's flood defences that is:

- Compatible with natural estuary processes;
- Compatible with adjacent developments, including preferred options for adjoining lengths of frontage;
- Sustainable taking into account future changes in the environment (human, built or natural), in sea level and in the climate.

To ensure that all proposals are:

- Technically feasible;
- Economically viable;
- Environmentally appropriate;
- Socially acceptable.

The detailed objectives given in table 1 demonstrate the wide range of issues that are being investigated.

Flood defence • To reduce the risk to people, property and the environment from flooding and erosion. Land use and planning • To provide standards of protection that are consistent with existing land use while permitting future development where appropriate. • To encourage the recognition of flood risk as an issue in regional planning guidance and in structure and local plans. Industry and commerce • To provide appropriate protection for industry and commerce and encourage future industrial and commercial development in suitable locations. Navigation and port development • To avoid adversely affecting navigation in the estuary or opportunities for its development. Fisheries • To avoid adversely affecting fisheries (inland or in the estuary) or the fishing industry. Agriculture • To protect, where appropriate high quality agricultural land by the estuary. • To minimise the impact of natural processes on land drainage to the estuary. Community interests • To protect the overall interests of people living near the estuary. Tourism • To allow for the importance of tourism to the local economy	Sport, recreation and access • To maintain and, where possible, improve the provision of sporting and recreational facilities by the estuary. Heritage and cultural resources • To avoid actions that adversely affect the estuary's heritage and cultural resources. • To comply with all statutory obligations arising from national and local designations and related legislation. • To protect, where necessary, the estuary's heritage and cultural resources against erosion. Landscape • To protect and, where possible, enhance the estuary's existing landscape character. • To complement the objectives of the Heritage Coast Management Strategies. Estuary processes • To build an understanding of the natural processes taking place within the estuary and work with these processes. • To respond to future climate change and sea level rise. Nature conservation • To comply with all statutory obligations arising from national and international designations and related legislation • To encourage habitat development that contributes to the UK Biodiversity Action Plan. • To create areas of new habitat in compensation for any habitat lost.

Table 1 Detailed objectives for the Humber Estuary Shoreline Management Plan

A shoreline management plan is a document which sets out the long-term, sustainable strategy for coastal defence for a specific length of coast taking account of natural processes, and human and other environmental influences (Ministry of Agriculture Fisheries and Food 1995). The purpose is to enable decisions to be made on flood defence investments in an integrated manner. Initially plans were prepared for 11 open coast "sediment cells" around England and Wales. The large size and complexity of the Humber's tidal environment led to the extension of the shoreline management concept to the estuary. An essential part of the project is to ensure that any flood defence works are compatible with the Humber's navigation and its designated habitats in compliance with the requirements of the European Habitats Directives and the UK Habitats Regulations 1994.

The approach to developing the Plan is:

- Gather information on:
  - The state of the defences;
  - Tides, channels and the physical state;
  - The ecological condition;
  - The human environment including heritage features;
  - Policy and legislation.

- Gain understanding of:
  - The estuary physical and biological processes;
  - Human issues.

- Wide and detailed consultation including partnership working with the principal organisations with an interest in the Humber.

The Agency works with many Humber stakeholders – Government departments, local authorities, other public agencies, ports, industry, fisheries, agriculture, landowners and tenants, developers, sporting interests, and heritage and conservation bodies. It is advised by the Humber Estuary Shoreline Management Plan Steering Group, on which such organisations are represented.

The plan is based on innovative studies and modelling of the physical behaviour of the Humber and its sediment movement, which an international consultancy consortium is undertaking for the Agency. This work began in 1998 (Environment Agency 2000b) and is now in a further phase. It has centred on historic analysis of changes in the estuary, long-term 'regime' modelling designed to deliver predictions of the general shape of the estuary and the extent of inter-tidal habitat loss over the next 50 years, and three dimensional modelling. The latter work will provide indications of both the local and more remote changes resulting from large-scale realignment of defences.

A key document, which informs HESMP, is the Humber Coastal Management Plan (CHaMP) (Environment Agency 2001). This provides an estimate of the potential gains and losses of habitat resulting from flood defence works and 'coastal squeeze'. The latter is the result of the rise in sea level being confined by embankments, floodwalls and other hard defences.

The broad strategy set out in the first HESMP is to maintain a line of defence around the Humber and to assess if the alignment can be improved in some places. In front of the urban areas of Hull, Cleethorpes and Grimsby there is no alternative to raising the defences on their existing line. Floodwalls were recently constructed along the dock frontages in Hull and new lock gates were also fitted. The need for these schemes was to bring the standard of defence for the City up to the indicative standard of protection of a 0.5% chance of over-topping in a year (the 1:200 year criteria). The design took into account the best estimate of the rise in sea level to 2050 of around 300mm.

The first HESMP (Environment Agency 2000a) is currently being refined, particularly by assessing the measures to counter coastal 'squeeze' by setting back some lengths of embankment to return farmland to wetland (Environment Agency 2002b). The modelling work indicates that coastal squeeze is not an issue for the tidal rivers but there could be substantial habitat losses around the Humber itself. There is a statutory duty to compensate for these losses in the inner, middle and outer parts of the estuary. The indication is that approximately 1000 ha of washland and new wetland habitat should be created. Managed realignment can, thus, have the following objectives:

- To achieve greater stability of embankments by having foreshore in front of them to dissipate the erosive energy.
- To reduce flood defence maintenance and capital costs
- To lower the highest water levels in the tidal rivers by creating washlands
- To compensate for the loss of protected habitat by engineering works
- To compensate for loss of protected habitat by sea level rise – "coastal squeeze". Very valuable new inter-tidal habitat for plants and birds can be created.

There is also some indications that setback can benefit fisheries, and provide a sink for carbon and nutrients (Jickells et al 2000).

The first two Humber managed realignment sites (Thorngumbald and Alkborough) are under development (see below). Another eleven potential sites (Figure 3) are being evaluated – four are possible washlands beside the Ouse and Trent, while the others are to counter coastal squeeze. The inundation of the washlands will probably be infrequent and agriculture should be possible most of the time. The Agency is likely to purchase the "coastal squeeze" sites. The concerns of the farming community are appreciated and being addressed by detailed consideration of the location and boundaries of the land required. Land purchase will probably be over about 15 years, which provides some flexibility.

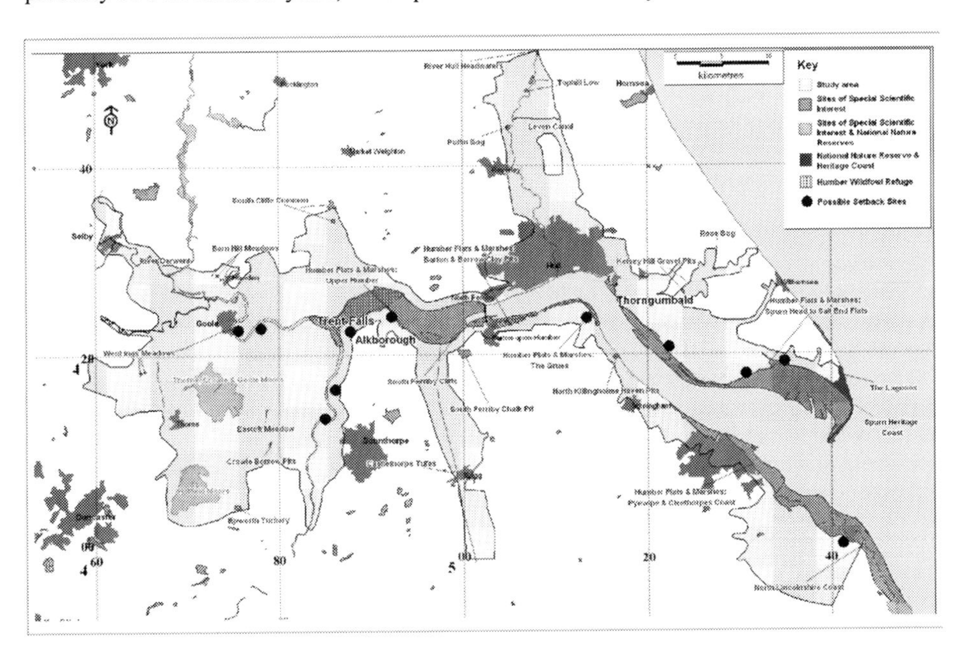

Figure 3 Humber Estuary Shoreline Management Plan and potential realignment sites

The refinement of HESMP will be subject to a strategic environmental assessment and sustainability appraisal, before publication at the end of 2003. These assessments will link with economic analysis, which is to demonstrate the robustness of the approach and to justify expenditure on the projects to implement the plan. The Agency is not required to carry out a strategic environmental assessment or prove the sustainability case, but both are likely to become standard in the near future.

The plan will be implemented by the long-term Humber Flood Defence Strategy, which is likely to require an investment of £250m - £300m over a bout 30 years.

Thorngumbald and Alkborough Managed Realignments

At the start of the work on HESMP there were pressing flood defence problems in various parts of the estuary, which required action before the plan was completed. These works have been undertaken in line with the principles of the plan and informed by the modelling and other technical studies. One such problem was the low bank at Thorngumbald (to the east of Hull on the north side of the estuary). Another was on the South Bank at Pyewipe between Immingham and Grimsby where foreshore lowering was in places undermining the sea wall in front of the industrial zone. There was no room to do work behind the existing defence because of the presence of chemicals works. There were objections to encroachment on the seaward side because a small area of the Special Protection Area for Birds (SPA) would be lost.

The solution produced was to set back the defence at Thorngumbald to creating 75 ha of new inter-tidal habitat, much of which will be initially, mudflat. Part of this area will compensate for encroachment into the SPA at Pyewipe and to offset the impact of future flood defence schemes. The Secretary of State for the Department of the Environment Food and Rural Affairs gave "over riding public interest" approval for work affecting the SPA, which enabled planning permission to be granted.

The Pyewipe scheme was completed in the summer of 2002 and the existing floodbanks at Thorngumbald are due to breached in the autumn of 2003. The transformation of the site to saltmarsh and mudflat will be closely monitored. These schemes highlight the need to look at the complex and extensively used estuary like the Humber as one entity and seek novel solutions.

A consortium formed of the Environment Agency, English Nature, the Countryside Agency and North Lincolnshire Council is purchasing the 440ha Alkborough site at the confluence of the Ouse and Trent. The aim is to return farmland back to the mudflat and saltmarsh, which existed prior to reclamation. This project will compensate for coastal squeeze in the inner estuary, should reduce high water levels in the tidal rivers, and will have other benefits being for wildlife conservation, flood defence and the economy of the local rural community. Extensive modelling is being undertaken to determine the configuration of the breach or breaches, and to ensure that the design does not adversely affect the navigation. The Alkborough project has received funding from Yorkshire Forward via its Humber Trade Zone initiative and from Europe.

Conclusions

The Humber is of vital importance to the UK's economy and environment. Its international wildlife conservation status influences all uses of the estuary's waters and activities around its banks.

The achievement of the goals of economic development and regeneration, flood protection, and environmental enhancement requires the use of novel solutions, such as waste minimisation and managed realignment. Partnership working is well developed and action is in progress to ensure that the numerous plans, strategies and management schemes and compatible and fully integrated. The Humber is a testbed for practical sustainable management.

Acknowledgements

We are grateful to colleagues who have provided information. The views expressed are the authors and are not necessarily those of the Environment Agency.

References

Department of the Environment Food and Rural Affairs (2001) *Development and flood risk.* Planning Policy Guidance No. 25, The Stationery Office, London, 52pp.

Department of Environment, Transport and the Regions (1999) *A better quality of life: a strategy for sustainable development for the UK.* The Stationery Office, London, 96 pp.

Edwards, A.M.C. (2001) *River and estuary management in the Humber catchment.* in Huntley, D., G. Leeks and D Walling (eds.) A land-ocean interaction study: measuring and modelling fluxes from rivers to the coastal ocean. IWA Publishing, 9 – 32.

Edwards, A.M.C and Johnson, N. (1996) Water and Wastewater minimisation: the Aire and Calder project. *J.C. Inst. Water Env. Manag.*, 10, 227 – 234.

English Nature (2002) *The Humber Estuary European Marine Site: English Nature's advice given under Regulation 33(2) of the Conservation (Natural Habitats etc.) Regulations 1994.* 134pp.

Environment Agency (1998) *Humber Estuary action plan: our commitment.* 58 pp.

Environment Agency (1999) *Humber Estuary - state of the environment report 1998.* 47 pp.

Environment Agency (2000a) *Planning for the rising tides: the Humber Estuary Shoreline Management Plan.* 58 pp.

Environment Agency, (2000b) Humber Estuary geomorphological studies: Phase 2 - final report, volume 1. 104pp.

Environment Agency (2001) *Scoping report and brief for the Humber CHaMP.* 31pp.

Environment Agency (2002a) *Our vision for the environment: making it happen – draft corporate strategy.* Bristol, 59pp.

Environment Agency (2002b) *Planning for the rising tides, the Humber Estuary - consultation on managed realignment: information for landowners and tenants.* 14pp.

Government Office for Yorkshire and the Humber (2001) *Regional Planning Guidance for Yorkshire and the Humber (RPG12).* The Stationery Office, London, 155pp.

Humber Forum (2003) *The Humber Economic Development Action Plan for the period April 2003 – March 2006.* Hull, 44pp.

Jickells, T., Andrews, J., Samways, G., Malcolm, S., Sivyer, D., Parker, R.., Nedwell, D., Trimmer, M. and Ridgeway, J. (2000) Nutrient fluxes through the Humber Estuary – Past, present and future. *Ambio* 29, 130-135.

Ministry of Agriculture, Fisheries and Food (1995) *Shoreline management plans, a guide for coastal defence authorities.* MAFF Publication PB2197, London, 24pp.

National Rivers Authority (1995) Contaminants entering the sea : a report on contaminant loads entering the seas around England and Wales for the years 1990 – 1993. *NRA Water Quality Series,* No 24, HMSO, London, 94 pp.

Uncles, R.J. and Stephans, J.A. (1999) Suspended sediment fluxes in the Tidal Ouse, UK. *Hydrol. Process,* 13, 1167 – 1179.

World Commission on the Environment (1987) *Our common future.* Oxford University Press, Oxford, 383 pp.

Yorkshire Forward (2001) *Humber Trade Zone – integrated development plan.* Leeds, 108pp.
Yorkshire Forward (2003) R*egional economic strategy: ten year strategy for Yorkshire and Humber 2003-12*. Leeds, 62 pp
Yorkshire and Humber Assembly (2001) *Advancing Together: towards a sustainable Region – the Regional Sustainable Development Framework for Yorkshire and Humber.* Wakefield, 36pp.

Challenges of implementing urgent works in the coastal and estuarine environment

Sarah Sinclair, Principal Engineer, Black and Veatch (Principal Author)
Tom Matthewson, Principal Environmental Scientist, Black and Veatch
David Barton, Engineer, Black and Veatch
Loreta Adams, Project Manager, Environment Agency

TOPIC 5/3

Coastal Management involves integrating the management of assets necessary to our wellbeing, such as flood defence or land drainage structures, with the natural, human and built environment. Engineering and Environmental practitioners are aware that coasts and estuaries often contain scarce environmental resources, whether natural or managed, which contribute to the local, national and international environmental assets of a country. Coasts and estuaries also offer visual amenity and resources which can be harvested or used by people, making them attractive places to live and work, both in present times and historically. The management of engineering works at the water's edge needs to ensure that all the general and specific issues relating to the environment are understood, and resolved as necessary. This is a particular issue when urgent works are needed because existing structures are in imminent danger of failure.

The authors are involved in implementing works for the Environment Agency, seeking government funding in the form of grant aid to capital works costs. Any works undertaken have to satisfy national and international regulations with regard to maintaining the natural environment. Statutory undertakers with interests in the area must be consulted to seek their approval. Landowners, local businesses and others with interests in the area must all be consulted, and agreements sought where necessary. Any necessary licences must also be obtained. In some cases, partners must be sought who would jointly benefit from any works, and would contribute to the costs of those works or take over longer term management of the asset.

Consultation is a long process, involving negotiation relating to the design of any works, construction methods, time-scale, access and any impacts on the environment in the short or long term, constraints and mitigation. Consultation is also nearly always an iterative process.

Recent studies by DEFRA have indicated that many of the coastal and estuarine flood defence assets are deteriorating in condition more quickly than money is being spent to maintain them. In this context, at least in the short term, much of the construction work carried out in the coastal and estuarine environment is likely to be urgent in nature. This paper examines the specific issues relating to undertaking urgent or emergency works in an environmentally sensitive area, using examples that were constructed in Spring 2003.

Coastal Management 2003, Thomas Telford, London, 2003.

Introduction

The British Isles has a population of approximately 57 million people. Many towns and cities are established on low lying land at the coast or along river valleys, where there is risk of flooding. Where areas have been subject to flooding historically, the remaining older parts of the town or village have tended to be outside the floodplain. However, development pressures and population growth have led to development on these prime sites, and over the past 30 years many have been developed. Recent Environment Agency figures estimate that within England and Wales, over 4 million people either live or work in flood prone areas.[1]. Coastal erosion is also increasing, partly due to the effects of sea level rise.

The majority of coastal flood and erosion protection in Britain is publicly funded, either through central or local government. The last severe coastal flood in Britain was in 1953, which led to an investment programme to improve coastal flood defence that lasted at least into the 1970's. Since then, although there have been some significant localised storms, there have been no events which have raised public awareness of the need for ongoing coastal defence investment.

A recent study of coastal flood defence and erosion protection assets [2] indicated that although £140 billion of land and property was at risk in England through coastal or tidal flooding or erosion, there was currently a shortfall in the level of investment needed to maintain current standards of protection, or to improve to meet the indicative standards of protection. At the time these findings were published, it was proposed to use this information to justify long term increases in the Government's funding programme. This will take time to be implemented – in the interim, many of the defences implemented from the 1950's to 1970's are starting to reach the end of their working life. In many cases, there has been historic under-investment in maintenance of these assets, and existing maintenance budgets are not sufficient to counter the deterioration.

The historic approach to flood and coastal defence has been piecemeal, often causing or exacerbating problems downdrift. In 1993 the Government published a Strategy for Flood and Coastal Defence [3] which led to a hierarchical approach to defence management, from Shoreline Management Plans (SMPs), through Strategy Plans to specific schemes. Although the approach has, in the Authors' view, been vital to maximise the effectiveness of government investment in flood and coastal defence, the process of undertaking SMPs and Strategy Plans has in itself been paced and prioritised such that many Strategy Plans are still currently being undertaken to assess the future justification for capital investment in defences. Whilst the prioritisation of these studies will have ensured that defence failure risks are managed, this process has deferred some of the capital spend as major schemes cannot be implemented without a Strategy in place.

The net effect of the lack of historic investment, and the requirements for the development of approved strategies, are that many assets are in poor condition. For some of these assets, the need for works to reduce flood or erosion risk, or manage localised failures, leads to urgent or emergency works before the strategy can be implemented.

Urgent Works

It is important to differentiate between 'urgent' and 'emergency' works as the approach to each and the funding mechanism for each is different.

Works to repair flood defence or erosion protection will only be deemed 'emergency works' if there is an immediate (in the event of a storm) risk to people or property. The emergency works undertaken will only be those necessary to alleviate the risk. For example, if a section of flood defence fails, exposing the previously defended area to an unacceptable level of flood risk, works are generally undertaken to minimise that risk by installing a temporary defence such as earth bund or sandbag wall behind the defence. The works are purely to manage short term risk, and are often of a temporary nature, to prevent damage whilst a capital scheme or strategy is developed.

There is no standard definition of 'urgent works'. However, urgent works are generally taken to be needed if the condition of a defence has deteriorated such that it would be likely to fail in a relatively minor storm, putting people or property, or significant environmental assets, at risk. The defence does not need to have failed.

Given the scale of risk to be alleviated with emergency works, works tend to be initiated immediately the problem is identified. Funding and permissions are sought retrospectively, although relevant statutory bodies are informed prior to any works occurring. For urgent works, the 'standard' approvals process needs to be followed, but usually over a very limited timescale. This will usually entail development of a justification report for the works, to be submitted to Defra for funding approval, followed by procurement of design and construction through a variety of routes. The time frame for urgent works is usually between one year to eighteen months from problem identification to completion of construction. Permissions and consents will be needed from statutory authorities. To achieve this, a process of consultation is needed, all within the reduced timescale.

An indication of the range of legislation under which consents are likely to be required for works in coastal and estuarine situations is provided below:

- Conservation (Natural Habitats &c.) Regulations, 1994 (for works in or near SACs and SPAs);
- Town and Country Planning Act 1988 (for works above Low Water);
- Food and Environment Protection Act 1985 (for works below Mean High Water Springs);
- Coast Protection Act 1949 & Harbour Works Regulations (ditto);
- Land Drainage Regulations 1991 (for works near a main river);
- Countryside and Rights of Way Act 2000 (for works in SSSIs);
- Controls on disposal of dredged material (under FEPA 1985).

In implementing urgent works at the coastline, the authors have encountered some general issues. These relate to the natural environment and the use of the local area, and significantly affect both access to the works and the timing of any works, as detailed below.

The Natural Environment

Almost all of the English coastline now has some form of environmental designation protecting it from potentially damaging development. At the international level, Special Protection Areas (SPAs) and Special Areas of Conservation (SACs), are designated under EU Legislation for the conservation of birds and other key habitats/species respectively. Several of these sites extend into tidal and coastal waters, although there are currently no sites designated offshore. At the national level, Sites of Special Scientific Interest (SSSIs) are the main conservation designation although these do not extend below the low water mark. The nature of the coastal or estuarine environment means that it provides some features or habitats of considerable conservation interest, such as brackish lagoons, saltmarsh, vegetated shingle, mudflats, coastal heathland, and dunes. Many areas are particularly important for birds, especially if there are freshwater roosting sites behind the defences. Furthermore, coastal areas can contain features such as cliffs that are designated as SSSIs for their geological interest.

Much of the coastline and estuaries have been the sites of human occupation for hundreds if not thousands of years. As well as living in the area and using its resources, people have often implemented defences either to defend against the incursion of the sea, or against foreign invaders. Former inhabitants have left a variable source of archaeological artifacts, most of which have probably not yet been uncovered. Those that are known about include Roman defences, bronze age workings and so on, up to more modern Napoleonic or second world war defences. Eroding cliffs may yield fossils; eroding foreshores may uncover drowned woodland or even 'Woodhenge'.

The coastline remains a valuable asset to humans today. It is used commercially for fisheries and harbours, and is used more and more as a recreational asset. Coastal footpaths and bridleways are used throughout the year. Recreational boating is increasingly popular. Other water sports such as surfing, windsurfing, waterskiing, jetskiing are undertaken in some areas. Recreational fishing and wild-fowling remain popular. Many coastal communities rely heavily on tourism for their income, and have developed tourist amenities such as caravan parks very close to the sea.

The environment, and the use of it, will always impose constraints on waters-edge construction projects. The waters-edge environment is much used by people and wildlife. In addition, it is not always easily accessible, particularly in less developed, more environmentally sensitive areas. These constraints will affect the design of the works; materials used, the aesthetics, the timing of construction, access routes (whether by land or sea), working hours, working methods, location of site compounds. Mitigation of adverse effects may be required.

In terms of mitigation, analysis and consultation with environmental bodies will be needed to determine whether effects are short-term construction-related impacts or longer term impacts of undertaking the works.

Particular considerations apply for defence schemes located in or near SPAs and SACs owing to the requirements of the Habitats Regulations. The developer for schemes in such areas is required to demonstrate *either* that the works are directly connected with management of the site; *or* that no significant adverse effects on the conservation site are likely. If there is a risk of such effects, it is necessary to provide the information to assist with the preparation of an 'Appropriate Assessment' of the scheme. This will establish, in consultation with English Nature, whether or not the defence scheme will result in an adverse effect on the integrity of the conservation site. If the assessment indicates the latter, the scheme will only be permitted to proceed if there are grounds of overriding public interest. In such circumstances, compensation habitat for that lost or damaged would need to be provided. Land purchase may be necessary to provide the replacement habitat and there is a general presumption that this habitat should be as close as possible to the site of loss.

In some circumstances, the compensation site may have been already identified through the CHaMP process. CHaMPs (Coastal Habitat Management Plans) are intended to identify among other things the habitat compensation required due to the implementation of flood and coastal defence works in SACs or SPAs. There is a risk of considerable delay whilst a suitable site is identified and developed, which may be reduced if a CHaMP is in place. In the cases where compensatory habitat is required it will be necessary to agree a pragmatic approach with English Nature that enables the urgent works to proceed in advance of the procurement or development of the compensation site.

Consultation

Consultation must be held with statutory authorities and others representing the interests of coastal users or managers. However, given that relatively few public objections are needed to stop a project from receiving some form of statutory consent it is important to consult the general public as well. This is time-consuming, and takes up a considerable part of the programme for urgent works; however it is necessary.

Generally, there may be an initial consultation exercise to scope the project and seek data, and to inform people of the project's existence. This is the start of developing a dialogue with affected parties and also informing other interested parties. Once options have been developed for the solution of the problem, these can be consulted on to gain an understanding of possible areas of concern relating to the proposals, which feed into the impact assessment and mitigation analysis within the option justification process. This process also identifies which long or short-term effects can be mitigated where appropriate. There is generally a further stage of consultation to inform interested parties of the preferred option as a result of the analysis process.

Once a preferred option is selected and detailed design commences, discussions will need to be held to develop any access agreements, agree the methodologies to allow for any constraints, fully establish likely effects and any mitigation or compensation requirements, finalise design details that are acceptable to consultees, and often to inform local bodies such as parish councils of the progress of the works and likely timing of construction.

Summary

As can be seen above, there is a significant amount of time consuming work, and often an iterative design process involved in implementing urgent works at the coastline. The environmental sensitivity of sites requires early consideration of issues. Early and ongoing consultation is also necessary to ensure that an appropriate solution can be reached in the restricted time frame. Figure 1 demonstrates the level of consultation required

Figure 1: Example flowchart of communication requirements

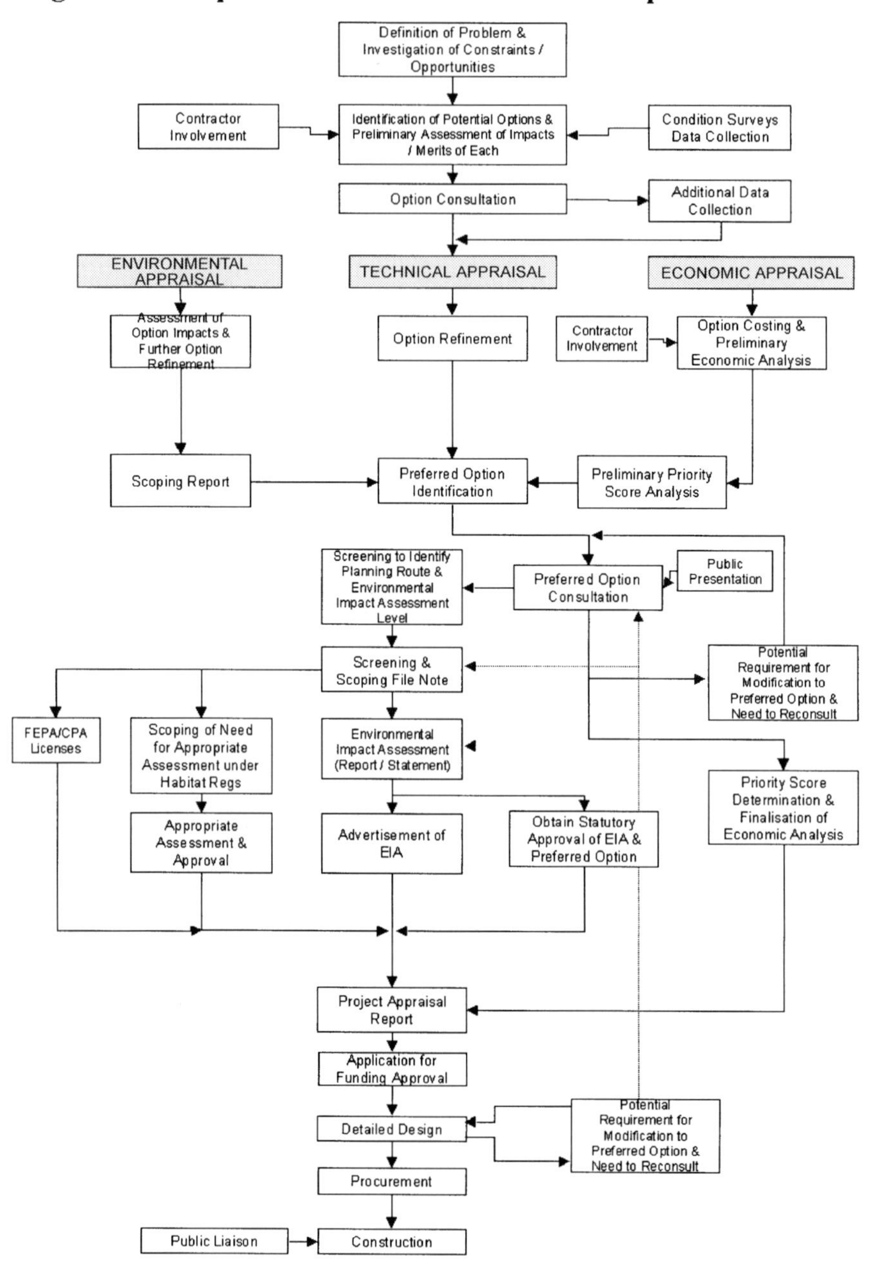

However, in some cases, the environmental sensitivity of the site, particularly if it is a European designated site, can help to focus the process. The Authors cite as a case study Minsmere Tidal Sluice Outfall in Suffolk.

Example: Minsmere Tidal Sluice Outfall

Introduction

The following case study describes the process of undertaking works to provide urgent erosion protection to an Environment Agency tidal sluice outfall. Collectively the Minsmere tidal sluice and outfall are located midway within the north Minsmere to Sizewell frontage of the Suffolk Coastline in an area of significant conservation value. See Figure 2 The detailed process of appraising options commenced in September 2001, DEFRA funding approval was obtained in November 2002 and the project was completed with construction of the preferred option between February and March 2003.

Background

The tidal sluice chamber lies on the landward side of an artificial clay-cored dune embankment, which provides defence against coastal inundation between the Minsmere Cliffs in the north, and the town of Thorpness in the south. Seaward of the embankment lies a relatively narrow sand and shingle beach that slopes to the waterline. The purpose of the sluice is to combine the discharges from the three freshwater watercourses (The Minsmere New Cut, Minsmere Old River and Leiston Channel) which drain the surrounding hinterland. The defended area includes the renowned Minsmere Nature Reserve, which is owned and operated by the Royal Society for the Protection of Birds. The sluice discharges river flow by means of two large diameter pipes that run beneath the dune embankment and beach, emerging at around the High Water Mark within a concrete headwall inspection chamber. From this point onwards the discharge is transferred within a piled rectangular concrete culvert which discharges the river water to sea approximately 50m from shore. The structure was first constructed in the late 19th Century and much of this is still use today, although the outfall culvert was reconstructed in 1967.

The sluice structure lies inland of coastal dunes that form part of the Minsmere to Walberswick Heaths and Marshes Special Area of Conservation (cSAC). The sluice drains freshwater to the sea by means of gravity drainage, underneath the dune embankment. This embankment forms the sea defence protecting the adjacent RSPB reserve, Minsmere to Walberswick SSSI, SPA and Ramsar sites and the surrounding area from marine inundation. The Environment Agency has a legal obligation to maintain the favourable conservation status of the designated habitats and species.

Figure 2 Location Plan

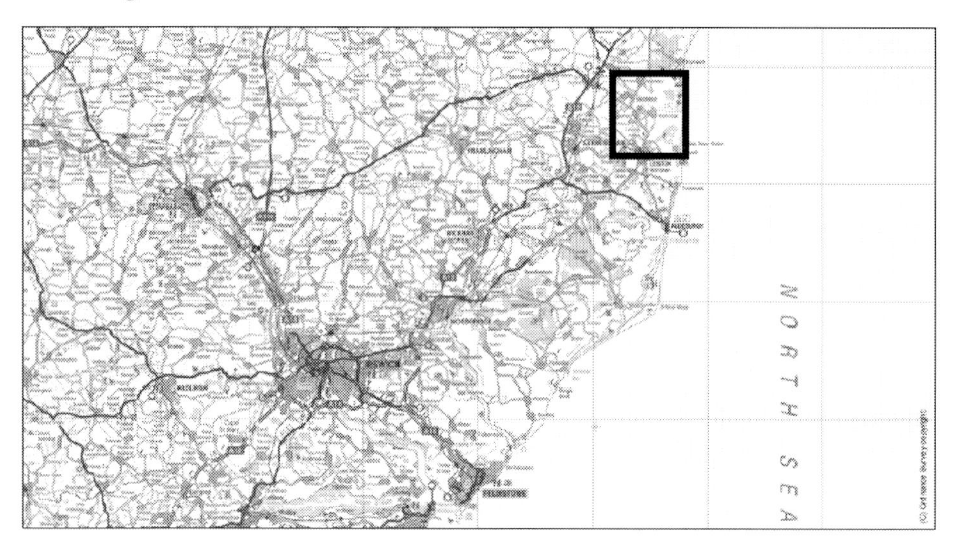

The designated area is shown on Figure 3 below. The large areas of nationally and internationally important nature conservation, including both brackish and freshwater habitats are all highly vulnerable to coastal change, should the current regime fail.

The remaining area is predominantly rural and settlements on the hinterland are confined to small towns, villages and isolated properties. Several residential properties, a farm and a public house lie within the flood plain, as do sewage treatment works at Leiston and Middleton. Should flooding at Minsmere occur, settlements stretching from the coast to upstream of the three watercourses will be affected.

Minsmere sluice is approximately 3km to the north of Sizewell Power Station. This is a major industrial development that comprises two nuclear power generators. The stability of the coastal defences protecting the power stations have been linked to the presence of Minsmere sluice which acts as a hard point to fix the eroding frontage.

Currently much of the area is used as a public amenity and many people, locals and visitors, use the area for bird watching, walking, fishing and other leisure / tourism activities.

The Problem

In 2001 a site inspection revealed that local scouring of the beach was occurring at the northern side of the outfall pipes, such that the pipes were exposed to the elements and potentially to undermining by the sea. Failure of these pipes may have the following implications:

- Blockage or partial blockage of the outfall, causing freshwater flooding along the three watercourses draining into the Minsmere RSPB reserve. This could have serious implications on the grounds of nature conservation interests within the Minsmere to Walberswick Heaths and Marshes cSAC and the Minsmere to Walberswick SPA, SSSI and Ramsar site.

- Breach of the outfall pipe higher up the beach. An event like this may contribute to local scour of the beach, which may in turn lead to faster erosion of the sea defence dune system, increasing its vulnerability in large storm events.

The risk of the outfall failing is highest over the winter months as the beach level appears to drop most during these periods. The degree of erosion of the beach material is dependent not only on the severity and duration of the prevailing sea conditions but also on the wave direction as it approaches the shoreline. The conditions which result in the erosion of the beach have not been fully determined due to the very localised coastal processes involved. It has been identified that the beach profile can vary significantly with each tide, and the summer to winter differences have been identified anecdotally as up to 2m. Given the multi-causal risk, and the unknown process effects it was not possible to identify the severity of storm event which could result in the failure of the outfall feed pipes, or indeed when this event may occur, although previous exposure of the pipes had occurred in relatively benign conditions.

Previous Studies

A Shoreline Management Plan[5] (SMP) undertaken for Lowestoft to Harwich was completed in May 1998. This includes the Minsmere frontage. The responsible coastal authorities have adopted the generic coastal defence policy options identified within this plan. The policy option adopted for the Minsmere frontage was to retreat the shoreline.

Following the adoption of the SMP, the Environment Agency in conjunction with Waveney District Council and Suffolk Coastal District Council commissioned a Strategy Plan[6] for the coast between Lowestoft and Thorpeness. The Strategy Plan, the final draft of which was issued in September 2001, developed the SMP policies in more detail in line with more recent guidance. These strategies were recommended for the medium term (up to 20 years). The proposed policies from the strategy are for a managed realignment of the shoreline north of the sluice, a hold the line policy in the vicinity of the sluice, with a policy of no intervention south of the sluice. This is in contrast to the SMP recommendations, in part.

More recently, the Environment Agency commissioned pre-feasibility studies with regard to the sea defences at Minsmere, and the tidal sluice outfall. The pre-feasibility study by Babtie Brown and Root[4] (June 2001) identified that the condition of the outfall structure in general was sound. However the study identified that erosion of the beach had resulted in exposure of the normally buried concrete feed pipes. It was concluded that the degree of exposure of the concrete feed pipes would require urgent action to prevent further erosion leading to failure. The pre-feasibility report considered the outfall to have a residual life of less than 5 years and speculated that failure of these pipes would effectively close the discharge from the tidal sluice and consequently cause flooding of the hinterland, with major impacts on the environmentally designated sites and on the adjacent agricultural land. The report recommended that protection work to the concrete outfall pipes was undertaken urgently.

The wind and wave conditions leading to the erosion of the beach and exposing the outfall feed pipes were unknown, although no significant storms had occurred in the period leading up to the outfall inspection at which the erosion problem had been identified. This, combined with the beneficial effect of the outfall on the Sizewell coastline, and the flood protection and drainage functions to people, property and the environment, led the Environment Agency to seek a short-term solution to the threat of outfall failure, ideally to be constructed before the next winter. Therefore, it was proposed that the works were treated as 'Urgent Works'.

The Project Appraisal Process

The works, as urgent works, would need an appraisal to be undertaken of the various options available to remediate the problem. The Strategy for the frontage, already in place, proposed a preferred option for management of the frontage in the longer term. This showed that cost-effective intervention to improve the flood defences and manage the coastline in the area would not be justified for between 5 and 10 years. On review of the Strategy findings, therefore, only short-term solutions were considered for the urgent works, to maintain the defence and outfall until the Strategy could be implemented. However, the works still needed to be justified in terms of economic, environmental and technical suitability.

The urgent works still needed to follow the standard approach of:

- Definition of the 'problem' and any constraints or opportunities
- Identification of options and impacts
- Analysis of factors relating to each option to reach a preferred option (including consultation)
- Seeking consents and licences to construct the preferred option
- Seeking funding to design and construct the preferred option
- Design of the preferred option
- Procurement of the preferred option
- Construction and handover

The Environment Agency commissioned Black and Veatch Consulting (BV) to undertake an appraisal of short-term options for the sluice outfall. BV were appointed in Winter 2001 with a view to undertaking the scheme justification (a Project Appraisal Report) by Summer 2002, so that funding could be sought from Defra and the project could be constructed during Winter 2002-2003.

Definition of the problem

In order to effectively develop potential options it was important to review existing data relating to the outfall and surrounding area. The data included technical data relating to the outfall and coastal processes, and data relating to the environment and land uses and ownership. Data gathering can be time consuming, as it depends on responses from other parties, who are invariably busy.

A first stage of consultation (scoping) occurred at an early stage in the life of the project. At this stage the consultation contained only a very brief discussion on the possible options and primarily described the background of why works to the outfall were necessary. This consultation was used to scope the general issues that would be needed to be addressed in the remainder of the study.

The review of existing data identified that there was a large amount of useful environmental information on the surrounding area but a shortfall on technical and construction information relating to the actual outfall structure. An understanding of the condition and likely deterioration of the outfall was needed to ensure that the full extent of the problem was being addressed. The project team combined primary data gathering in the form of a low water condition survey[7] in November 2001, and a dive survey[8] in January 2002 with detective work on the original construction, including approaching the now retired Contractor who was present on site during the 1967 modification works.

From the condition survey[7] it was concluded that the outfall structure itself is in relatively good condition for its age, and is not at risk of immediate failure (severe storms notwithstanding). The data gathering confirmed that the only problem to be addressed was that of the pipes feeding the outfall being exposed intermittently, due to localised scour around the structure, leading to the risk of failure if the pipes are unsupported by the beach when the beach material erodes, as the joints are likely to move.

Initial identification of constraints and opportunities

The initial scoping data gathering was used to develop an understanding of the environmental assets either at risk of flooding or at risk if construction works were undertaken. These assets include those used for amenity or commercial purposes.

The area defended is environmentally important. The Strategy Study[6] stated that continued provision of the sluice would help to fulfil the following environmental objectives:

- Protect freshwater habitats in RSPB reserve, including fresh and brackish lagoons, reed bed and unimproved basic grazing marsh (cited Ramsar habitats, together supporting ten qualifying SPA species);
- Avoid any interference with beach systems that would adversely affect vegetated shingle (qualifying cSAC habitat);
- Manage salinity within the RSPB reserve and provide protection in the event of a major breach of the dune ridge (RSPB);
- Maintain facilities at Minsmere for recreational and tourist use.

Recent guidance from Defra when dealing with this scale of environmental importance was to consult English Nature in the first instance to establish whether any works to maintain flood defences would be necessary for the maintenance of the features of the site.

Figure 3 – Environmental Designations

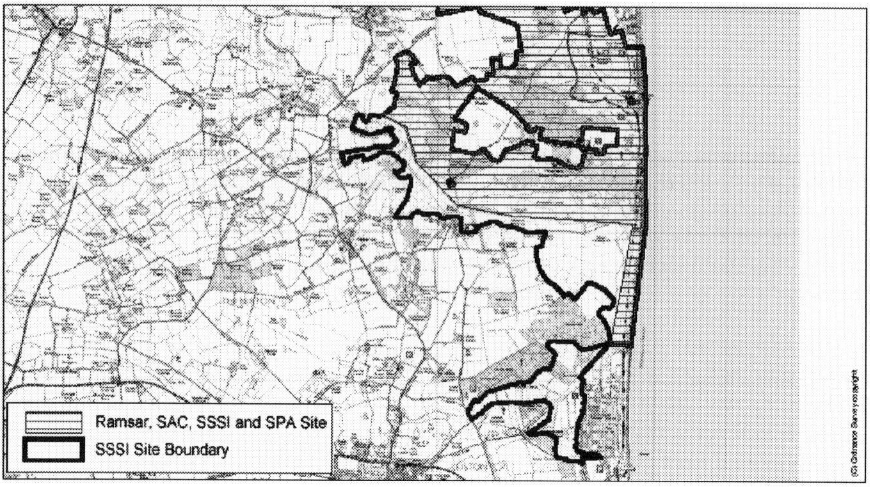

BV considered the implications of adopting the 'no active intervention' (Do Nothing) option at the sluice outfall on the adjacent SPA, cSAC and Ramsar Site. A document titled 'Implications of the "Do Nothing" Option'[10] was prepared by BV for English Nature, which identified the main implications of failure of the sluice pipes. The information in this document was comprised largely from correspondence and information provided by RSPB in June 2002 describing the impacts that periodic flood events have upon the management of the wetland habitats.

The evaluation process is similar in principle to that for an appropriate assessment of such an option under the Conservation Regulations 1994. As such, the evaluation method adopted was based on the Conservation Objectives for the site, and relating the physical changes arising from "Do Nothing" to features of European interest within the designated sites as detailed in English Nature's Favourable Condition Table (FCT). In that way, a judgement could be made concerning the favourable conservation status and therefore likely site integrity under the "Do Nothing" scenario. This approach was developed in accordance with an early draft guidance note produced by English Nature in April 2002 intended to outline procedures in such cases. The consideration of the "Do Nothing" implications is 'Stage 1' under that guidance.

The above report identified two main implications of a failure of the sluice pipes on the Minsmere to Walberswick Heaths and Marshes SPA, cSAC and Ramsar Site:

- Flooding of the hinterland would have potentially serious implications on nature conservation interests on habitats within the Minsmere to Walberswick Heaths SPA and Ramsar site, including reedbeds, standing waters and the coastal grazing marshes.

- Localised scouring of foreshore could lead to the loss of cSAC habitat. There is a risk that the scour could impact upon the sea defence embankment and if a breach occurred there would be a resulting increased vulnerability of saline intrusion of the SPA. This would again impact upon reedbeds, standing waters and the coastal grazing marshes to the detriment of the SPA and the Ramsar site interest.

In September 2002, English Nature[11] confirmed its agreement with the findings of the 'Implications of Do Nothing' report. This identified that no active intervention would have serious implications on the ability to maintain the designated features of the European site in favourable condition. English Nature gave its conditional support and approval that a scheme is required to conserve the international interest of the site. On these grounds it was agreed that some "active intervention", adopting the "Do Something" option was necessary to conserve the interest of the site.

The implication of the acceptance of the urgent works being necessary to the maintenance of the site was that under the European Habitats regulations, the Environment Agency would be legally obliged to undertake these works. This legal obligation enables the justification to be presented to Defra on environmental rather than economic grounds. For this reason, although the scheme should still have a positive ratio of benefits to costs, it is not subject to the prioritisation system of allocating funding, and therefore does not necessarily entail such extensive economic justification as may otherwise be required.

Do Something Options and impacts

Any options to improve the defence were shown within the Strategy[6] not to be viable until year 10. Therefore, a range of short term options to minimise the risk of pipe failure until the Strategy is implemented were examined. These included:

- Localised beach management
- Erosion protection using either sheet piling or rock revetment
- Monitor the structure and repair as necessary.

Access to the site was found to be a major contributor to the short-term construction impacts. The tidal sluice outfall is situated in an area internationally recognised for its aesthetic, environmental and historic value. It is situated in area that is not easily accessible by any existing infrastructure and it has been recognised that the chosen access route to the construction site would have a considerable effect on the likely impacts of the scheme on the human and natural environment. Routes would be required which would be able to transport heavy equipment and large quantities of materials. The constraints included any potential disturbance to the RSPB bird reserve, inadequate strength of culverts in the RSPB reserve for plant crossing, vegetated shingle on the foreshore, the Dunwich Cliffs' archaeological content, a Scheduled Ancient Monument at the Dunwich beach access, Sizewell nuclear power station's security constraints and amenity use of the beach for fishing to the south. The sea was also fished commercially locally, with boats launching from the beach at Sizewell.

The access options that were considered as potentially appropriate for this scheme were:

- Approach from the beach to the north, past Dunwich SAM and cliffs

- Approach from the south past Sizewell power station
- Sea-based delivery of large plant and materials
- Access for small vehicles though the RSPB reserve (all shown on Figure 4).

The process of technical, economic and environmental appraisal of these options was undertaken though a second stage of scoping consultation undertaken by BV in July 2002 with Statutory and Non- Statutory consultees. However, given the potential access constraints, the access proposals were also consulted on.

Figure 4 Access Options

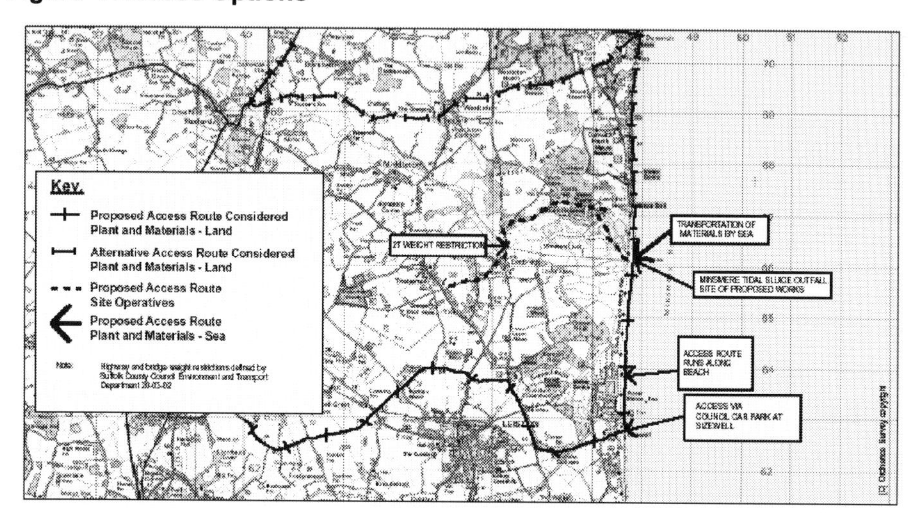

Analysis of options to reach a preferred option

In terms of option analysis, localised beach management was excluded given the very dynamic nature of the beach and the need for greater understanding of the processes. Monitor and repair was considered to pose at least a threat of fluvial flooding to the European sites. Erosion protection was considered in more detail.

The feedback from consultation included:

- Scheme is to be designed with view to potential alterations to the management of sluice following the larger study of sea defences;
- Scheme has regard for requirements of Sizewell Nuclear Power Station;
- Scheme has regard for coastal processes and alterations to topography of foreshore;
- Recognised requirement for site compound in vicinity of working area.
- Scheme has regard for the preservation of features of conservation interest within designated European sites;
- Scheme has regard for surrounding landscape in the types of materials used;
- Consideration of long and short term effects on Protected species.

Environmentally, erosion protection using rock armour option was considered more appropriate than sheet piling and concrete. The option offered less impact on landscaping . This coastal area is of national importance for its landscape value, and the majority of the rock was likely to be beneath the normal beach level and would therefore not normally be visible.

The coastal processes along this frontage and the effects of any works upon them are of paramount importance in defining the preferred option. In particular the effects of each option on the longshore transportation of beach material were considered. The protection of this frontage, which includes Sizewell Nuclear Power Station, is reliant upon any works not impacting upon these coastal processes. Following consultation with the Sizewell Shoreline Management Group the height of the erosion protection was reduced to the height of the existing structure to minimise the effects on longshore transport along the upper beach.

The preferred access option was, at appraisal stage, to deliver the rock armour, which comprises the majority of the construction materials, directly by sea. This would reduce the volume of traffic on the foreshore, and therefore minimise the potential impact upon the cSAC. The remaining construction materials and plant were to be brought to the construction site via the car park access at Sizewell Beach some 3km south of the outfall structure (the preferred access route by land). The access route along the beach was to be below Mean High Water Springs (MHWS) and clearly marked to ensure that vehicles do not encroach on to the vegetated strandline or shingle. The timing of the works meant that any potential impact on beach nesting birds would be avoided. A small secondary site compound was proposed immediately inland of the dunes at Minsmere to reduce the need for plant movement along the site

The tidal sluice and outfall lie in close proximity to a European site designated for its nature conservation interest. Therefore, this scheme is subject to Regulation 48 of The Conservation (Natural Habitats &c) Regulations, 1994 implementing Article 6(3) of the Habitats Directive (92/43/EEC). As it was determined in consultation with English Nature that the works would not have a significant effect on the European site the Agency was able to proceed on the basis of permitted development rights. Consequently the scheme was covered by the land drainage improvement EIA Regulations (SI 99/1783).

Once a preferred scheme was established, as the 'drainage body' under the Regulations, the Agency determined in consultation with English Nature through submission of an Appendix 11 form that the proposed works would not give rise to significant effects on the environment and so no statutory Environmental Statement (ES) would be required. However, to fulfil the Agency's own internal requirements for environmental assessment a non-statutory Environmental Report (ER) and Environmental Action Plan (EAP) were prepared. The preferred option was then consulted on.

Seeking Funding, consents and licences for the preferred Option

BV prepared a standard format report for the Environment Agency's Project Appraisal Board, subsequently to be submitted to Defra. The early involvement of Defra's Regional Engineer's staff and Environmental Advisor in the project, plus English Nature's acceptance of works being necessary for the conservation of the European site facilitated the approvals process. Marine consents to deliver the rock by sea, and to place material below the high water mark, were sought at the end of the appraisal process.

Design and procurement of the scheme

This project was carried out as part of a 5-year framework between the Environment Agency, 4 major consultants and 7 major contractors to provide engineering and environmental services. The nature of the framework encourages the early selection of the Contractor, and the early development of an integrated project team. Van Oord ACZ was appointed at option appraisal stage to advise on buildability and costings. The project team at design stage was extended to include a cost consultant from Arup.

The initial adverse consultation responses had slowed the consultation and hence the appraisal process, such that approval for the scheme was not received from the Agency to progress until November. The detailed design and development of the Works information was undertaken concurrently for speed, and more consideration was given to the construction of the option over the winter. Further concerns about the stability of the works prompted the alteration of the preferred option to an amalgamation of the 2 erosion protection options previously consulted on. The design solution was two lines of steel sheet piling to be constructed between the existing headwall and the high water line to support the pipes and surrounding beach material (See Figures 5). Rock armour would be placed to the outside of the structure to provide additional protection and prevent scouring from wave action. Following construction, the existing beach profile to be reinstated thus covering much of the new erosion protection structure.

Figure 5 – Preferred Option at Design Stage

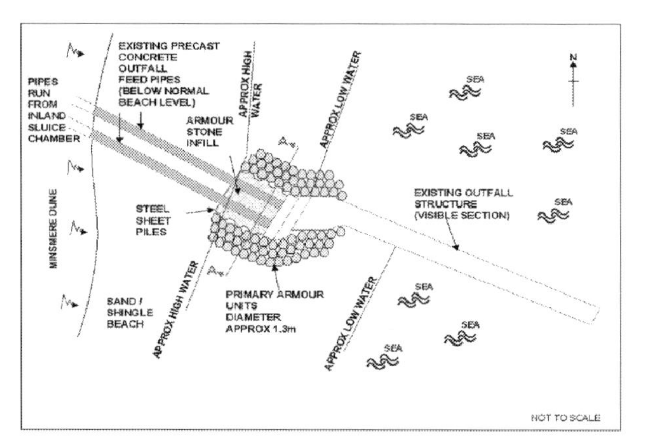

Although the initial proposal had been sea delivery of the rock, the reduction in rock quantities from the composite option, and the high risks associated with winter sea delivery made it preferable that all construction plant and material would be delivered to the site along the intertidal zone of the beach via Sizewell Beach car park some 3km south of the outfall. This change had to be re-consulted on with English Nature and other consultees to ensure they still accepted that there would be no adverse short-term impacts from this change.

Mitigation measures

A number of mitigation measures were agreed to minimise the potential impacts of the works. Amongst others these included that;

- Works to be completed by the end of March 2003 to minimise disturbance to nesting birds returning to the beach.
- The working area around the outfall was limited to 30m on either side.
- Materials and plant to be transported to site at periods of low tide to minimise any impact upon the strandline vegetation.
- Construction plant not to be allowed within 5m of the sand dunes.

These timings imposed quite severe constraints on construction working periods and methods, and meant that further arrangements had to be put in place for safety requirements such as emergency access or evacuation in the event of a storm. An emergency dune crossing was agreed in the location where RSPB vehicles accessed the beach, so that plant could be stored behind the dune in the event of storms, and appropriate dune protection measures such as protective matting were agreed in advance with materials for this to be stored at the location.

Construction

Works commenced on site in February 2003, after a final public consultation was held at the RSPB reserve to publicise the works. Effective working and good weather enabled the scheme to be built by mid March. The work was delivered in advance of programme and under budget, offering a saving both to the Agency and the Contractor from the Target Cost NEC Option C Contract. As well as an Environmental Action Plan, an Environmental Officer was appointed for the project. His role included an initial 'toolbox talk' to the Contractor's staff to highlight the environmental sensitivity of the site. There were no environmental incidents on site.

Conclusions and Management Implications

The current investment process for capital flood defence or erosion protection funding, combined with the decreasing residual life of many existing defences, and delays in agreement making to access some extremely environmentally sensitive sites makes it likely that more works will be undertaken as urgent works, at least in the near future.

The unforeseen need for urgent works is likely to have budget implications among funding authorities. The method of prioritising funding through Defra and the need for economic justification will be the main criteria in all cases except those where doing nothing could adversely affect the integrity of a European site. If capital works have been postponed pending the development of a strategic management plan, it will generally be necessary to ensure that any urgent works do not prejudge the outcome of the Strategy. This will be a difficult issue to manage in places where the Strategy may alter the existing standard of defence provided, and a pragmatic approach will be needed.

The nature of urgent works schemes and the short time scale to develop, justify and consult on options can place a significant short-term workload on consultants. Any changes such as changes to options because of consultation responses can necessitate another round of consultation, which can have a significant impact on the programme. There are several key elements which, in the opinion of the authors, contribute to the successful delivery of the project on programme.

- An established and experienced team – for urgent works, there is very little time to scale a learning curve, either for technical experience or for team development.
- Existing relationships between the project team and English Nature and Defra are particularly important for urgent works. If there is an existing level of trust and mutual understanding it is far easier to resolve any issues which arise.
- A flexible approach must be adopted by all parties to manage changes.
- Partnering arrangements, such as the Environment Agency's framework agreements, enable integrated teams to work together to seek the best solutions. The existence of the frameworks and the dedicated teams within consultants and contractors to work in these partnerships enables the short term peaks in workload occasioned by mitigating programme delays to be managed more easily. The long duration of these frameworks also enables a genuine partnering approach to develop in the teams, fostering everyone's participation in achieving the goals.

There is a lack of existing data on many structures that are now reaching the end of their useful life. That which is available is often design drawings rather than as-builts, and is micro-fiched to the point of unreadability. Therefore, early budget and programme setting should allow for a level of investigation appropriate to managing the risks in option development for the urgent works.

Authors' note:
English Nature's view on works necessary to the conservation of the site appear to be changing currently, and Defra's approach to funding flood defence in rural areas is also changing– this may affect the way urgent works in environmentally sensitive areas would be undertaken in the future.

REFERENCES

[1]. Environment Agency, 2000, **Our Contribution towards Sustainable Development** 1999-2000 Annual Review

[2] Defra, 2001, **National Appraisal of Assets at Risk of Flooding and Coastal Erosion, including the Impact of Climate Change** .

[3] MAFF, 1993, **Strategy for Flood and Coastal Defence in England and Wales,**

[4]Babtie Brown & Root, June 2001. **Pre-feasibility study for Minsmere Tidal Sluice.**

[5]Halcrow, May 1998. **Shoreline management plan for sediment sub-cell 3C Lowestoft to Harwich.**

[6]Halcrow, September 2001. **Lowestoft to Thorpeness Coastal Process and Strategy Study.**

[7]Binnie Black & Veatch, November 2002. **Minsmere Tidal Sluice Outfall Environmental Report.**

[8]Binnie Black & Veatch, December 2001. **Minsmere Tidal Sluice Outfall. Condition Survey.**

[9]Binnie Black & Veatch, January 2002. **Minsmere Tidal Sluice Outfall. Dive Survey Inspection Report.**

[10]Sizewell Shoreline Management Steering Group, February 2002. **Sizewell Flood Defences and Coastal Stability Beach Management Plan.**

[11]Binnie Black & Veatch, July 2002. Minsmere. **Implications of the 'Do Nothing' Option.**

A comparison of approaches to sustainable development of tidal areas in UK and Korea

Jonathan Simm, HR Wallingford, Wallingford, UK
Dr Sang-hyan Park, Rural Research Institute, Ansan, South Korea
Dr Dae-su Eo, Rural Research Institute, Ansan, South Korea

Introduction

This paper presents a summary and comparison of the situations in between Korea and the UK with regard to development of tidal areas. The paper reflects the authors' experience on projects in the Korea and the UK. It is also written in the context of the authors' work in an international working group on "Sustainable development of tidal areas" which has been set up under the auspices of the International Commission on Irrigation and Drainage (ICID) and which will be producing a guidance handbook.

International Context

The coastal zone comprises only 3% of the earth's surface, but contains a disproportionately high amount of its assets. For example (Huntington, 2002):

- It accounts for 25% of global primary productivity
- It contains ports and harbours for international trade.
- It is the source of 90% of the world's fish catch
- It contains a major portion of the world's prime agricultural land
- It accommodates 60% of the world's population – a figure set to increase to 80% by 2050
- It contains 2/3 of those cities in the world with a population of more than 1.6 million people
- It offers recreation and tourism
- It comprises a most valuable environment and heritage resource.

For the purpose of this paper, "tidal areas" are taken to be all those coastal areas where the tidal processes are capable of affecting man's activity or of being influenced by man. This roughly extends tidal areas between the following limits:

- On the seaward side up to the limit of conventional construction or dredging activity (typically of the order of 30m water depth);
- On the landward side up to the limit of the action of the sea, including all those areas that might be subject to flooding by seawater and up all estuaries and rivers to the tidal limit (the point where water levels are no longer influenced by tidal propagation).

Throughout the world, tidal areas have been and are being developed. The initial development is generally for agriculture, but gradually the development focus moves to ports, harbours, transportation, urbanisation, leisure etc. Tidal areas are also a valuable natural resource, which needs protection, and hence the role of agriculture in initiating development is critical.

Coastal Management 2003, Thomas Telford, London, 2003.

By 2025, it is likely that three quarters of the world's population will live within 100 km of the sea, placing huge strains on coastal ecosystems (Shultz, 2001). On this basis, the burden of guaranteeing sustainable development must be shared locally and globally for the wise use and conservation of tidal areas.

South Korean Practice

Geographical and climatic context

South Korea forms the southern part of the north-south oriented Korean peninsula, lying broadly between 34 and 38 degrees North, and between 126 and 130 degrees East. It is surrounded by sea on three sides (see Figure 1), with the Yellow Sea to the west (bordered by China, North Korea and South Korea this semi-enclosed shallow sea has an average depth across the whole of only 44 m), the South Sea or "Tsushima Strait" to the south, and the deep East Sea (or Sea of Japan) to the east. Due to these seas' ameliorating effects, South Korea has elements of both continental and maritime-temperate climates. Winters are generally dry and cold (Moores, 2003) with periods of prolonged subzero temperatures in the north, especially away from the coast, but often milder, wetter conditions in the south, leading to inland rivers freezing in the north but largely unfrozen tidal-flats and coastal reclamation lakes in the southwest and southeast. Summers are generally hot and humid (with daily maxima typically above 30C in the warmest areas), and the heaviest rains usually fall between June and September, either as part of summer monsoons or carried by typhoons. Drought conditions are reasonably frequent, further stretching very limited water resources.

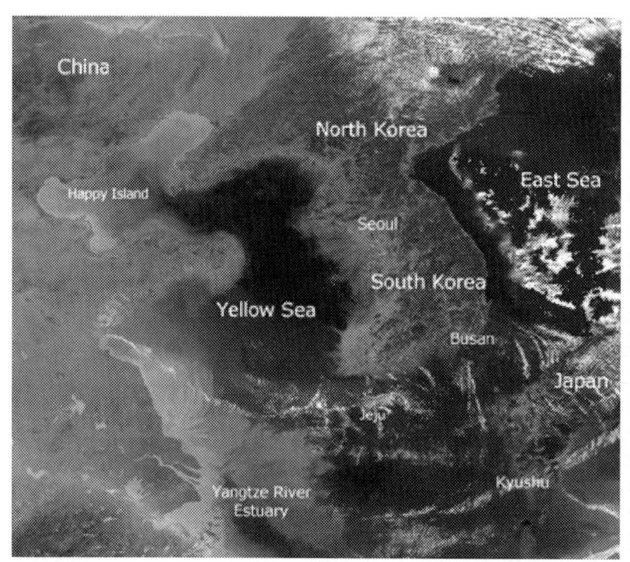

Figure 1. Regional geography around South Korea

Tides and morphology

Along the mainly rocky and sand beach east coast the tidal range reaches only 0.3 m. The tidal range, however, increases gradually along the south coast, from an average 1 m in the Nakdong estuary in the far southeast to 5 m during spring tides in Mokpo in the southwest, and northward into the Yellow Sea, where it peaks at 9-10 m (one of the world's highest tidal ranges) in the vast Gyeonggi Bay in the northwest. As a result of this tidal range, combined with a shallow coastal gradient and heavy sediment deposition (by Korean rivers and also by the Yellow and Yangtze rivers across the other side of the Yellow Sea), extensive and wide tidal-flats stretch southward from Gyeonggi Bay along much of the west coast and in bays and small estuaries along the south coast. Mud and silt flats tend to predominate in the north.

Land reclamation in Korea

The earliest known tideland reclamations in Korea took place after the Mongolian invasion of 1232. A seadike was constructed in the northeastern coast of Ganghwa island located 50km west of Seoul. The objective was to reclaim paddy fields for food supply of about 10,000 people who went on to resist the invasion for more than 30 years. With a spring tide range at Ganghwa of more than 9 metres and extensive tidal flats, reclamations here were easy to carry out using dikes and low tide drainage to create polders in small bays in the island coast. Further small-scale reclamations were carried out at Ganghwa until the 18th century when the use of new crane machines was initiated by a famous engineer, Sir Cheong, Yak-Yong to move stones for dumping. Then from 1917 to 1938 during the Japanese colonial period, some 40,000 hectares of paddy fields were developed.

In the period 1945 to 2000, a mixture of Korean government and private companies reclaimed some 76,000 hectares of land (see Figure 2.) The major push for land reclamation started in the 1960s with surveys of tidal flats carried out by the Korean government with the assistance of UN special funds. Many large-scale tideland reclamation areas were then developed from 1971 in the western and southern coasts with the aim of increasing crop production with increased arable land. (Only 30% of Korea is usable for agriculture, being a very mountainous country.) These tideland reclamations contributed greatly to rice production and other national objectives such as expansion of national territory, water resources development, convenient transportation network, promotion of tourism, etc.

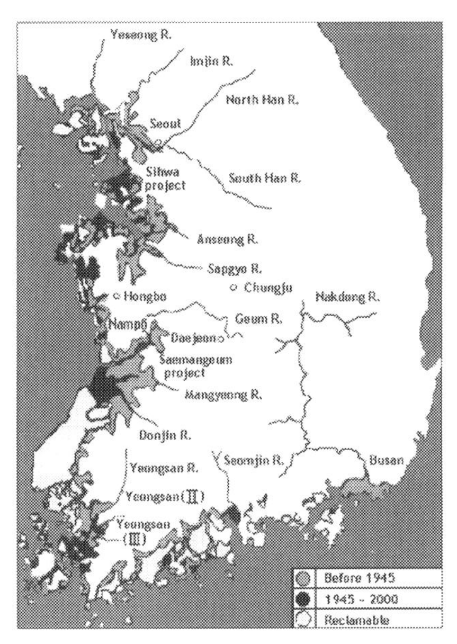

Figure 2 Tideland development projects in Korea

Now at the peak of their technical development, agriculturally based tideland reclamations in Korea are now becoming difficult to justify because of environmental pressures and the overcoming of rice shortage problems in recent years. The average size of the reclamations had increased from 15 hectares per project during the 1960s to 1,100 hectares during the 1990s. The most recent agricultural development in Korea, the Saemangeum project (see Case Study 2), is envisaged to retain some 40,000 hectares (400 km^2) with a 33km sea dike. This was delayed for 3 years in a half-completed state whilst a government review took place, but is now proceeding again amidst careful review and environmental monitoring and mitigation measures. It is now likely that there will be no further major land reclamations after Saemangeum. Long term plans for tideland reclamation were reduced from 402,00 to 208,000 hectares in 1995, further reduced to 157,000 hectares in 1998 and down to 21,000 hectares in 2000.

In contemplating future tidal reclamation developments, the Korean government is tending to move the focus away from pure agriculture to multi-functionality. The additional functions taken into account include water resources, flood protection by providing flood storage, tourist resort areas and new roads. The largest 5 of the most recent tideland development projects off the West coast of Korea which total some 137,000 hectares offer (1998 figures)

- 1,395 million m3 of water resource per year
- 480 million m3 of flood storage
- 35 million car journeys/year along the sea dikes at a national benefit of US$ 165 million
- Flood control and amenity benefits valued at US$ 462 million

The major environmental sensitivities primarily relate to loss of mudflat habitat for birds. South Korea lacks endemic bird species, and is rather poor in terms of avian "dry land" diversity compared with other East Asian countries, but its wetlands and waterways are extremely important for the future conservation of migratory waterbird species, around 13 of which are globally threatened. The tidal flats are also of importance to fisheries industries based around seaweeds, shellfish, crabs and fish, which are concerned about the loss of egg-laying and nursery areas. The government is taking environmental sensitivities into account to some extent by seeking mitigation measures in terms of fish passes, groynes to create some mitigatory tidal flats and creation of nature reserves within the reclaimed areas. Environmental NGOs remain very vociferously opposed to any further tidal reclamations at all, believing that irreparable damage has already occurred.

Recent changes in coastal management in Korea

Korea has experienced a fundamental change in managing coastal areas over the last five years (Lee, 1999). The change is characterised by the changes in the goals of coastal development and coastal governance, and in the approaches to coastal resources allocation. The goals of coastal development in Korea have been readjusted from economic growth and land-oriented perspectives which have driven the land reclamation activities toward sustainable use and environment conservation. Coastal governance has been restructured from a sectoral to an integrated system. In the past, coastal activities have been managed by 9 government agencies according to about 45 laws without an appropriate coordinating mechanism. The creation in 1996 of an integrated ocean governance system, the Ministry of Maritime Affairs and Fisheries, provided an opportunity to manage coastal uses and conservation in a coordinated manner. As a legal framework for integrated management of coastal areas, the Korean government enacted the Coastal Management Law in December 1998. According to this law, the allocation of coastal resources in the coming 21st century will be made on the principles of integrated management by implementing 'integrated coastal management plans' and 'coastal zone enhancement programs' both at national and local levels.

It remains unclear how the previous national policy aims for Korean agriculture to secure food and water resources are going to be addressed given this policy shift in coastal management. The food self-sufficiency ratio remains at best 30%. Rice paddy accounts for about 61% of Korea's entire farmland, and most of the water for agriculture is used for rice farming. Unfortunately, the seasonal and spatial uneven distribution of precipitation means that there is a need to develop water resources to meet future demand, it being estimated that there will be a national water shortage of 1.2 billion m^3 by 2011. Even so, it now seems unlikely that the scheduled remaining 21,000 hectares of reclamation will proceed and, of current projects, the nationally significant Saemangeum project (see below) may still be cancelled or modified even at this relatively late stage due to environmental arguments.

Case example 1: Inchon Airport, Seoul

Inchon International Airport was planned and conceived as a new hub airport for Seoul, Korea and to rival Kansai and Hong Kong in south-east Asia. An offshore inter-island site was identified to give maximum room for development and to permit 24 hour access without noise restrictions by over-water flight paths.

The airport site (see Figure 3) was created by a land reclamation project between two islands: Yong-Jong and Yong-Ju in a location where the spring tide range is 9.5m. A total of 17.3 km of dikes was necessary for the project, with a bottom width of 120m and a height of 9m. An average height of 5m of reclamation was made over 1,174 hectares (11.4 km^2) of the airport area, with soil dredged from the nearby sea bottom or excavated out of the surrounding mountains. Fill used for the site preparation amounted to 96.5 million m^3.

Figure 3 Aerial view of Inchon Airport under construction

Ground preparation was carried out on all 3.5 million m2 of aircraft movement areas. As the sea bed soils underlying the site reclamation included an upper layer of silty clay 5 m thick, some 610,000 sand drains or absorption paper drains were driven an average depth of 13m into the seabed. The ground was then preloaded with 5 to 6 m of fill to drive out the pore water and deliver some 0.5m of consolidation at depth. The dredged fill was then placed and then triple hammered with 10-ton weights to force a further 0.4 to 0.5m of settlement. Residual subsidence is expected to be less than 25mm over 20 years.

The airport was opened to international traffic in the spring of 2001 (in advance of the FIFA World Cup.) A total of US$5.6 billion was invested in the Construction Project (cheaper than Kansai at US$13.5 billion and Hong Kong at US$9 billion)

Case example 2: Saemangeum project

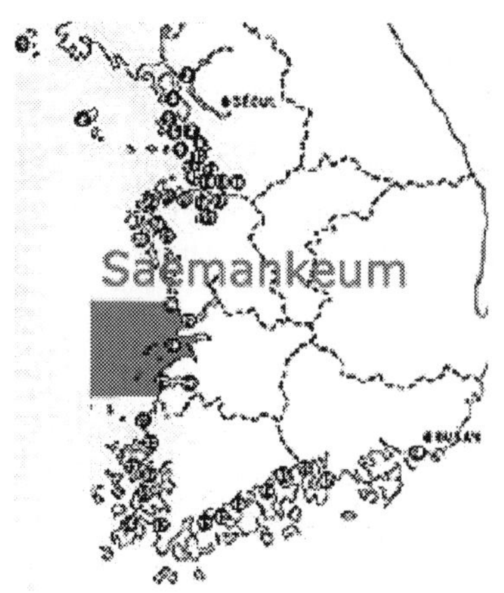

Figure 4 Location of Saemangeum

The Saemangeum (See Figure 4) reclamation project was first conceived in the 1970s with the aims of increasing useable national territory and of creating more land for rice culture, thus assisting in national food self-sufficiency.

A full economic feasibility study and environmental impact assessment performed from 1986 to 1991 led to the start of construction works for the Saemangeum Project in November 1991. The first phase of the project involved the construction of the outer seadikes. As proposed (see Figure 6) the wall will eventually dam off two estuaries in order to create 28 300 ha of agricultural and related rural areas, and 11 800 ha of freshwater lake. Subject to ongoing legal action (see below) final closure of the sea dykes is now expected to be completed by 2006, with the conversion of the tidal-flats to rice-fields following a few years later.

Construction costs of this outer wall (see example cross-section in Figure 5) which includes a 2 lane highway, were estimated at US$2.3 billion dollars in 2001, with 75% of the 33 km long wall completed by 2002.

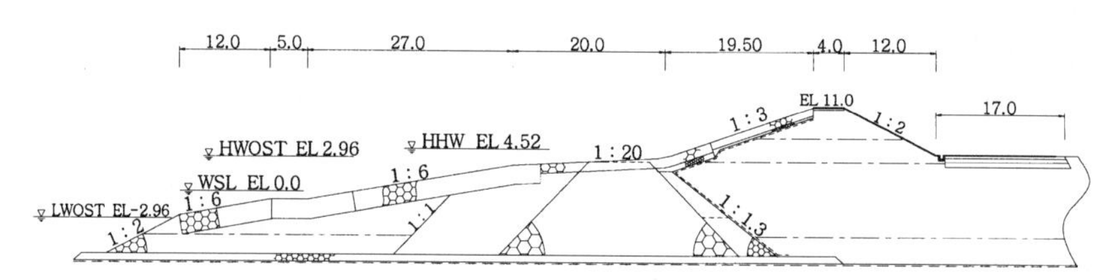

Figure 5 Cross-section through part of Seadike no.4, Saemangeum

In late 1998, in spite of significant regional support for an agricultural reclamation project in the area, concerns were raised by the public, led by non-governmental environmental organizations, about the negative effects of the project on environment, water quality deterioration and its low economic feasibility.

In April, 1999, a Joint Committee of Public and Government under the Office of the Prime Minister was set up to survey and re-evaluate the Saemangeum Project. Their terms of reference was to re-evaluate environmental impact; economic feasibility; water quality preservation. The committee had 30 members, 20 specialists from public sector including NGO, 9 from the related Ministries and 1 Chairperson. The Committee submitted a re-evaluation report of Saemangeum to the Government on August, 2000. Their final recommendations may be summarized as:

- For water quality preservation, the long-term water quality of Saemangeum lake would meet the quality standard for freshwater lakes, so long as certain river water mixing and pollution control measures were introduced.

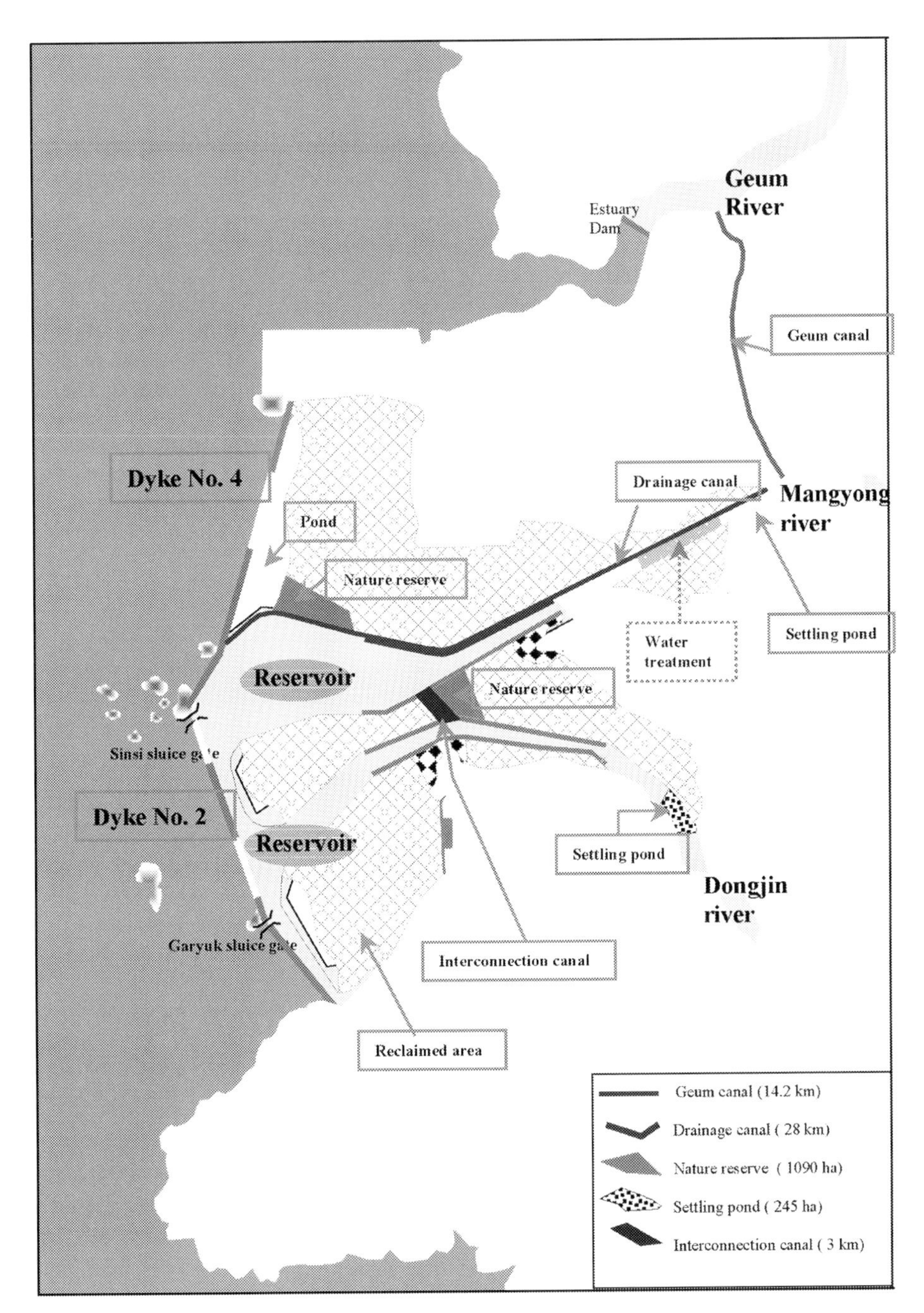

Figure 6 Layout of Saemangeum project

- For environmental impact, they suggested that new technologies be developed to construct artificial wetlands and settling basins to provide compensatory habitats for native and migratory birds.
- The Committee concluded that the Project was feasible with benefit/cost ratio 1.25, although several committee members submitted a minority report.

Based on the recommendations of the Joint Committee of Public and Government, several public hearings and extended consultation with experts, the Office for Government Policy Co-ordination under Office of Prime Minister decided on 25 May 2001 to continue with the Saemangeum Project on the following terms:

- The dike construction works will be continued to completion by the year 2006. The Dongjin land (South side) will be developed first (for agriculture.) The development of the Mangyeong land (North side) will be postponed until the water quality of Saemangeum lake meets the control standard.
- The Project will be implemented in an environmentally-friendly manner introducing various technologies such as polluted water treatment plant, settling basin, nature reserves, polluted water drainage culvert, fish ways, bird sanctuary, eco-parks as well as leisure facilities for the public. Environmentally sound and sustainable practices for agriculture should be adopted.

As a result construction works recommenced but recently (Summer 2003) the Seoul Administrative Court ordered the suspension of all construction work on the Saemangeum sea dikes. This order was in response to a request to halt construction by environmentalists and a group of residents. The Ministry of Agriculture and Forestry said it could not accept the ruling and the Minister resigned in protest at the decision. An appeal to a higher court is in progress and at present it is unclear whether these legal moves will just mean a further delay or reconfiguration in the project or whether it will finally be abandoned.

UK practice

The issues affecting the UK's socio-economic use of tidal areas include demands for flood defence and coastal protection, ports and harbours, mineral resources, urban regeneration, renewable energy production, waste disposal and leisure. These have to be set against the demands for environmental conservation of habitats, geological exposure, cultural heritage and landscape. In comparison with Korea the UK has advanced much further down the road of sustainable practice and environmental protection. Perhaps the UK now needs to re-emphasise the potential of sympathetic development of tidal areas for socio-economic gain. Sustainability is about choosing a development path that will not disadvantage the social or economic welfare of present or future generations as well as not adversely affecting the environment.

Figure 7 Multiple uses of tidal areas – Christchurch Harbour, UK

To achieve this, significant technical challenges remain, including finding adaptable solutions to climate change, appropriate materials resources and being able to predict long-term morphological change. However, in the international context, the UK is well-advanced in developing holistic GIS systems-based approaches to integrated coastal zone management. The remaining challenge is to ensure comprehensive and balanced implementation involving all stakeholders.

Drivers for development or management of tidal areas in the UK

Flood defence and coastal protection

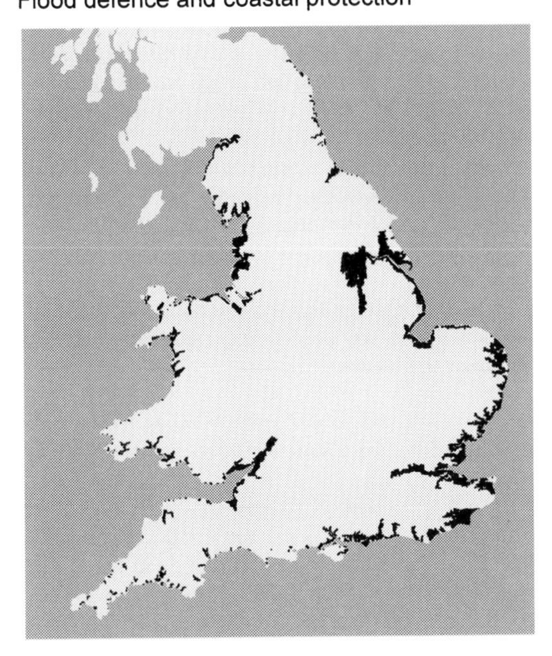

Figure 8 Coastal areas at risk from flooding in England and Wales

The UK government has a policy aim of reducing the risk to the natural and developed environment from flooding and coastal erosion by adopting technically sound and economically, socially and environmentally sustainable solutions. Climate change is leading to sea level rise and increased or changing patterns of storminess. This, combined with isostatic land adjustment, is leading to significant "coastal squeeze" resulting in eroding coastlines and the narrowing and steepening of intertidal areas, such as beaches and salt marshes. Despite this, there are still pressures (from industry, housing) to develop land that is already at risk of tidal flooding, and to "reclaim" areas presently within the inter-tidal zone.

The recently completed revision of the Flood Risk Assessment for England and Wales suggests that some 545,000 hectares (5,450 km^2) are at risk from coastal flooding (see Figure 8), with the Annual Average Damage due to flooding of these areas estimated at some £276 million pounds per year.

Ports and harbours

Increasing sizes of ships and changing trade patterns are demanding significant developments of port infrastructure. Major debates between interested stakeholders are underway at present in relation to planned new port facilities at Southampton and in the Thames Estuary. Even where existing facilities are satisfactory, significant deepening of navigation access channels may be envisaged. In the context of a privatised ports industry, a National Ports Strategy may be required to ensure that appropriate, but not excessive provision is made.

Leisure usage of tidal areas, particularly sailing, is also increasing leading to a rapid increase in the number of coastal marinas, often accompanied by other associated developments such as housing and retail outlets.

Airports

Major land reclamations, previously carried out at places such as Osaka, Hong Kong and Seoul, and discussed for London in the 1960s at the Maplin Sands in the Thames Estuary are now being contemplated again in the UK. Specifically plans have been put forward, but not approved for airports at Cliffe and Severnside:

- Cliffe airport is a proposed new multi-runway airport to be located in the area of the mud flats to the south of the Thames estuary known as Cliffe, Halstow, and St Mary's Marshes opposite Canvey Island on the north of the estuary. The location offers over-water flight paths, rail links using the new channel Tunnel Rail link to Kings Cross and road access by way of extensions to the M2/A2 linking with the M25.
- The Severnside airport is a proposed international airport serving Wales and the west of England, which would have runways on a manmade island in the Severn Estuary, near Newport, south Wales.{PRIVATE "TYPE=PICT;ALT=Severnside airport/PA"}

Exploitation of coastal and seabed resources

Demand for aggregates for construction is leading to significant exploitation of offshore areas. The effects of such dredging on tidal areas are of concern and the scope and procedures for environmental assessment continue to be modified.

In some areas of the UK where sand and gravel is still mined from beaches as the only "economically viable source." The challenge here is to make more use of re-cycled materials.

Barriers and barrages

Moveable barriers have been constructed in a number of UK estuaries for flood protection purposes, most notable in the Thames. As sea levels continue to rise, modification/replacement of such barriers and new barriers in vulnerable areas have to be considered. Elsewhere in Europe, flood defence barriers are continuing to be built in locations such as the Netherlands (Rotterdam) and Italy (Venice.)

Fixed barrages have been less common, but have been built on a small scale at a number of locations. The largest barrier constructed so far in the UK is that in Cardiff Bay – with the primary purpose of urban regeneration after a period of industrial decline. Smaller urban regeneration barrages have also been built on the Tees and Tawe.

Tidal power barrages have been considered in the past for the UK. Britains largest tide range is in the Severn Estuary – at 14m it is also nearly the largest in the world. Here proposals for a Tidal Power barrage may be revived after more that 10 years under pressure to meet targets for reducing carbon emissions under the Kyoto treaty. Other changes including the privatisation of electricity may also make the project more financially attractive when combined with the possibility of the barrage providing flood protection against storm surges.

Large water supply reservoirs behind barrages have been evaluated (e.g. within The Wash and Morecambe Bay) but never built. However, small barrages have been constructed to improve water quality e.g. on the Lagan at Belfast.

Guidelines for the assessment and planning of estuarine barrages have recently been compiled by HR Wallingford (Burt & Rees, 2001)

Renewable energy

Another growing use of the coastal seabed (see Thorpe, 2001) relates to the use for wind farms and in areas of high tidal currents, for underwater propeller farms and for wave power in Scotland.

Industrial uses of water

Coastal power stations and sewage treatment outfalls are a significant feature of the coast. Management of water quality is a major issue for sustainable development of tidal areas.

Waste disposal

Dumping at sea has been controlled for some time via the London convention and other international and European agreements. Meanwhile pressures to find landfill sites for domestic, commercial and industrial waste remain strong. In Guernsey (one of the channel islands between England and France), a land reclamation bund has been created with the specific purpose of retaining landfill waste – with the eventual plan that the land so created could be used for industrial or port purposes.

In the meantime, we have a substantial legacy of disposal sites around the coastline of the UK, containing industrial and domestic waste products. Often these areas have been subsequently built over, often for car parks, storage areas or even residential housing. Many of the sites are now eroding and causing pollution problems.

Leisure activity

There is a continuing demand for marinas and provision of other leisure related facilities. Much of this has to be provided within the context of redevelopment of existing small fishing and other harbours (Watchet is a recent good example.) At the same time conservation and tourism demands that as much as possible of the UK's coastal areas are left without further human intervention.

Engineering challenges

In meeting these demands, there are significant engineering challenges for the UK:

Adaptability to change

Solutions need to be easily adaptable in the future for natural or socio-economic changes. For example, both flood defences and barriers/barrages need to be able to meet the challenge of rising sea levels. Unadaptable structures, or structures which are difficult to decommission, will have to be rejected in favour of more adaptable solutions.

Materials resources

The resources required to build, manage and maintain coastal defences may be very significant. A particular challenge here is to find adequate supplies of coarse material to recharge our gravel beaches to the extent required to maintain them as defences. At the same time production of such resources have their own harmful effects on the environment. Sustainable resource solutions (Masters, 2001) need to avoid such effects and be based on supply and use of as wide a range of materials as possible, maximising the re-use and recycling of both primary and waste materials.

Long-term prediction of morphological change

To produce sustainable developments, the capacity to predict long term changes in morphology is essential. Conventional process-based computational modelling tools can show instability when run over long time scales. There is a need to develop more pragmatic tools, which are based on long term morphological concepts and which take account of the physical constraints to morphological change resulting from geological features and man-made structures. Although it may be difficult to construct such "models" in a completely reliable way, there remains a need for some sort of assessment of the potential for significant change to seabeds and coastlines.

Socio-economic pressures

Added to this, social changes in the UK are leading to demand for more housing through a combination of immigration and the trend for smaller family units. The population density in the UK is relatively high and the amount of land available for such new housing is limited, because much potential candidate land is already developed, or has been designated as prime agricultural land or for conservation for habitat or landscape reasons. As a result land that may be subject to either coastal or fluvial flooding may offer the only remaining possibility.

Coastal regions remain very popular with the British, particularly with people who are no longer working. As the UK population age balance shifts towards the older generations, demands for such property can only increase, in spite of well-publicised erosional losses of homes in isolated areas e.g. on top of coastal cliffs and of flooding/ risk of flooding to homes in urban areas.

Environmental considerations

After a period of more informal environmental control, Europe made a major step moved towards a statutory conservation with the "Directive on the conservation of natural habitats of wild flora and fauna" (92/43/ECC), generally known as the "Habitats Directive." This set out the necessary requirements to preserve the habitat, and in particular those of favourable conservation state. Particular areas are designated under this directive as SPAs (Special Protection Areas) or SACs (Special Areas of Conservation.) These designations are starting to severely restrict development, particularly in the UK's tidal estuaries.

In some areas, unsuitable or undesirable for development, it has been recognised that the best way of dealing with coastal squeeze and improving habitats at the same time may be to set back the traditional lines of flood defence. This allows the land so released seaward of the defence to revert to marshland and thereby create new or compensatory habitat.

Other aspects of the environment are also significant. The coast is rich in geological exposures, landscape and heritage and all these factors are valuable, not only in their own right, but also as a significant attraction for economic activity in terms of tourism and leisure.

Strategic planning in the United Kingdom

In determining the way the UK should move forward in its management of the coastal zone, it is faced with significant uncertainty:

- There are spatial and temporal variations in natural hazards
- There is incomplete knowledge of physical processes and such descriptions as are available are based on incomplete data.
- There are differing and conflicting views on what is a desirable outcome, because different coastal attributes are valued differently by a range of stakeholders. For example,

there are some 37 consultees that might have a legitimate interest in a Coastal Habitats Management Plan (CHaMP)

The only way to deal with these issues is by a whole systems approach with tiered or nested analysis and decision making, supporting national strategy and feeding down via regional planning (Shoreline Management Plans and the nested more detailed Strategy Plans) to specific schemes. The building blocks of such a system are currently under development and include:

- Data with associated metadata, plus a reliability trail
- Knowledge bases – to guide decision makers through key issues and processes
- Tiered risk assessment methods
- Techniques for representing and propagating uncertainty
- Open modelling support systems
- GIS interfaces to present risk and other information to decision makers and stakeholders

The aim is take account of the widest possible set of management actions (including major schemes and structures) that may impact on the coast along with climatic, demographic and socio-economic changes.

The approach emerging for the planning for aggregate dredging licences is of a similar top-down nature with inputs envisaged at national and regional levels.

Other papers at this conference will address these issues in more detail.

The ICID Handbook on Sustainable Development of Tidal Areas

Both Korean and UK experience, along with that from many other countries will be captured in a handbook being planned by a working group on Sustainable Development of Tidal Areas set up under the auspices of the ICID (International Commission on Irrigation and Drainage.)

The ICID Handbook is being designed to provide guidelines for planning, design and management of irrigation and drainage, and flood control for sustainable development in tidal areas, focussing primarily on agricultural requirements. The handbook will serve mainly to provide design support for engineers. It is intended to publish the handbook by the year 2006. Drafts of the document will be made available on the ICID website (www.icid.org) as the work progresses and there will be a system available for commenting on the documents.

Provisional chapter headings/subheadings for handbook

1. Introduction: philosophy: historical context of agricultural land development practices in tidal areas; objectives, readership and structure of guidelines.
2. Aspects of managing development of tidal areas, discussing agricultural needs in the context of other demands for freshwater storage, aquaculture, industrial development, urban regeneration, tidal power generation, river navigation, coastal ports and harbours, tidal power generation, water-based leisure activities, resource abstraction. The chapter will touch on:
    - Conservation (habitat/bio-diversity) planning.
    - Irrigation planning requirements.
    - Aquaculture planning requirements.
    - Catchment flood management planning.
    - Coastal zone management planning.

3. Description of tidal and lowland features, briefly explaining their geological and sedimentological origin and their significance for development of tidal areas. Features discussed will include peatlands, mangrove areas, salt marshes, dune-beach systems and other coastal features.
4. Existing natural processes: climatic, isostatic, hydrological, hydrodynamic, morphological.
5. Tidal reclamations and their impact on natural processes (physical chemical and biological) and on human systems.
6. Engineering for sustainable development of tidal areas, including design, construction, operation and maintenance of: Dikes and water retention structures. Water management systems and structures. Dredging and channel improvements. Land fill in reclaiming areas and artificial nourishment in the tidal areas.
7. Integrated decision support framework drawing together:
   - Institutional aspects.
   - Community participation and motivation.
   - Natural processes.
   - Engineering feasibility and cost of proposals.
   - Methods of assessing acceptability of proposed changes

Case studies

An key part of the handbook will be the Appendix containing case study descriptions of historically significant monumental tidal reclamation projects from member countries of ICID and related organisations.

These case studies will illustrate a variety of stages of development requiring different engineering solutions. In developing countries, the development of tidal areas is often mainly aiming at providing agricultural development to feed the growing population. At the other end of the scale, the demands in the industrial world mean that developments focus on flood defence and coastal protection, ports and harbours, exploitation of seabed resources, barriers and barrages and increased use of water by industry and for power generation, waste disposal and leisure.

However, all developments have in common that, once started, the development will continue forever to face challenges like falling land levels, often in combination with a rise in sea levels due to climate changes (changing rainfall pattern and intensity), population growth and movement. The case study descriptions will distinguish between addressing natural changes (sea level rise, rainfall pattern, etc) and addressing man-made changes (subsidence, change in land use. etc).

Taking account of its place in the relevant national development cycle, the case study review of each tidal reclamation project will review:

- Physical, chemical and biotic characteristics, including climate, soils, habitat, etc.;
- Water management: irrigation, drainage and flood protection, including water quality, sedimentation, flood mitigation, strategic storage, etc.;
- Management and institutional aspects, including performance and risk analysis;
- Socio-economic impacts, including cost/benefit analysis.

Conclusion

Balancing the engineering challenges, socio-economic pressures and environmental constraints associated with sustainable development of tidal areas is not easy. The nature and justification for tideland developments varies from country to country depending on the level of their technical and socio-economic development and of their environmental awareness. In developing countries, the demands for agricultural production may drive developments that in other locations would be viewed as being environmentally undesirable. However, whereas in Korea, the need is for increased environmental sensitivity, in the UK and Europe, we may need to rediscover a vision of the socio-economic benefits of pressing forward with appropriate developments in tidal areas, whilst seeking to maximise benefits for all stakeholders.

References

Burt, T.N. and Rees, A.W. (2001) "Guidelines for the assessment and planning of estuarine barrages." London: Thomas Telford.

Crossman, MP and Simm, JD (2002). "Sustainable coastal defences – the use of timber and other materials." Proceedings of the Institution of Civil Engineers - Municipal Engineer 151, Issue 3.

European Commission (2001) "EU focus on coastal zones." Luxembourg: Office for Official Publications of the European Communities

House of Commons Environment Committee (1992) "Coastal zone protection and planning." London: HMSO

HR Wallingford (2003) "National Flood Risk Assessment." Bristol: Environment Agency

Huntington, SW (2002) Keynote address. Proc. Int. Conf. Coastal Engng. ASCE.

Lee, J (1999) "Application of Integrated Management for the Protection of Water Resources in Coastal Areas: Korean Example." Proc. Mayors' Asia Pacific Environmental Summit Honolulu, Hawaii

http://www.csis.org/e4e/Mayor32Lee.html

Masters, N (2001) "Sustainable use of new and recycled materials in coastal and fluvial construction." London: Thomas Telford.

Moores, N (2003) "Wetlands: Korea's most-threatened habitat." http://www.wbkenglish.com/wco.asp

Moores, N (2003) Saemangeum: Internationally Significant Wetlands to be 100% Reclaimed." http://www.wbkenglish.com/saemank.asp

Schultz, B (2001) "Development of tidal swamps and estuaries." Keynote address at 1st Asian Regional Conference at 52nd IEC meeting of International Commission on Irrigation and Drainage (ICID), Seoul, Korea.

Thorpe, TW (2001) "The UK market for marine renewables." All-energy futures conference. Aberdeen, February.

Scarweather Sands Offshore Wind Farm: Coastal Process Investigation

N. J. Cooper, J. M. Harris, W. S. Cooper and J. R. Drummond
ABP Marine Environmental Research Ltd., Southampton, UK

Introduction

The UK government is committed to reducing carbon dioxide emissions in accordance with the arrangements set out in the Kyoto Protocol. This is to be achieved through a framework of measures intended to ensure that by 2003 some 5% of UK electricity requirements will be met by renewable sources, rising to 10% by 2010. Consequently, much attention has focused on offshore wind as a potential source of renewable energy.

The UK is considered to be one of the windiest countries in Europe, and it has been estimated that present electricity needs, potentially, could be met more than threefold by offshore wind farms. Despite this, only one small offshore wind farm (OWF) has been constructed to date: the two 2MW turbine development located 1km off the Northumberland coast at Blyth.

In December 2000, the Department for Trade and Industry (DTI) announced 'Round 1' of OWF development, resulting in the substantial growth of the OWF industry in the UK and the successful pre-qualification of 18 companies for site development at a number of pre-defined locations (Figure 1). Of the resulting development proposals, six to date have gained the necessary consents to enable construction: Scroby Sands, Rhyl Flats, North Hoyle, Barrow, Kentish Flats and Solway Firth. Throughout 2002, the DTI undertook a strategic review of future leasing arrangements and announced 'Round 2' of OWF development in March 2003; due to this, the growth of the OWF industry is set to accelerate in the UK.

No.	Name	Turbines
1	Teeside	30
2	Lynn / Inner Dowsing	60
3	Cromer	30
4	Scroby Sands	30
5	Gunfleet Sands	30
6	Kentish Flats	30
7	Scarweather Sands	30
8	North Hoyle / Rhyl Flats	60
9	Burbo Bank	30
10	Southport	30
11	Shell Flats	90
12	Barrow	30
13	Ormonde	30
14	Solway Firth	60
15	Tunes Plateau	30

Figure 1. Offshore Wind Farm Sites around the UK ('Round 1').

Coastal Management 2003, Thomas Telford, London, 2003.

Coastal Process Investigations

In addition to the technical engineering challenges of constructing and operating wind turbines in the marine environment, such structures must be proven to demonstrate no significant adverse impact on the existing coastal process regime in order to ensure that ongoing coastal management activities, such as flood and coastal erosion management or nature conservation management, remain unaffected.

The generic scope of the coastal process investigation necessarily associated with any OWF development proposal has been defined by both central government (DTI) and CEFAS guidance. There is an additional requirement for this generic scope to be further enhanced by specific issues raised during a consultation process associated with a Scoping Phase of the development. Essentially, the aim of the coastal process investigation is to assess the baseline coastal process regime, and to quantify the magnitude and significance of any changes to this brought about by the OWF development. The coastal process regime is defined according to:

- Hydrodynamics – water levels, current velocities, waves;
- Sediments – sediment composition, sediment transport pathways and erosion/accretion patterns, suspended sediment concentrations.

Assessments of the above factors are required to be made for existing baseline conditions, together with construction, operational and decommissioning phases of the proposed development, and over the following spatial scales:

- 'Near-field' – defined as the footprint of the whole development site that resides in the marine environment, including turbine support structures, foundations and cable route(s);
- 'Far-field' – defined as the coastal area surrounding the development site over which remote effects may occur.

In this paper, an applied case study of a proposed OWF development at Scarweather Sands in Swansea Bay has been used to demonstrate a variety of techniques that can be applied to assess the location, type and scale of potential coastal process impacts associated with such developments (see ABPmer, 2002 for full details).

Scarweather Sands Offshore Wind Farm

Scheme Details

United Utilities Green Energy Ltd. has pre-qualified for a 25-year lease of a 10km^2 area of the seabed from the Crown Estate for the development of an OWF. The allocated site is in the lee of a sandbank known as Scarweather Sands, located towards the south-east of Swansea Bay in the Bristol Channel (Figure 2). The development site also extends across parts of the adjacent Hugo Bank.

The lease arrangements permit the installation of up to 30 turbines within the allocated site, along with associated inter-connecting cables and main export cables to shore landing to the north of Margam Moors. There will also be an anemometer mast and an offshore substation on the development site. Each turbine will be c. 85m in height, 5.6m in diameter and mounted either on gravity foundations, monopile foundations driven into the seabed, or tripod multipile foundations (to be confirmed subject to further engineering design assessments). Cabling to shore and inter-connecting cabling will be buried under surface sediments. With the proposed layout there is a distance of approximately 6km between the adjacent coastline near Porthcawl and the location of the closest turbine.

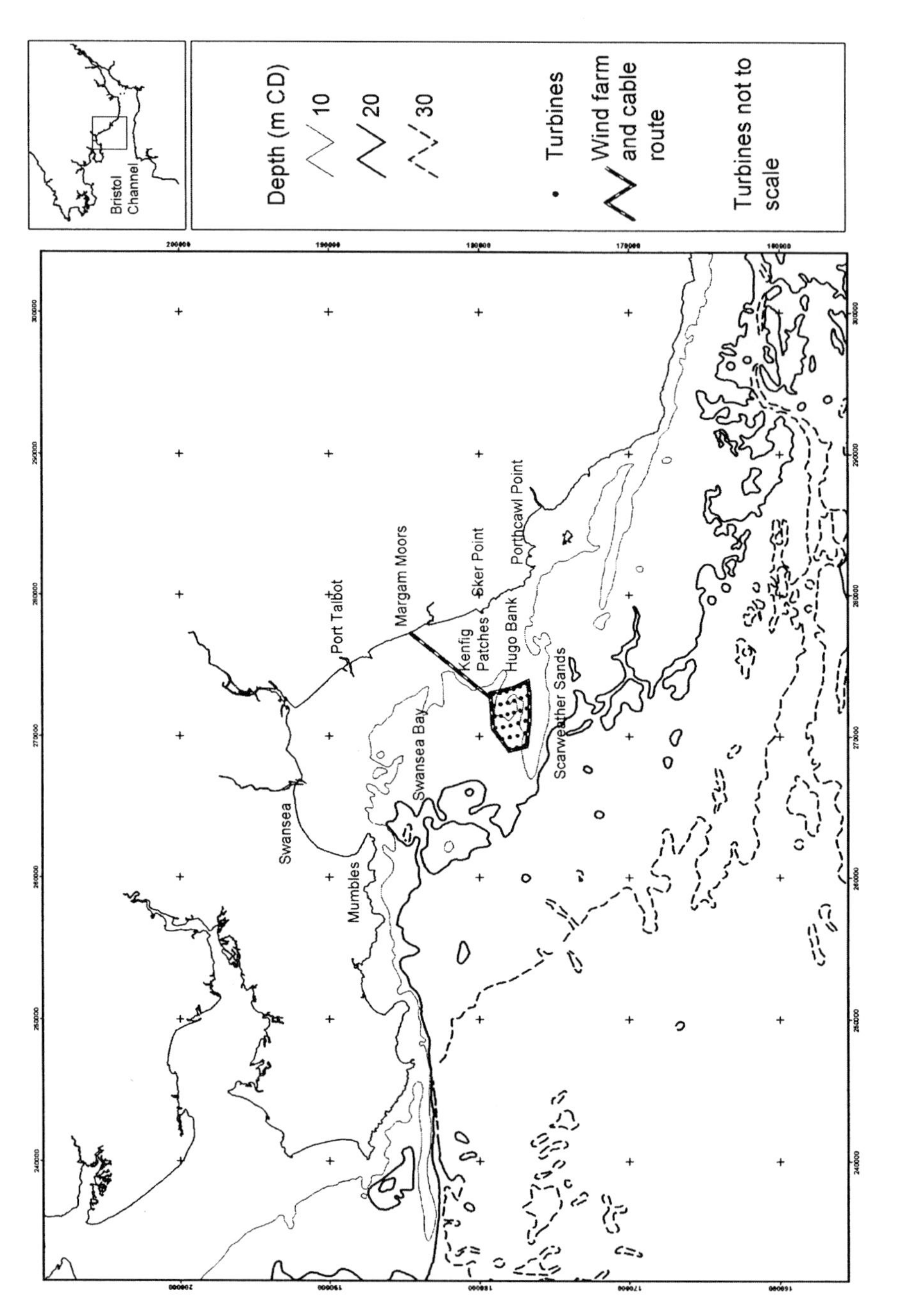

Figure 2 - Scarweather Sands Development Site, Swansea Bay.

Two consenting processes potentially were applicable to the development proposal, namely the Transport and Works Act (TWA) Order and Section 36 of the Electricity Act. The TWA was identified as the most preferable route because of potential navigation issues raised during initial consultation (Swansea Bay has several ports) and because it enabled the decision-making process to be devolved entirely to the Welsh Assembly Government (WAG). Under this route there remained a requirement for consents under the Food and Environmental Protection Act (FEPA) 1985 and the Coast Protection Act (CPA) 1949, and an Environmental Impact Assessment (EIA).

Whilst the focus of this paper is on the coastal process investigation undertaken as an integral component of the EIA, due consideration was also given by appropriately skilled consultants to a wide range of other relevant issues, including: human environment; terrestrial and marine ecology; landscape, seascape and visual aesthetics; noise and vibration; electromagnetic effects; navigation and fisheries; archaeology; hydrology and geology; and contamination. Following consideration of all of the above factors, an application and accompanying Environmental Statement was submitted to WAG in January 2003 (United Utilities Scarweather Sands Ltd., 2003).

Methodology for Coastal Process Investigation

The coastal regime is a dynamic environment that responds to the periodic effects of tides, the episodic effects of winds, waves and river discharges and the longer-term trends associated with climate change issues (e.g. sea level rise). The form of the response may influence rates of sediment transport or lead to changes in the local morphology of the seabed and coastline. Therefore, a baseline description of each of these processes is necessary, along with a description of changes to these processes with the scheme in place. However, it is important to recognise that this baseline condition is not a fixed state, but varies over timescales ranging from seconds/minutes/hours (different water levels and current velocities throughout the tidal cycle) to years/decades/centuries (effects of sea level rise).

The method of coastal process investigation adopted in this study was based on a combination of assessment of site data, empirical evaluation, detailed numerical modelling and engineering judgement to define the magnitude and significance of any changes to the existing coastal process regime. Through application of a suite of numerical modelling tools it was possible to provide a detailed description of the relevant coastal processes at an appropriate range of temporal and spatial scales (Tables 1 and 2).

Table 1. Suite of numerical modelling tools

Process	Regional Scale (Far-field)	Local Scale (Near-field)
Hydrodynamics	DELFT3D-HD • 4 layers in the vertical	DELFT3D-HD • nested into Regional model
Waves	DELFT3D-WAVES (HISWA) • Refraction (due to both depth and current) • Shoaling • Breaking	MIKE21-BW • Refraction • Shoaling • Back-scattering • Reflection • Diffraction
Sediments	DELFT3D-SED DELFT3D-PART	Empirical scour method

Table 2. Range of events modelled

Events		Baseline	Post-Development
Tides	Lunar cycle (spring to neap range)	Yes	Yes
	Future sea-level scenario	Yes	Yes
Waves	100:1 year offshore return period	Yes	Yes
	10:1 year offshore return period	Yes	Yes
	1:1 year offshore return period	Yes	Yes
	1:50 year offshore return period	Yes	Yes
Sediments	Plume dispersion from turbine scour	N/A	Yes
	Erosion / accretion patterns (potential transport)	Yes	Yes

A comparison between baseline and post-development scenarios has enabled an assessment to be made of the potential impacts of the scheme on coastal processes.

At a local scale (near-field) modelling information has supported an assessment of the following issues:

- Local scour around the base of the wind turbines and the cumulative effect of multiple structures;
- Effects on the local wave climate (wave reflection off turbine structures and diffraction around them) and how this might influence sediment mobility;
- Quantification of the volume and type of fine sediments released into the water column as a result of construction / operation and decommissioning activities.

At a regional scale (far-field), the following issues have been considered:

- Changes to the overall tidal flow regime, and how this may alter sediment pathways.
- Changes to the wave regime, caused by refraction and focusing effects, and how these may impact upon sediment transport and the coastline.
- Fate of any sediment dispersion arising from the development either during construction / decommissioning or scour development in the operational phase.

Morphological Development

Swansea Bay is a shallow, crenulated bay extending between the Limestone headlands of Mumbles and Porthcawl (Figure 2). It was formed by the inundation of the lower reaches of the rivers Tawe, Neath and Afan and the erosion of the less resistant Millstone Grits and Coal Measures located between the more resistant headlands. The Bay is situated out of the main tidal stream of the Bristol Channel and acts as a depositional centre for fine sands and muds. Kenfig Patches, Hugo Bank and Scarweather Sands exist to the south-east of Swansea Bay, developed from sand transport moving off Sker Point and Porthcawl Point. Although there remains a supply of contemporary sediment input to Swansea Bay from the rivers and from shore and seabed erosion, this is relatively small when compared with the volume of sediment presently within the Bay and supplied during the Holocene. Over the last 100 years or so, the morphology of this whole area has been altered by dredging, construction of training embankments along the Neath Estuary and coastal defence activities.

Scarweather Sands and the smaller Hugo Bank and Kenfig Patches are located within the 15m Chart Datum (CD) isobath. Scarweather Sands is believed to have formed in the Flandrian glacial stage during a period of sea level rise. It is a linear sandbank orientated with the dominant tidal streams and separated from the adjacent Hugo Bank by the Shord Channel. The initial accumulation of sand is thought to have been the result of a small eddy created by the presence of the protruding headland of Sker Point. As further sand accumulation developed, interactions between the tidal currents and the bank itself further enhanced the eddy, encouraging the growth of the bank (Pattiaratchi and Collins, 1987).

Scarweather is 9.4km long and has a maximum width of 1.9km (at the 10m CD contour). It protrudes to a maximum height of 20m above the seabed (Pattiaratchi and Collins, 1987). The western section of the bank is higher, flatter and wider than the eastern end. The breadth of the western end can be attributed to swell wave action from the southwest. The bank is asymmetrical with a steeper south facing slope in the west and a steeper northerly slope in the east. Megaripples of up to 2m in height and with wavelengths of 14-25m are present over the western and central section of Scarweather Sands. Asymmetrical sandwaves occur on the eastern end of the bank. These features are up to 5m high and 200m long (Pattiaratchi and Collins, 1987). An extensive gravel region occurs along the southern flank of the bank and over parts of the Shord Channel. There are also some sand waves evident in the Shord Channel (Stride and Belderson, 1985).

Hugo Bank is located to the north of Scarweather Sands and is approximately 5km in length and 1km wide. Further to the north, off Sker Point, is Kenfig Patches, a more diffuse sand body with limbs of sand extending from the main body in a south-easterly direction.

Charts between the following dates were used to undertake a quantitative assessment of the changes over time in the vicinity of Scarweather Sands: 1939, 1949, 1979 and 1994. This assessment was brought up to date using contemporary bathymetric survey data (Titan Environmental Surveys, 2002). Based on analysis of these charts it has been shown that the development site is part of a dynamic region that has experienced considerable changes in the form of the seabed, particularly in relation to the sandbanks. Figure 3 indicates that Scarweather Sands appears to have undergone considerable reshaping with the sandbank at times exhibiting a quite sinusoidal crest alignment (e.g. 1938, 1994), and at other times a much more linear crest alignment (e.g. 1979). This variability in the sand bank has implications for the natural variability of wave climate in its lee.

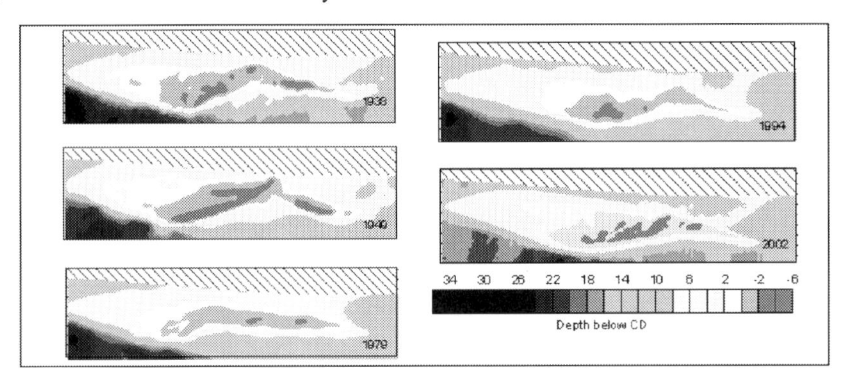

Figure 3. Historic development of Scarweather Sands.

Tidal Regime

The tidal regime is defined here as the behaviour of bulk water movements driven by the action of tides and non-tidal influences such as river flows and meteorological conditions (e.g. winds, atmospheric pressure and storm events). The influence of Scarweather Sands OWF on the tidal regime has been assessed through the application of a calibrated hydrodynamic model. The model includes two levels of detail, a regional description across Swansea Bay and a higher resolution nested grid of the area around the development site. The models have examined the potential implications the scheme may have on the tidal regime by comparing a standard set of baseline tidal scenarios to equivalent scenarios including the offshore wind farm.

The Bristol Channel has one of the largest tidal ranges in the world, with a mean spring tidal range of around 6m in the Outer Bristol Channel and around 12m in the Severn Estuary. In Swansea Bay the tidal range is 8.5m on spring tides and 4.1m on neap tides. The principal implication of such a large tidal range is that large volumes of water are moved in and out of the area on each tidal cycle, which in turn leads to strong tidal currents. Non-tidal meteorologically driven effects may also contribute to the hydrodynamic regime in the form of surge contributions to water elevations. The surge effect may provide up to an additional 1.6m on tidal levels in extreme cases.

The main axis for flood and ebb tides within the Bristol Channel is aligned west-east, although these main channel flows get deflected into the various embayments along the Welsh and English coasts. Flows deflected into and out of Swansea Bay are slower than the main Bristol Channel flows, thereby creating a relative depositional area within Swansea Bay. Scarweather Sands creates an obstruction to flows, and the residual circulation around the bank is clockwise. Flows across Swansea Bay vary temporally as a function of the tide and tidal range, and spatially as they move across the bathymetry of the seabed.

Figure 4a presents a spatial distribution of predicted surface flows across Swansea Bay during the time of peak flood on a spring tide, with Figure 4b indicating the equivalent pattern for peak ebb flows. These figures indicate that, in general, peak ebb speeds are marginally greater than peak flood speeds within Swansea Bay. Periods of slack water occur around high and low waters.

Influence of Scarweather Sands Offshore Wind Farm on the Tidal Regime

The presence of an installed turbine tower provides a local obstruction to flows that otherwise would not occur in the baseline scenario. The effect of the obstruction is to increase local turbulence in the flow regime. The head-on flow first slows down in front of the obstacle before bifurcating to find an alternative passage around the turbine tower. At this point the diverted flows join with the adjacent flow to lead to locally increased speeds, before meeting-up behind the obstacle to form a wake in a region where the flows have again been slowed. This effect continues through the tidal cycle and becomes most prominent at times of peak flow (i.e. mid-flood and mid-ebb periods on a spring tide).

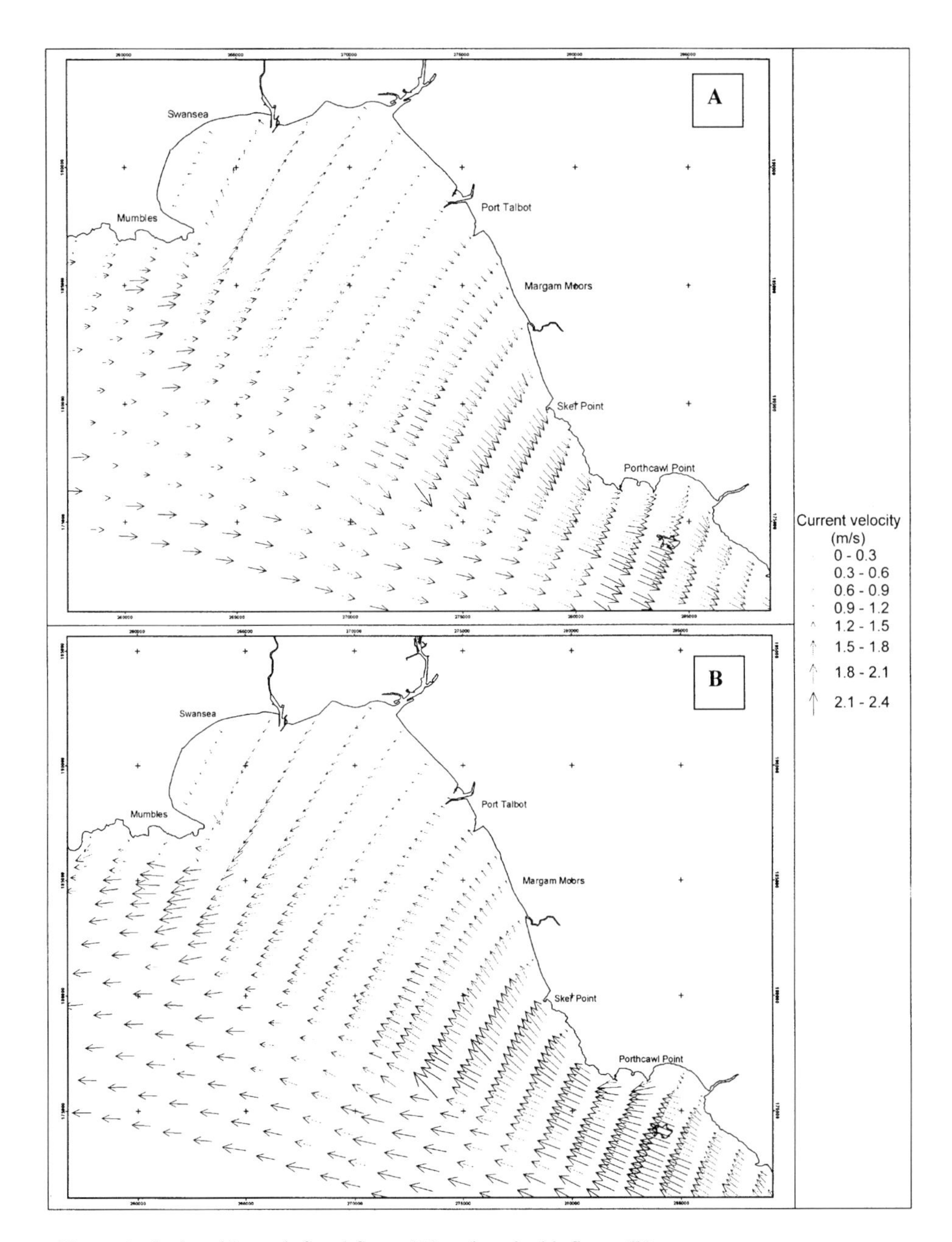

Figure 4. Spring tide peak flood flows (A) and peak ebb flows (B).

Figure 5 illustrates an indicative pattern of predicted differences in the flow regime around an individual structure. The results from a baseline condition have been subtracted from equivalent results including the OWF to produce a set of predicted changes in the local flows, with negative values representing reduced flow speeds and positive values representing locally increased flow speeds. The figure illustrates the extent of change occurring at the peak of the ebb flows for near bed flows. A prominent feature of these flows is the locally increased flow speeds around the turbine unit that will be the primary initiator for local scouring. Scour will initiate in situations where locally increased flow speeds exceed the local threshold for sediment erosion.

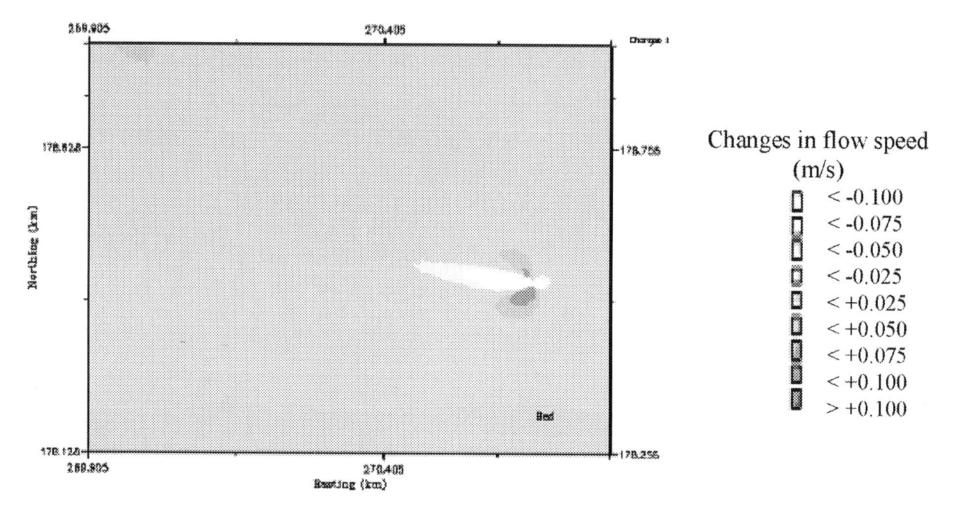

Figure 5. 'Near field' changes in flow due to turbine tower during peak ebb conditions.

To determine whether any changes in flow regime would be observed at the regional scale, as well as these localised changes, a number of representative locations around the perimeter of the development site, along the cable route and within the wider environment of Swansea Bay were investigated for baseline, operational and predicted sea level rise scenarios (assuming a sea level rise of 2mm per year for the 22 years operational life of the development).

For water levels, the changes demonstrated between the baseline event and the scenario including the OWF were very small, with a change at most sites being at the level of 0 to ±0.003m (i.e. 0-3mm). This level of change is negligible when compared to the natural variability in the system and is outside the accuracy of the model's numerical scheme. Furthermore, the changes observed under the climate change scenario clearly indicate that the magnitude of natural change in water levels due to sea level rise over 22 years is considerably greater than the changes induced by the offshore wind farm.

For flow speeds, the maximum change modelled between the baseline event and the scenario including the OWF was identified to be located within the development site during peak ebb flow and -0.18m/s in magnitude (i.e. flow speed decreases). During peak floods, the maximum decreases within the development site were -0.07m/s. However, the changes in flow speeds at sites beyond the immediate boundaries of the development site were very small, being within the range ±0.02m/s. This demonstrates that the effect of the proposed development on the flow speeds is very localised. Under the climate change scenario, an increase in flow speeds was observed at all observation sites, ranging from +0.01 to +0.11m/s at peak flood on spring tides and from 0 to +0.24m/s at peak ebb on spring tides.

For flow directions, the changes modelled between the baseline event and the scenario including the OWF were within the range ±4°, with zero change observed along the cable route and coast. Under the climate change scenario, results indicate that the magnitude of natural change in flow directions over 22 years is greater than the changes induced by the offshore wind farm.

One further area of investigation related to the potential for the OWF to influence the hydrodynamics in the vicinity of the sand bank at Scarweather Sands, hence inducing a change in its form and/or volume. The implications of any such change could be a reduction in the degree of natural wave protection offered to the shoreline by the bank. This issue is of principal concern to the Swansea Bay Coastal Engineering Group and Countryside Council for Wales who respectively have flood and coastal management responsibilities and nature conservation responsibilities within Swansea Bay, and would not wish to observe an increase in wave energy at the shoreline threatening the integrity of the natural and structural coastal defences, or the designated nature conservation area of Kenfig candidate Special Area of Conservation (cSAC). In order to make further assessment of this issue, changes in bed shear stress at various locations along the sand bank were modelled. Results indicated that the development does not increase the potential for a change in bed shear stress and consequently will not induce a change in bank morphology. In some locations along the sand bank, the potential for change in bank morphology is actually reduced, based on an assessment of localised reductions in bed shear stresses.

Summary of Impacts on Tidal Regime

Near-field and far-field changes in the tidal regime were assessed using a detailed modelling approach that first was proven capable of describing the baseline conditions. Predictions of the flow regime with a representation of an OWF included were then compared with the baseline conditions in order to quantify the magnitude of the effects both in spatial and temporal scales. In general, the observed effects were related to moderate reductions in flow speeds in the lee of individual structures, with moderate increases in flow speeds around the base of the structures. The potential for the changes in tidal regime to influence Scarweather Sands and accelerate changes in its morphological form and/or sediment volume was considered to be insignificant, based upon interpretation of changes to flow speeds and bed shear stresses in the vicinity of the bank. Additionally, the changes observed due to the scheme were considerably less than that variability in the 'baseline' natural coastal system due to sea level rise of 2mm per annum over the period of the operational life of the development. Consequently, the development is considered to have no significant adverse impact on the tidal regime.

Wave Regime

The wave regime can be regarded as the combination of swell waves moving into the area (having been generated remotely from the area) and locally generated wind-waves. To the west of the development site, data from a deep-water directional wave prediction point was obtained from the Meteorological Office wave model at a site considered suitable for use as an offshore boundary condition for the wave modelling studies. The location of the offshore data is 51.28°N, 4.92°W and in a water depth of 63m.

Analysis of these offshore wave data led to the derivation of a set of extreme wave events for selected return periods and wave directions. These data also provided the definition of offshore boundaries for the wave model.

Table 3. Extreme offshore wave conditions under different directional sectors

Direction Sector (°N)	Wave height (in m) under various return periods (in years)					
	0.01	0.1	1	10	20	50
0-30	3.06	4.55	5.95	7.27	7.66	8.17
30-60	2.27	3.2	4.03	4.8	5.02	5.31
60-90	2.55	3.36	4.03	4.63	4.8	5.02
90-120	2.77	3.8	4.69	5.5	5.73	6.03
120-150	2.71	3.67	4.5	5.25	5.46	5.73
150-180	2.76	3.58	4.26	4.86	5.03	5.25
180-210	3.92	5.36	6.6	7.72	8.04	8.45
210-240	4.85	6.77	8.44	9.96	10.39	10.96
240-270	4.96	7.06	8.9	10.58	11.06	11.68
270-300	4.23	5.93	7.42	8.78	9.17	9.68
300-330	3.15	4.1	4.9	5.59	5.79	6.04
330-360	2.65	3.56	4.33	5.02	5.22	5.47
All	4.96	7.06	8.9	10.58	11.06	11.68

The largest offshore wave condition is present with waves travelling from the 240-270°N sector, from which waves are duration-limited. The impact of the OWF on the wave regime has been assessed for the following conditions:

- From low water to high water on a spring tide to determine tidal influences on the wave regime;
- For the principal offshore direction (255°N) to determine the main exposure condition;
- For wave events with offshore return periods of 100 in 1 year, 10 in 1 year, 1 in 1 year and 1 in 50 years to determine changes in wave heights occurring across the site.

Influence of Scarweather Sands Offshore Wind Farm on the Wave Regime

The influence of Scarweather Sands OWF on the wave regime was assessed through the application of a calibrated wave model. The model included varying levels of detail, including both a regional description of the wave regime across Swansea Bay and a higher resolution nested grid of the local wave conditions around the development site. Representative wave events were selected to define the variability in the wave regime from typical conditions through to extreme events. Predictions with and without the wind farm in place were then compared to determine the potential implications of the scheme on the wave regime.

The presence of an installed wind turbine tower provides a local obstruction to waves which otherwise would not occur in the baseline scenario. The effect of the obstruction is to reflect waves off each structure and scatter them radially as a consequence of the circular shape of the structure. The reflected wave interacts with the oncoming waves to create wave-wave interaction with a local small increase in wave heights in-front of each unit, and a shadow effect in the lee of each unit. The reflected waves and shadow effects off one structure appear to interact with adjacent structures but at a level that is small in comparison to the incident wave.

It is commonly accepted that diffraction effects around a cylinder become important when the ratio D/L becomes greater than 0.2 (Isaacson, 1979); where D is the diameter of the pile and L is the wavelength. Therefore, with turbine towers of 5.6m diameter;

$$\frac{D}{L} = \frac{5.6}{L} > 0.2,$$

and diffraction becomes important when the wavelength, $L < 28$m.

Due to the fact that the wavelengths of most waves propagating within Swansea Bay are considerably greater than this value, diffraction effects are unlikely to be important for the size of structure and the type of wavelengths involved at this site. Consequently, the effect of the OWF structures on the long period waves that characterise this area is expected to be minimal, as the waves will pass each turbine tower and re-group, with only very localised influences and negligible wider-reaching effects. However, to assess the far-field effects further modelling was undertaken using a highly conservative approach. To achieve this, the turbine towers, sub-station and anemometer masts were represented as 'obstacles' across the entire edge of a model grid cell. To account for the fact that the turbine diameter (5.6m) is very much less than the width of one grid cell (100m), a transmission coefficient was applied to each obstacle to allow partial propagation of waves. This value was selected to represent a conservative scenario after using the Boussinesq model to simulate near-field changes across the development site. However, it should be remembered that the model conservatively assesses effects over a much wider area of influence than will be observed in reality, by virtue of the fact that each model obstacle is 100m long. Simulation of such a large obstacle in the model created artificially exaggerated shadow effects because the obstacle length is unrealistically large with respect to wavelength. Despite this, the approach was adopted as a sensitivity test in order to assess the effects of a highly conservative scenario on the 'far-field' environment. Under this conservative scenario, the general pattern of change was for a small reduction in wave heights in a 'down-wind' direction that created a small wave shadow between the development site and the adjacent coastline. At the shoreline, changes in wave height under this conservative scenario were identified as both negative (small wave height reductions) and positive (small wave height increases). The magnitude of the changes at the shoreline was generally less than 1% (and sometimes less than 0.1%) of baseline wave height conditions and consequently of insignificant impact on the overall wave regime.

Summary of Impacts on Wave Regime

Near-field and far-field changes in the wave regime were assessed using a detailed modelling approach that first was proven capable of describing the baseline conditions. Predictions of the wave regime with a highly conservative representation of the OWF development were compared with the baseline to quantify the magnitude of the effects both in spatial and temporal scales. In general, the effects under this conservative modelling approach were related to small changes in wave heights immediately in the lee of individual structures with no significant far-field implications. Indeed, natural changes in the morphology of Scarweather Sands (bank height and shape) are likely to cause greater degrees of variation in the far-field wave regime as wave refraction across the bank is affected.

Sediment Regime

The contemporary sediment regime across Swansea Bay comprises populations of surface seabed deposits, mobile bedforms, suspended sediments and various sources and sinks for material. The behaviour of these sediments depends on their respective response to the applied hydrodynamic forces of waves and tides.

Mobilisation of sediments can be expected when the shear stress effects from the applied hydrodynamic conditions exceed a certain threshold relevant to the specific material type. Transport then occurs in the direction of the sustained flow regime until shear stress levels drop below a further threshold and the material is deposited. The combination of wave-current interaction provides the most significant influence to the initiation of sediment transport. Fine sediments (e.g. silts and muds) tend to be transported as a suspended sediment load and may remain in suspension for long periods throughout the tidal cycle, whereas coarser sediments (e.g. sands) may only be transported at time of peak flow.

To investigate the potential consequence of the proposed OWF development on the sediment regime, appropriate modelling tools were applied. An initial baseline description of sediment movements was defined, adopting tidal and wave scenarios for the baseline case, and compared to results from model runs with the OWF introduced. Further tests considered scenarios where, hypothetically, scour protection would not be included to investigate the release of sediment plumes. These scenarios served to indicate the potential dispersion patterns of materials released into the water column as a consequence of construction and decommissioning activities. It should be noted, however, that these investigations must be viewed as 'worst case' scenarios since it is intended in practice that scour protection will be provided during the construction phase, thereby preventing such scour from occurring (i.e. mitigating potential effects in design and construction).

Near-field and far-field changes in the sediment regime have been assessed using a combination of detailed modelling approaches and empirical relationships. Predictions of changes to existing sediment transport pathways, and hence areas of erosion and deposition, were proven through modelling to be insignificant. Consequently, the development was not considered likely to adversely affect existing coastal defences or designated nature conservation areas, such as the Kenfig cSAC. Modelling was also used to assess the fate of any sediments released into the water column through processes of scour at the base of each turbine tower (which will occur only in the absence of scour protection). Under this scenario, scour material will increase suspended sediment concentrations temporarily whilst the scour holes are being developed. However, the increased levels of concentration will remain within thresholds of natural variability and the strong tidal currents will disperse the suspended material widely, meaning that no significant impact is predicted on existing seabed uses, such as shell fisheries. It should also be noted that because in practice scour protection will be provided at the base of each turbine tower, the volume of sediment released into the water column will be considerably lower than the predictions made in the modelling for the scenario with no scour protection, and predominantly confined to during construction and decommissioning activities.

Many of the sediments within Swansea Bay comprise large proportions of sands, which do not inherently contain high levels of toxic metals. This observation has been confirmed by grab samples retrieved from the study area, where levels of contamination for hydrocarbons and various heavy metals were identified to be lower at sites located within or very close to the development site than those sites located more remotely. Since metals and organic contaminants tend to most-readily bind to clay particles, the sediments in the development area are unlikely to contain any persistent contaminants that pose a risk of adverse impact if dispersed. Therefore, the recorded concentrations are deemed to be representative of typical background concentration rates and remain below industry-standard critical thresholds for probable effects.

The development area is within a seabed zone where sediments are expected to be very mobile because of the strong currents and waves creating a 'dispersive' environment. Since there has been no long-term build-up of pollutants originally derived from spoil dumping, this evidence further suggests it is unlikely that there are any unexpected reserves of contaminated fine sediments within the development area.

Conclusions

There are many advantages in the promotion of offshore wind farms as a source of renewable energy in the UK. However, it remains necessary to ensure that any such development does not have an adverse effect on the coastal process regime. The uses and management issues within the coastal zone are complex and varied, ranging from flood and coastal defence, through nature conservation and seascape, navigation dredging and material disposal, tourism and shell fisheries. Consequently, it is imperative that the potential location and magnitude of any changes brought about by the introduction of an OWF development are robustly assessed.

In this paper a comprehensive investigation of the coastal process regime in Swansea Bay has been presented to characterise the baseline conditions and determine the potential effects that the installation of an offshore wind farm at Scarweather Sands may have on these conditions. Consideration has been given to issues related to the immediate area of the proposed development site (known as the near-field) as well as the regional context across Swansea Bay (known as the far-field).

Studies have been based on the best available historic and contemporary site data, which has included a suite of on-site monitoring to quantify wave, tidal and sediment variations. The data has been collated and analysed to derive direct measures of key parameters, as well as supporting the configuration and calibration of a comprehensive suite of numerical models used to evaluate potential impacts of the wind farm. Various scenarios have been evaluated using conservative assumptions and sensitivity tests to draw out the maximum potential impact that the development could have on coastal processes. The findings from the investigation reveal that the proposed offshore wind farm development generally will cause localised changes to the 'near-field' area, for example through local bifurcation of flows around turbine towers, but without significant 'far-field' effects on the wider coastal process regime.

Many of the approaches presented in this paper have been developed and applied by ABP Marine Environmental Research Ltd. during coastal process investigations associated with proposed offshore wind farm developments at Burbo Bank, Gunfleet Sands, Scarweather Sands and Teeside. It is anticipated that the methods presented have wider applicability as 'Round 2' of OWF development proceeds in the UK in order to meet the requirements of existing CEFAS and DTI guidance on the type and scale of coastal process investigations required.

Acknowledgements

The authors would like to thank Hyder Consulting Limited (Gill Lock, Steve Scone, Julian Galloway) and United Utilities Green Energy Limited (Chris Williams, Eleri Owen, Bob Ayres) for their assistance during the course of the Scarweather Sands Coastal Process Investigation and their permission to publish results in this paper.

References

ABP Marine Environmental Research, 2002. *Scarweather Sands Offshore Wind Farm: Coastal Process Investigation*. Report R.985. December 2002.

CEFAS, 2001. *Offshore Wind Farms: Guidance note for Environmental Impact Assessment in respect of FEPA and CPA requirements*.

DTI, 2001. *Guidance Notes: Offshore Windfarm Consents Process for England and Wales*.

Isaacson M., 1979. Wave induced forces in the diffraction regime. In: *Mechanics of Wave-Induced Forces on Cylinders*.

Pattiaratchi, C.B and Collins, M.B, 1987. Mechanisms for linear sandbank formation and maintenance. *Prog. Oceanog.* 19, 117-176.

Stride, A.H and R.H. Belderson 1990. A re-assessment of sand transport paths in the Bristol Channel. *Marine Geology* 92, 227-236.

Titan Environmental Surveys, 2002. *Scarweather Geophysical Survey*. Report No. CS0040/D1/1. Report to United Utilities, July 2002.

United Utilities Scarweather Sands Ltd., 2003. Environmental Statement. Submitted to Welsh Assembly Government. For further information, please see www.scarweathersands.com

Developing Procedural Guidance for SMPs

R. Spencer, Arun District Council, Littlehampton, UK
A. Hosking, Halcrow Group, Swindon, UK
P. Frew, North Norfolk District Council, Cromer, UK
R. Young, Shepway District Council, Folkestone, UK

Introduction

Shoreline Management Plans [SMPs] provide a large scale assessment of the risks associated with coastal processes and present a policy framework for coastal defence. The concept was introduced by the Department of Environment, Food and Rural Affairs [Defra] (then MAFF) in 1993, with full guidance issued in 1995.

It is generally recognised that the first round SMPs, despite the guidance, lacked uniformity and consistency (not least in scope and depth); output quality and format also varied. Defra issued revised guidance in 2001 [2001 Guide][1], in order that the SMP could be used as universal coastal planning tool. Interim procedural guidance [PG][2] has recently been developed to help coastal groups deliver more consistent and usable results in the reviews of their SMPs.

In providing a background to the subject, this paper calls upon information in a number of papers presented at the recent Defra Conference[3] on the subject of SMPs and the interim guidance. It is both intended to give an introduction to the subject and to recount the experiences of the coastal groups that have acted as test frontages during the development of the guidance.

The experiences of the coastal groups in using the Interim PG in the review of their Plans and the ways in which Plans, in particular their procurement and pre-planning, should be approached are highlighted. The review work is ongoing and it is anticipated that further experience will be gained and presented to the Conference, as appropriate

Background

49 SMPs have been prepared for the coast of England and Wales since the concept was first introduced by Defra in 1993. Formal guidance was issued in 1995 but despite this it was considered that there was scope for improvement if the whole process was to become one with which to manage the whole coastline on a consistent basis, using the most current information and tools available.

The Universities of Newcastle and Portsmouth undertook a review of a selection of the first round SMPs in 2000; this review[4] highlighted a number of areas where particular attention was needed:-

- Timescale of Plans
- Long-term implications of coastal evolution (including climate change)
- Using the Plan to inform the Planning system
- Implications of policies on European conservation sites and biodiversity
- Uncertainty of predicting future management and its change over time
- Consultation
- Identification of funding requirements of schemes
- Boundaries of the SMP especially within estuaries
- Standard formats and dissemination methods

Recognising that the existing SMPs would soon be due for review (generally, a 5 yearly review of each SMP was envisaged), Defra issued the 2001 Guide.

The 2001 Guide did not deal in any great detail (due to timing of production) with the strategic and research work that was undertaken at around that time. The following areas of contemporary work were seen as important in the development of the next generation of SMPs and would need to be incorporated in the review of the first SMPs

- Futurecoast - provides predictions of coastal evolutionary tendencies over the next century and helps in determining broad scale future coastal defence policy throughout the open coast shorelines of England and Wales,
- CHaMPS - Coastal Habitat Management Action Plans seek a sustainable future for coastlines designated under European Directives or as Ramsar sites and are intended to fulfil obligations under Article 6.2 of the Habitats Directive to avoid deterioration of natural habitats,
- RASP/MDSF - RASP is intended to develop and demonstrate supporting methods for dealing with risk to large floodplain areas; whilst MDSF consists of a set of procedures and a GIS based software tools to help automate parts of the process and make Plans easier and quicker to prepare, (MDSF was developed as a fluvial tool but has since been extended for use on the open coast).

There were also concerns over the interpretation of critical words and phrases in the 2001 Guide and the fact that little reference had been made to the rapidly evolving local and regional beach monitoring programmes was also an issue.

The culmination of these concerns and issues was the commissioning, by Defra, of Halcrow Group in Autumn 2002, to prepare procedural guidance to augment the 2001 Guide. The new PG was intended to be a "how to" guide and as such it was considered that 'live' testing would be beneficial. This testing was carried out using three of the first SMPs due for review and is dealt with in more detail below.

The development of the 2001 Guide was project managed by Defra together with input from a steering group composed of representatives from Defra, National Assembly for Wales, Environment Agency, English Nature and the 3 trial SMP coastal groups.

The development of the guide itself involved many of the industry's consultants already involved in SMPs, so that 'best practice' guidance could be gained. Some 17 consultants, as well as Halcrow, were directly involved in the development of the Interim PG, addressing the issues through a number of 'Core' and 'Linked' studies. Further consultants and groups were involved through a process of e-mail consultation.

Core Studies	Linked Studies
Shoreline Processes	Estuaries
Issues and Objectives	Open Coast Boundaries
Policy Appraisal	Risk & Uncertainty
Format & Presentation	Consultation & Stakeholder Involvement
	Conflict Avoidance & Resolution
	Economics
	Data Access and Management
	Modelling and Decision Support Framework [MDSF]

Having a number of consultants, rather than solely Halcrow, involved in the development of the Interim PG gave the industry an opportunity to understand fully the requirements for the second round of SMPs, while at the same time providing Defra with best practice experience for the PG document.

KEY CHANGES FROM THE 2001 SMP GUIDANCE

Whilst the Interim PG is supplementary to the 2001 Guide, there have been a number of clarifications in its development that have resulted in changes to the framework set out in the 2001 document. The main changes to those requirements are identified below:

- In line with changes to Treasury Guidance in 2003, the appraisal of SMP policies is to consider a 100-year timeframe;
- 'Limited Intervention' is no longer a policy option;
- The term 'Management Unit' is to be replaced by 'Policy Unit';
- Some of the work items previously specified as being carried out in Stage 3 will now be undertaken during Stage 1.

Objectives, policy setting, and management requirements should be considered for 3 main epochs; 0-20 years, 20-50 years, and 50-100 years. Whilst interim periods may be considered, output should reflect the shoreline status as a result of these policies for years 2025, 2050 and 2100

Management Units are replaced by "Policy Units", with these defining a length of shoreline with a uniquely defined policy. A further deviation from the 2001 Guide is that these should be established towards the end of the SMP process, after determining draft final policy, not at the outset

The existing 'Management Unit' boundaries may need to be retained in some instances, as they have been used as reference points in monitoring and inspection schemes. This does not, however, mean that if they are retained they should be over-riding in the SMP Reviews; a distinction should be made in the purpose of their retention.

The PG is currently provided as 'interim' guidance and is in two parts:

- Part 1 An overview of the approaches to undertaking an SMP.
- Part 2 Technical appendices on the methods to employ.

A third part (example outputs) will be provided with the final Procedural Guidance in Spring 2004. This will be once the rigorous testing (and amendment if necessary), currently underway with the 3 trials, is complete.

Trial SMPs

The first three SMPs to come forward for review were invited by Defra to become test frontages for the various aspects of PG that were being developed by Halcrow. These were:

- Sub-Cell 3b – Sheringham to Lowestoft (ACAG) North Norfolk DC
- Sub-Cell 4c – South Foreland to Beachy Head (SEGC) Shepway DC and
- Sub-Cell 4d – Beachy Head to Selsey Bill (SDCG) Arun DC

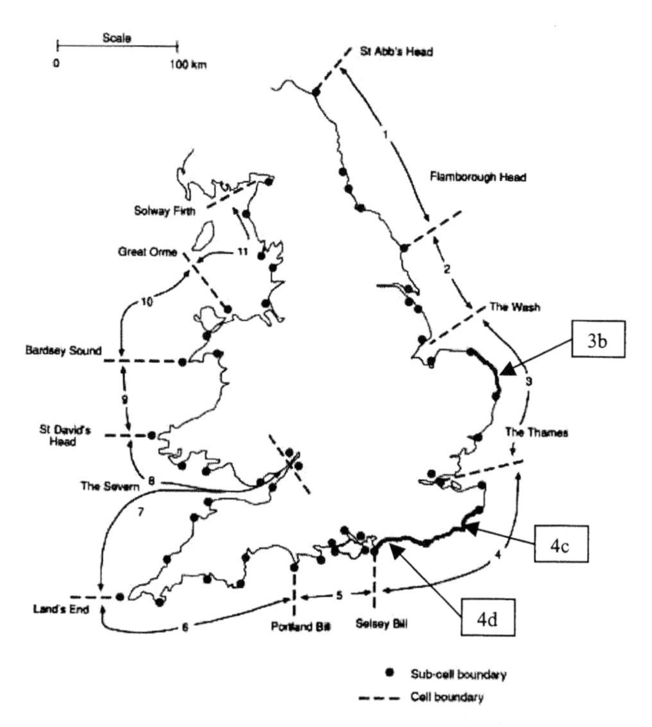

Procurement of the Trial SMP Reviews

In developing a procurement strategy for these three reviews, it was initially envisaged that a normal competitive tendering process would be used. Once a consultant had been appointed, and as work proceeded, Halcrow would suggest work elements on each test frontage to be addressed in a certain manner. The results would be fed back for comparison with the other two test areas.

However, as the scope and extent of the PG work became clear, it was decided that this method would introduce too many risks, notably in terms of timescale and budget. It was decided to separate the production of each of the three 'trial' SMPs into Parts A & B.

Part A being the use of each respective frontage to test and assess the merits of various methodologies, to aid the formulation of the interim guidance and,

Part B being the bringing together of the initial work and the completion of the SMP in the manner most appropriate from the Part A work – some elements would need to be redone if it was found that another method (used on one of the other trial areas) was more advantageous.

All the Part A work was commissioned through the Halcrow Group in November 2002, with the PG work being carried out in tandem; all of the studies were substantially complete by April 2003. Final comments were made by the Steering Group and an interim draft of the guidance made available to Defra for release at its Keele Conference for River and Coastal Engineers in July 2003.

Towards the end of the Part A work, consideration was given to the procurement of the Part B works for the three trial SMPs. It was decided that, whilst there were a number of concerns about awarding further work to Halcrow under a 'single tender action', there was an over-riding need to ensure that three consistent examples would be available for the remaining 46 reviews.

Many of the other Coastal Groups had expressed an interest in proceeding with their reviews at the earliest opportunity. If the National Steering Group aim of having consistency of methodology, output and quality in SMPs around the whole coast was to be achieved, then some relaxation of the lead authorities' normally strict procurement rules would be needed. Accordingly, Variation Orders were issued to Halcrow to undertake the Part B work for all three 'trial' SMPs.

The Defra grant rate awarded to the three lead authorities had been set at the maximum of 65% and this was extended to the whole process, to ensure that the individual authorities (and/or Group members) were not financially disadvantaged locally for the national benefit.

Key Consideration for Coastal Groups

Coastal Groups have been set up for some time – indeed many Groups were set-up at the time of the first SMPs and have continued to meet; not only for undertaking the further studies and Coastal Defence Strategies that came from the SMP but also for meeting Government High Level Targets and realising the benefit of a more formalised flow of information regarding the management of the coast.

Coastal Group Inputs

A number of areas were identified as requiring specific attention of the Coastal Groups, both prior to and during the Plan review. Section 1.4 of the Interim PG sets out a number of criteria that need to be defined before the procurement process is initiated.

Pre-procurement action

It is usual for one authority to act at 'lead' in submitting applications to Defra and to procure the competitive tender, contractual arrangements with a consultant; etc. Whilst some Plans have been developed in-house, the majority have been commissioned through consultants. Given the specialist nature of some of the new elements of the guidance, this route is more than likely to continue.

Procurement Strategy

The Coastal Group should firstly establish a Project Group or Board to manage the progress of the Plan development and to act on behalf of the Coastal Group. Whether the financing of the review is covered by Group funds, lead authority budgets or is shared in some way between the Coastal Group members, the Project Board should be authorised to act, within pre-defined bounds, without need for constant reference back.

The first question that the Client Group should ask is 'is it the correct time to undertake a review and have all the necessary studies that may have been previously highlighted been carried out satisfactorily'? Progression to Defra and contract should not be considered if there is a risk that delay and further expenditure may be incurred by the consultant in gathering data etc.

It is imperative that a defined conduit through which information is passed to and from the consultant is established and maintained; as the study progresses the need for this will only increase. The Plan must be seen to be the 'property' of the Coastal Group and its constituent members, rather than a study that has been produced by 'outsiders' and can somehow be disowned if the preferred policies are not to some members' liking. Early and informed 'buy-in' to the process should aid the Plan in this respect. Some control over the direction in which the study is progressed may be delegated to a wider Group, depending on the consultation model chosen (see below).

Physical Limits of Plan

The physical limits of the study should be set as part of the Brief. This relates both to the coastal limits (see example below) and the fluvial limit. Appendix F of the Interim PG relates to guidance on the establishment of the geographical extent of Plans i.e. how far the Plan should extend into an estuary.

The Brief should establish this boundary to ensure there is no overlap (or gap) with Catchment Flood Management Plans. The guidance is clear, with flow diagrams to aid the decision process, which is relatively straightforward and relies upon reference to tables within Futurecoast. The approach is qualitative and, like many other parts of the SMP process, utilises rankings rather than absolute scoring; an element of local knowledge is also required in setting these boundaries.

The open coast limits of each of the first round SMPs were also considered. Annex 4b of the Interim PG presents a series of questions to help in establishing whether or not the existing boundaries are appropriate. The questions are as follows:

- Does the boundary NEED to change?
- Should a boundary change be considered?
- Are there other boundaries to consider?
- Are there major inter-SMP considerations? and as a concluding statement:
- Should the SMP area be altered?

Consultation / Stakeholder Models

Chapter 2 and Appendix A of the Interim PG give examples of models that have been found to be appropriate. They include a range of stakeholder involvement levels and take account of the new constitutional arrangements now found in most authorities. Individual Groups should decide which is more appropriate to its particular frontage and use the experience gained from the previous SMP and Coastal Defence Strategies. These decisions should be built into a Stakeholder Engagement Plan, in order that the process is transparent and auditable - another feature and underlying tenet of the new PG.

Two further points which should be decided upon prior to the award of any contract are:
a) the level of precision of any studies and analysis - the existing data holding will guide this decision to a large extent and
b) how will the plan be presented upon completion - with the digital age comes a lesser need for hardcopy reports; web and CD based dissemination is encouraged. There will be a need for meetings and possibly 'flyers' and educational packs but the degree to which these are needed/desirable will need to be balanced with what is worthwhile.

Trial Frontage experience

The trial frontage work was carried out in tandem with the PG work and as such a number of the above issues have had to be addressed retrospectively.

The one which developed the most debate within the South Downs area was the Plan's coastal boundaries. In assessing the eastern boundary in line with the 5 questions (above), there was no doubt about the appropriateness of Beachy Head. The western boundary was not as clear cut. The physical feature of Selsey Bill was used as a boundary for the 1st South Downs Plan but the adjacent Eastern Solent Plan overlapped this, by using Pagham Harbour (to the east) as its boundary.

The situation had not been compromised, as similar policies had emerged in both Plans for the overlap area. However, it was suggested that the Medmerry frontage to the west of Selsey Bill, could have an effect on the risk within Pagham Harbour (and frontages to the east) if the existing shingle beach was allowed to breach. The likelihood of breach, inundation routes, erosion rates and the effect of the spinal road to Selsey may have on these issues was debated at length. In considering an appropriate boundary position, it was decided that little advance in the understanding of processes in this very complicated and dynamic area could be made within available time and budget constraints. There was a conflict between setting policy, based on coastal processes in one area, and the risk that those policies may have on another area outside that immediate area.

By setting policy in the South Downs SMP for a portion of the western Selsey Bill, there could be a complication in relation to managing coastal process in the Witterings and Chichester Harbour area. The prime aim of setting policy for Medmerry was for that (western) frontage and the risk that existed to other areas, whilst real, was not over-riding. It was decided to make broad assumptions of what may or may not happen in the longer term for the whole Selsey Peninsula under a 'do nothing' scenario and to set policy from the Bill, for Pagham and beyond. Importantly it was necessary to make it clear in the Plan how this decision was reached.

Timscale and Scope of the Studies

Consultation is an important element of the SMP process and the 2001 Guide highlighted this. However, it is important to distinguish between consultation, stakeholder engagement (establishing aspirations etc.) and data gathering. Due time should be allowed for these various elements but it is not anticipated that new PG will have an impact on the overall timescale outlined in the 2001 Guide if the 'consultation' is managed in an appropriate manner.

Protracted correspondence can be avoided, for example, if the consultees are made aware of the way in which their comments will be considered and indeed they are made aware at the outset of the scope of a Shoreline Management Plan; i.e. comments will be welcomed on shoreline engineering/management matters but coastal zone management is outside the scope of the current work. Similarly, the wider inspirational issues and smaller parochial issues are best brought up in other arenas

Trial output as Examples

The Part B work is ongoing and it is anticipated that further experience will be gained prior to the presentation of this paper; this will be presented to the Conference as appropriate. It is anticipated that all three Plans will be available as examples by April 2004.

Coastal/Client Groups and Consultants drawing up briefs and proposals will be able to draw upon the experience gained in these three trials, through consideration of example output and discussion with the parties concerned. The National Steering Group is set to continue its meetings, albeit less frequently, to help make the transition to the next generation of SMP as smooth as possible.

Expected Outcome of the 2nd Generation SMPs

Following the development of this next round of SMPs, not only will a more robust, transparent and auditable coastal planning tool exist but also Coastal Groups, their constituent members, Planners and environmental groups should be able to look forward to a number of developments. These will include:

Longer timescales considered, including three general epochs to facilitate transition from existing management practice
The consideration of the environment will be an integral part of the SMP process
More standardised databases will be developed
There may be a fewer number of SMPs, as adjacent Plans become established and common approaches and formats are adopted
There will be a greater integration with other strategic processes
Greater input to the Planning (Town & Country) process, with the possibility of 'red lines' on plans and/or time limited planning consent.

Acknowledgements

The authors would like to acknowledge the efforts and contributions of the steering group, and all other consultants and individuals who contributed to the PG project and the development of the 3 SMP Reviews.

References

[1] Defra/NAW. *Shoreline Management Plans: A Guide for Coastal Defence Authorities*. 2001.
[2] Defra/NAW. *Procedural Guidance for Shoreline Management Plans*. 2003.
[3] Defra/NAW. *Proceedings of the 38th Conference of River and Coastal Engineers*. 2003.
[4] University of Newcastle/University of Portsmouth. *A review of existing Shoreline Management Plans around the coastline of England and Wales*. 2000.

Coastal change analysis: a quantitative approach using digital maps, aerial photographs and LiDAR

Dr R Moore and Dr P Fish, Halcrow Group Ltd, Birmingham, UK
A Koh, Geotechnologies, Bath Spa University College, Bath, UK
Dr D Trivedi and Dr A Lee, British Nuclear Fuels Plc, Risley, UK

Introduction

There is a need to make reliable projections of coastal change in support of coastal management strategies and decision-making. In most cases, projections of coastal change over the next hundred years or so are more than adequate to support shoreline management and coastal defence strategy plans. There are, however, a number of coastal sites that are used for nuclear power generation and low-level nuclear waste disposal which require projections of coastal change over much greater timescales in support of post-closure safety cases.

Projection of coastal change is not new, but until recently, the historical and baseline data on which they were based were largely qualitative and incorporated unknown errors. Probabilistic methods can be used with qualitative assessments to account for the errors with data and the wider uncertainties associated with the prediction of coastal change. The reliability of coastal change projections will ultimately reflect the quality of data and the judgements upon which they are based, whether they are short-term or long-term projections. Judgements are inevitably required when making projections of change and their reliability will be guided by knowledge of historical coastal behaviour and rates of change. Therefore, it is desirable to quantify and reduce the errors associated with data wherever possible given the nature of coastal hazards and risk and the consequences of making decisions on high levels of uncertainty.

With recent advances in survey and computer technology, it is now possible to make use of spatially referenced (i.e. map-accurate) aerial photographs and integrate these with digital mapping using desktop computers. The archive of historical mapping and aerial photography within the UK dates back to the 1860s and 1930s respectively, providing a space-time dataset of considerable value to coastal change studies. Use of specialist photogrammetric and geographical information system (GIS) software has allowed the integration of historical data with baseline surveys and state-of-the-art digital remote sensing data. A wide range of analysis and applications are possible with the digital approach.

This paper demonstrates the benefits of the digital approach for quantifying historical coastal change for a case study on the west Cumbria coastline, between St Bees Head and the Esk Estuary (Figure 1). The work presented is the world's first dataset of simultaneous and synchronous medium format digital aerial imagery and LiDAR acquisition for the purposes of orthophotomap production.

Coastal Studies Programme

To ensure the responsible management of its sites, British Nuclear Fuels Limited (BNFL) has commissioned an environmental programme of which coastal management and monitoring forms an integral part. This programme aims to support the development of future scenarios and projections of coastal change. The development of a coastal change projection model for these sites needs to consider the likely evolutionary scenarios for the coastline in the short (c50 years), medium (500 years) and long-term (c10,000+ years).

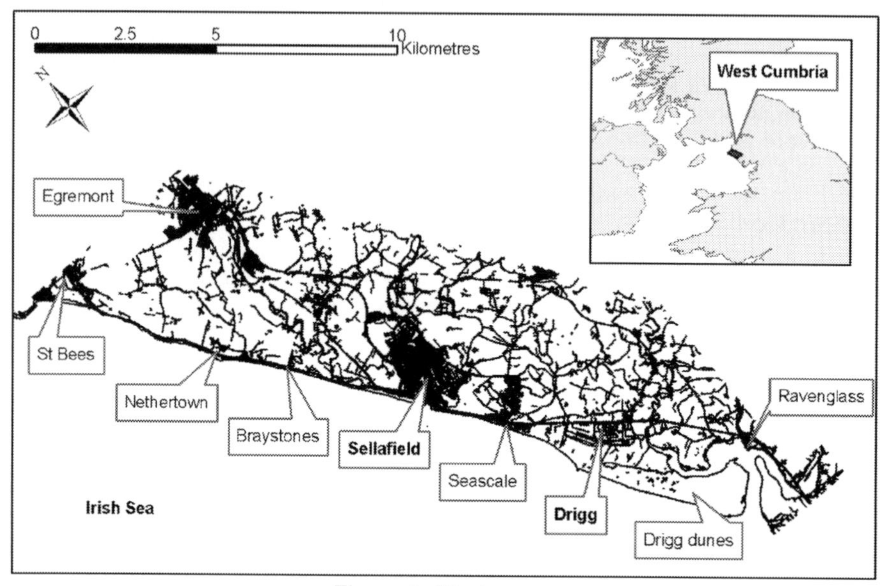

Figure 1. Study area

During 1999, Halcrow Group Ltd was commissioned to undertake a scoping study and review of the potential impacts of climate change and coastal evolution along the west Cumbria frontage. This scoping study involved a comprehensive review of the literature and identified potential sources of information and data. In 2000, a programme of data acquisition commenced comprising the procurement of historical maps and charts, aerial photographs, bathymetry and other data. In 2002, a baseline survey was completed, comprising an integrated state-of-the-art digital aerial imagery (DAI) and Light Detection and Ranging (LiDAR) survey, offshore bathymetry and shoreline surveys, and real-time monitoring of waves, currents and tidal conditions. The programme is supported by annual monitoring surveys and a range of other activities including ongoing collation and analysis of historical records and data and the development of a conceptual coastal change projection model.

Background and Approach

The coastal sites at Drigg and Sellafield fall within subcell 11d of the shoreline management plan (SMP) prepared by Bullen Consultants (1998), which extends from Walney Island to St Bees Head. The plan provides general information on the local coastal environment in support of strategic coastal defence policy and management. The SMP includes a projection of shoreline movement based on an assessment of historical maps. The errors with data are not specified and are probably unknown but would be usual for general assessments based on the interpretation of hardcopy maps and aerial photographs. The plan indicates potential shoreline retreat of 0.5 m to 1 m per year along the Drigg frontage, retreat of 0.2 m to 0.5 m at St Bees, and no change at Sellafield. The rates quoted are stated to be typical for the length of coast and to reflect average conditions.

BNFL recognised a need to obtain higher quality and reliable shoreline data that can be used to underpin future coastal change projections and decision-making. One of the main objectives of the programme has been to identify data deficiencies and to reduce uncertainty in the understanding of coastal change and its prediction. In order to do this, a quantitative approach has been adopted from the outset involving the procurement, processing, integration and analysis of historical, baseline and ongoing monitoring data.

Central to the approach has been the use of digital photogrammetry, which is a method of obtaining accurate measurements from aerial photographs and digital terrain models. The method allows quantification of terrain features in terms of their location, extent and surface topography. Digital photogrammetry makes use of scanned images or digital aerial photography captured 'on-the-fly', and references these to digital maps and other survey data to produce map-accurate orthophotographs or photomaps (Graham and Koh 2002).

Historical data procurement

Information sources of potential value to the coastal studies were identified and located (Table 1). For the analysis of coastal change, the more important of these are topographic maps and aerial photographs that date back to 1860 and span more than 140 years. These are considered below with regard to their availability and suitability for analysis.

Table 1. Information sources

Information	Date	Source
Topographic maps, 1:2,500; 1:10,000; 1:25,000 and 1:50,000 scales	Current	Ordnance Survey
Topographic maps, 1:2,500 scale past editions	1860-1970	Landmark
Topographic maps, 1:10,000 scale past editions	1867-1983	Landmark
Aerial Photographs, 1:10,000 scale	2000	Getmapping.com
Aerial Photographs, 1:5,000 to 1:25,000 scales	1964-present	Ordnance Survey
Aerial Photographs, 1:5,000 to 1:10,000 scales	1941-1958	Ministry of Defence
Admiralty Charts, 1:100,000 and 1:200,000 scales with local detail	1877-present	UK Hydrographic Office
Geological Data, 1:50,000 solid and drift, memoir, borehole and geotechnical data, seabed sediments	1870-present	British Geological Survey
Coastal Hydrodynamic data	1915-present	Various
Coastal Study/Engineering Design/Maintenance Reports	1995-1997	Copeland Borough Council

Topographic maps and heighting data

The first County Series of Ordnance Survey (OS) mapping date from the mid-1800s and were regularly updated to 1945 when the British National Grid system was introduced and a new series of maps published. Past editions of the OS 1:2,500 and 1:10,000 scale topographic maps are available in digital format from Landmark Information Group Ltd, under a joint venture agreement with the OS. Historical map data are provided as black and white raster data composed of a grid of pixels. Each map sheet has been 'warped' to the National Grid and referenced by the grid co-ordinates of the map corners. The source mapping for the scanning process was the original mapping published at the time of survey and held in the OS archive. Each map sheet has been scanned at a resolution of 300 dots per inch. Map coverage available for the Cumbria coastline for different time periods (epochs), comprised:

Epoch 1 – first County Series published between 1860-1893;
Epoch 2 – first revision County Series between 1891-1912;
Epoch 3 – second revision County Series between 1904-1939;
Epoch 4 – third revision County Series between 1919-1943; and
Epochs 5 and 6 – National Grid survey published since 1945.

The OS Land-Line (recently relaunched as MasterMap) 1:2,500 scale digital data is available for the Cumbria coastline. However, there is a paucity of heighting data for the study area that are contiguous in extent and capable of representing both the Digital Surface Model (DSM)[1] and the Digital Terrain Model (DTM)[2]. The available heighting data for the site are limited to OS Landform Profile data series with contour intervals at 10 m (±5 m).

The lack of good heighting data is set to improve with the NEXTMap Britain project. This project uses Interferometric Synthetic Aperture Radar technology and aims to produce data at 5 m postings and vertical accuracies of 1 m for DSMs and 1.5 m for DTMs across the UK. Greater accuracy and tighter postings can be achieved using laser scanning technology, such as LiDAR (see below), but these datasets generally cover small, specially commissioned areas. Data postings and vertical accuracies of 0.2 m can be achieved using laser scanning.

Aerial photography

Historical aerial photography has been obtained from various national archives and commercial aerial survey companies. Prior to the baseline survey in 2002, the most recent aerial photography of the Cumbria coastline was flown in 2000 for the Millennium Mapping Project.

The most comprehensive archives within the UK are those of the Ordnance Survey (OS) and the English Heritage National Monument Record Centre (NMRC). Of these, the NMRC have provided the best coverage of the study area with many sorties dating back to 1941. The archive comprises 1:5,000 to 1:10,000 scale photography flown by the MoD between 1941 and 1958, and 1:5,000 to 1:25,000 scale photography flown by the OS since 1964. Other aerial photograph archives and suppliers are available but coverage is limited. Details of the historical aerial photography procured for this study are provided in Table 3.

Suitability of data, relative accuracies and cost

Selection of appropriate aerial photographs and maps is required for specific applications. Some guidance on the choice of products for coastal change analysis is provided below. A key factor to consider is the scale of photography, which controls:

- the number of photographs to be purchased and processed;
- the amount of detail that can be seen in the image; and
- the accuracy of data extracted from the photograph.

---

1 Digital Surface Models represent the elevation of the ground including vegetation and built structures.

2 Digital Terrain Models represent the elevation of the ground excluding vegetation and built structures.

Together, these factors determine the suitability and cost of aerial photography and mapping for specific applications (Lawrence *et al.* 1993). At large scales (1:5,000 to 1:10,000), a wealth of detail can be seen and confidently interpreted, whilst at smaller scales only larger features will be visible and important detail may be lost. Coastal features that can be observed in aerial photography at various scales are indicated in Table 2.

Table 2. Coastal features observed in commonly found scales of aerial photography

Scale	Features recognised	Approximate area of image (km^2)
1: 60,000	Coastlines, rivers, towns and hinterland geomorphology	185.5
1: 20,000	Coastlines, large landslide systems, streams, road network and villages	20.6
1: 10,000	Coastlines, cliffs and shorelines. Discrete landform units, sand bars and houses	5.15
1: 5,000	Detail of coastal landforms, including scarps and benches, instability features, sediment types and transport pathways	1.28

Choosing an appropriate scale is essential to ensure both accurate recognition of features whilst avoiding too much detail and high costs. It is worth noting that doubling the photograph scale will quadruple the number of photographs needed, which has a significant impact on purchasing and processing costs.

Data on the positions of coastal features prior to the earliest aerial photography in the 1930s can be derived from historical maps, provided the uncertainties with the accuracy and depiction of features in the early surveys are taken into account. An issue to bear in mind with the latest Land-Line and MasterMap products is that they are sold as 'current' data and may show active features, such as clifflines, out of position as they will invariably be based on survey data several years out-of-date. The timeless nature of these products is a major setback for geomorphological studies as periodic updates of mapped features of interest are neither identified nor dated separately. This renders them of little value in the analysis of coastal change, emphasising the need for photogrammetric studies.

Baseline surveys

A baseline survey was carried out in 2002 to capture high-resolution spatial and temporal data across the entire study area, and to provide a datum to which future surveys and monitoring can be compared to quantify change. The baseline survey comprised the following:

- installation of a GPS active and passive control network;
- an integrated digital aerial photographic and LiDAR survey of the beach and coastline;
- a Real Time Kinematic (RTK) topographic survey of the beach; and
- a bathymetric survey of the seabed to 5 km offshore.

A key objective was to carry out the baseline surveys simultaneously so that quality checks could be conducted in areas of overlap between the surveys. In practice, this could not be achieved as the RTK and bathymetry surveys were conducted over several weeks. The aerial survey took place on the 5th May 2002 and the RTK and bathymetry surveys were completed on the 10th May 2002.

Digital aerial photographic and LiDAR survey

The objective of the aerial survey was to capture simultaneous digital elevation data using LiDAR and land cover data using vertical colour digital aerial imagery (DAI).

The camera system used for acquiring the DAI comprised integration of a Kodak DCS Proback with a Hasselblad 555ELD body and a calibrated 40mm Carl Zeiss lens. The system was based on a single charged coupled device image sensor with a recorded image array size of 4072 pixels along and 4072 pixels across track. The image information was stored in a digital format at the instant of image capture onto 1 Gigabyte Compact Flash cards which are changeable in-flight. Survey altitude was 850 m (2,800 feet) above mean ground level and aircraft ground speed was maintained below 120 knots. Average image frame rate for the duration of the sortie was 8.8 seconds.

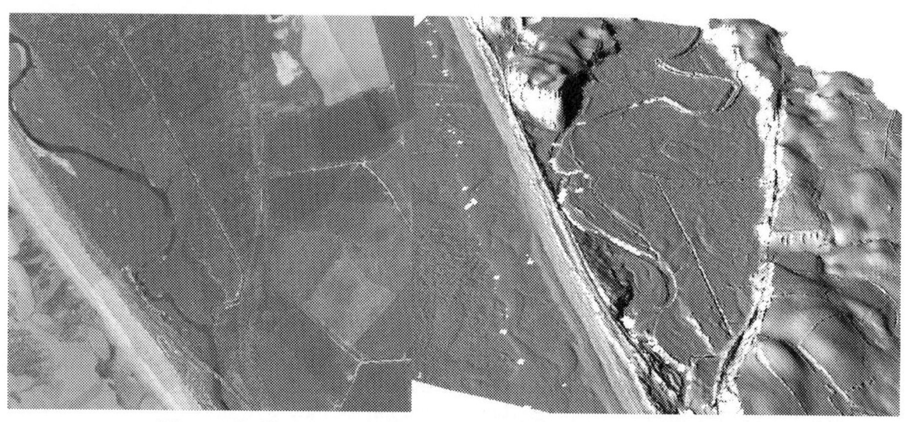

Figure 2. Example digital aerial image and LiDAR DSM

LiDAR or laser radar performs in a similar manner to radar systems, with the exception that narrow pulses of light are used instead of broad radio waves. The system has a receiver that times, counts and processes the returning light photon. Laser radar depends on knowing the speed of light to calculate how far a returning light photon has travelled to and from an object. The laser generates an optical pulse that is reflected off an object and returns to the system receiver. The high-speed counter measures the time of pulse from the start pulse to the return pulse and the time measurement is converted to a range where:

$$\text{Range} = (\text{Speed of Light X Time of Pulse}) / 2$$

The return pulse may be presented twice to the receiver. The first pulse measures the range to the first object encountered at a given location and the last pulse measures the range to the last object encountered for the same location. By acquiring first pulse and last pulse data simultaneously, LiDAR can measure both object-heights (i.e. DSM) and ground height (i.e. DTM) in a single scan. Measurements to targets can be made at any angle.

The LiDAR sensor used was an Optech 3033 ALTM configured to scan at 30Hz, with a repetition rate of 33kHz, a 19° scan angle and a swath width greater than 618 m.

The products of the aerial survey were:

- DAI at a nominal ground sampled distance of 19 cm, with each image frame covering an area of 770 m by 770 m;
- 305 image frames with a database size of 7.3 Gb in TIF format;
- 66.8 km^2 of LiDAR elevation data at 2 m postings with a height accuracy ±11 to 25 cm and a point plan accuracy of ±45 cm (RMS at 1 standard deviation); and
- 21.5 M LiDAR heightings with a database size of 354 Mb.

This is the world's first dataset of simultaneous and synchronous medium format DAI and LiDAR acquisition for the purposes of orthophotomap production.

Shoreline RTK

RTK GPS profiles of the beach and intertidal zone were carried out to link the offshore seabed bathymetry and LiDAR digital terrain models. The locations of all profiles were pre-determined and data were recorded within ±0.1 m of each profile line. Threshold settings for each data point were set to ±0.02 m in height and position. Survey data was collected ±2 hours of low water, to maximise the overlap between the LiDAR and bathymetry surveys.

Bathymetry

The seabed bathymetric survey extended 5 km offshore and overlaps with the LiDAR and shoreline RTK beach survey of the intertidal zone. Two boats were used for the survey. A manoeuvrable, shallow draft vessel, fitted with a Deso 14 echo sounder and a CSI Max GPS D-beacon system was used in the nearshore zone. A second vessel with deeper draft and fitted with a Leica 530 RTK GPS unit and a Deso 15 echo sounder was used for the offshore survey area. Tidal influences were corrected with reference to local tide gauge data.

Data processing

Integration of historical and baseline data into digital mapping solutions requires specialist software, GIS and analysts. There were various stages to the data processing chain, as follows:

1. assess 'fitness for purpose of use' of data;
2. image processing;
3. referencing of historical maps and aerial imagery;
4. 2002 DTM and orthophotomap production; and
5. error analysis.

The most important consideration was 'fitness for purpose of use'. This assessment was often made subjectively as traditionally analogue media are used (see suitability of data above). The method employed when assessing fitness for purpose for a specified application is based on the information content of the available datasets, and digitising at a level that does not exceed the information content of the least capable analogue data within the dataset. This ensures consistency in recognition, identification and analysis of historical datasets.

Image processing

The historical aerial photography is classed as derived digital imagery (DDI). Such imagery is captured by analogue cameras with film as the image sensor, where the film negative or diapositive is scanned, and the derived image is stored in a digital format. Each pixel of the digital format represents a pre-specified nominal ground sampled distance (Table 3).

Table 3. Details of raw image datasets

Source	Date	Image Type	Original Scale	Ground Sampled Distance (m)	No. of digital images & filesize
MoD	1941	DDI	5000	0.5	41 (206 Mb)
MoD	1957	DDI	5000	0.5	95 (335 Mb)
Ordnance Survey	1965	DDI	7500	0.5	4 (63 Mb)
Ordnance Survey	1966	DDI	7500	0.5	33 (377 Mb)
Getmapping	2000	DDI	10000	0.5	48 (2800 Mb)
Geotechnologies	2002	DAI	N/a	0.19	305 (14100 Mb)

The baseline aerial survey DAI data is captured using charged coupled device or complimentary metal oxide semiconductor image sensors, with the information stored in a digital format at the instant of image capture. As with DDI, each pixel of the digital format represents a pre-specified nominal ground sampled distance (Table 3).

For most historical aerial photography where the original negatives or diapositives are not available, the capability of the complete dataset is typically in the order of 20 to 30 microns (μm), and attempts to scan at higher resolutions to obtain smaller ground sampled distances will be of little value. An example of the capability of DDI at different scan resolutions and ground sampled distances is presented in Table 4.

Table 4. Influence of photo scale and scanning resolution of DDI capability

Scale of photography	Data resolution (μm)	GSD required (metre)	Scanning resolution required (μm)	File size for colour 9" % 9" photography
1:20 000	20	0.5	25	240 Megabytes
1:15 000	20	0.5	33	135 Megabytes
1:10 000	20	0.5	50	60 Megabytes
1:5 000	20	0.5	100	15 Megabytes
1:20 000	20	0.25	not possible	not applicable
1:15 000	20	0.25	not possible	not applicable
1:10 000	20	0.25	25	240 Megabytes
1:5 000	20	0.25	50	60 Megabytes

Most historical aerial photography is supplied without data on the calibrated camera focal length, flying height, radial lens distortions and locations of the principle points and fiducial marks. These photogrammetric parameters are essential for high accuracy and precision mapping. In these instances a series of additional processing steps are needed to attempt to re-construct the camera model and determine the data capture parameters at the time of image capture.

The major photometric issues associated with data quality include brightness, contrast, vignetting, and physical defects. These must be considered with reference to the entire dataset and a master data file created from the entire set. The data making up the entire dataset is then digitally image processed against the master data. This process is repeated for all epochs under consideration.

Referencing of historical maps and aerial imagery

Referencing of topographic maps and aerial imagery allows them to be viewed with other spatially referenced data in a GIS. The process of georectification is used to 'warp' topographic maps and aerial imagery to a large-scale digital map, such as the OS Land-Line or MasterMap products. This uses a polynomial approach, where the effects of terrain height are neglected.

Orthorectification is a more accurate method of referencing which fits aerial imagery to a DTM. In simple terms, this process 'warps' the image to the landsurface by applying co-ordinates and heighting data of features on the DTM to the same features seen in the aerial imagery. The process applies the collinearity principle, and transforms the photographic perspective into a 'map-like' orthogonal view, taking into account camera parameters, aircraft tilts and terrain heights.

The aerial imagery from this study was processed to derive usable coverage for each photo model. A dataset of ground control points was assembled, comprising x and y coordinates (from digital maps) for georectified photo models and x, y and z coordinates (from the 2002 DTM) for orthorectified photo models. The photo models were processed for interior, relative and absolute orientations and image matched to produce seamless orthophoto mosaics. For data management reasons the mosaics were then split into 1 km^2 OS grid tiles and index catalogues prepared.

2002 Digital terrain model and orthophotomap production

The integration of the digital terrain model (DTM) datasets derived from LiDAR, RTK and bathymetry surveys was conducted so that the integrity of the highest resolution dataset is preserved. The datasets had postings appropriate to the technology used for their acquisition:

- Raw LiDAR data is posted at 2 m centres oriented in x and y direction to the National Grid (OSGB36);
- RTK data is posted along transect lines oriented orthogonal to the shoreline at 200 m intervals; and
- Bathymetry data is posted along transect lines oriented orthogonal and parallel to the shoreline at 200 m and 1,000 m, respectively.

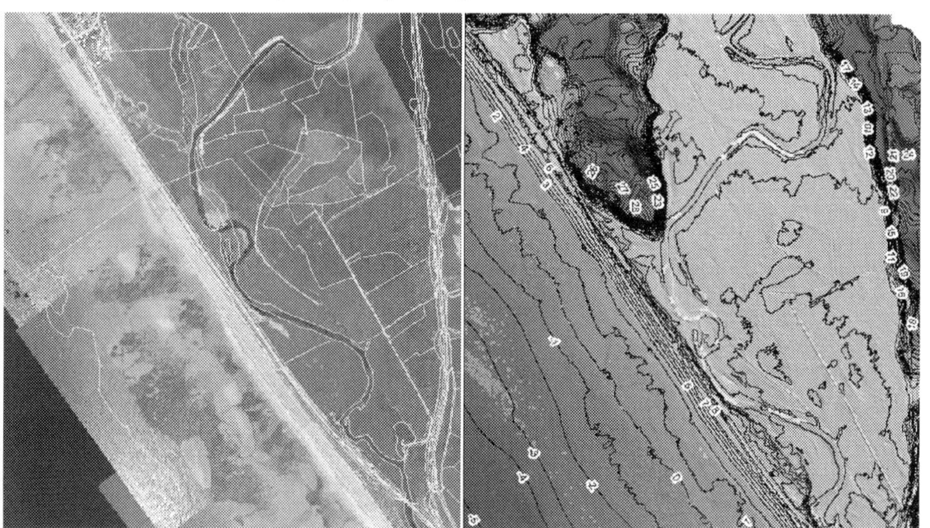

Figure 3. Orthophotomap and 2002 DTM derived from DAI and LiDAR

The DTM points from all three datasets were combined into a single dataset and the points triangulated to generate a new DTM surface with postings identical to the LiDAR data. This ensured that no detail was lost from the highest resolution dataset. Where datasets overlap, the DTM data associated with the higher order instrument was used. This was particularly important when processing data in the intertidal zone. Integration of the 2002 DTM with DAI allows orthophotomaps to be produced at any desired scale (Figure 3).

Error analysis

A key benefit in using GIS-based mapping and aerial imagery is the ability to quantify the errors associated with these dates. This allows the accuracy and error-bands of derived data to be determined. Errors are calculated using the Root Mean Square (RMS) statistic, which compares feature positions on historical maps and aerial imagery to the same feature positions on OS Land-Line mapping. These calculations are done in the GIS.

The errors associated with the georeferenced historical maps will include damage to the original paper maps, distortion from the scanning and rectification process and errors in the original survey. It is noted that OS Land-Line has two absolute RMSEs, one for rural areas and one for moorland and mountains; the latter applies for Cumbria. Calculated RMS errors for the historical maps are presented in Table 5.

Table 5. Accuracy of digital topographic maps

Data	Year	Sample size	Accuracy (x, y RMSE ±m)
Historical map*	1860	13	4.74
Historical map*	1880	7	6.26
Historical map*	1895	9	5.92
Historical map*	1899	22	4.11
Historical map*	1924	8	4.55
Historical map*	1963	8	1.29
Historical map*	1969	7	1.81
Historical map*	1971	7	1.11
OS Landline**	2002	N/a	4.8

* Relative to OS Landline   ** Absolute accuracy for study area

As the aerial photographs are referenced to topographic maps, the absolute accuracy of historical photography will only be as good as the map to which it is referenced. An accuracy assessment for all photomap tiles was conducted (Table 6) to quantify the relative accuracies and error-bands for all epochs. RMS values for the 2002 survey are relative to LiDAR survey co-ordinates and elevations. All RMS values for earlier epochs are relative to 2000 rectified images and have been calculated for common points occurring in all five surveys.

Specialist software and GIS

Production of orthophotomaps requires specialist software. Typical products are VirtuoZo, ERDAS Imagine or ER Mapper, that enable stereo viewing and warping of imagery over a DTM to generate 'map-accurate' orthophotographs. These applications can also extract height data from orthophotographs and allow digitising of features, such as cliff edges, using 3D visualisation.

The orthophotomap files can be saved as TIFFs with separate World files (.tfw) that contain the georeference data, or less preferably as geoTIFFs, which include the georeference data within the image file which cannot be easily viewed or edited. In both these examples, the resultant file sizes can be very large (typically over 10 Mb), posing difficulties with data storage, retrieval and analysis. This can be overcome by exporting the data to ER Mapper raster format (.ecw). These files are typically ten times smaller than the equivalent TIFFs and can be viewed in GIS packages using a free plug-in, or viewed using the free ER Viewer software.

Both MapInfo Professional 6 and ArcView 8 GIS systems have been used to view and analyse the orthophotomaps for this study. The ArcView '3D Analyst' plug-in was used to visualise the photography in 3D.

Table 6. Accuracy of orthorectified imagery

Data	**Year**	**Sample size**	**Accuracy (RMSE ±m)**		
			x	y	z
MoD	1941	54	1.59	1.46	N/a
MoD	1957	45	1.05	1.22	N/a
OS	1966	45	1.05	1.11	N/a
Getmapping	2000	55	1.26	1.28	N/a
Geotechnologies	2002	139	0.42	0.39	0.23

Analysis of historical coastal change

Geomorphological framework

A number of distinct coastal landforms can be recognised along the study coastline. The sensitivity, response or evolution of these landforms to past and future conditions will vary and needs to be taken into account when analysing historical coastal change. The presence of coastal protection works or structures that potentially constrain the free response of landforms also needs to be considered by such assessments.

Nine 'management units' were defined by the SMP as "lengths of shoreline with coherent characteristics in terms of both natural coastal processes and land use". More detailed geomorphological studies carried out for BNFL's coastal studies programme have identified up to sixteen key landforms (Table 8).

Each landform is linked (spatially and temporally) with others to varying degrees. The shoreface influences wave climate at the shoreline, controlling transport and accumulation of beach material, while the shoreline and beach affords protection to the backshore or hinterland and partly controls the extent of erosion or inundation that occurs. The hinterland may release sediments (e.g. cliffs) or accumulate them (e.g. estuaries) controlling sediment availability at the shoreline and shoreface, that in turn affects the morphology of these zones. A change in any element can potentially trigger readjustments in others to maintain the form and position of the whole system. Occasionally, a key threshold may be exceeded triggering major readjustments of form and position of other elements, themselves initiating further changes. Each coastal landform should be understood in terms of its longshore connectivity with adjoining landforms.

Feature recognition

The digital historical maps and aerial imagery were viewed in the GIS at an appropriate scale to allow identification and digitising of landform features. Care was required at this stage as certain landform features can be observed better than others, and significantly, landform features seen in aerial imagery may not be shown on historical maps (Table 7). In other cases, features needed to be interpreted before they can be digitised with confidence. Superimposition of mapping with aerial photography and other data can assist with recognition and interpretation of landform features and thereby improve confidence with analysis. If available, the use of LiDAR data greatly assists in feature recognition. Correct recognition of landform features is fundamental to understanding historical and future coastal change.

Table 7. Feature recognition on historical maps and aerial photographs

Feature	Historical Maps	Aerial Photographs
Cliff edge	Shown as a vector line on current OS Landline maps; historical maps generally indicate cliff line with the steep cliff face or coastal slope hachured or graphically portrayed. Cliff edge is distinguishable from the cliff base or upper beach limit.	Observed as a sharp break in slope when viewed in stereo or ortho mode, and also as a major break in vegetation cover.
Dune front	Cannot be detected from historical or current OS Land-Line maps.	Observed as a sharp or smooth break in slope when viewed in stereo or ortho mode and also as a major break in vegetation cover where dunes are densely vegetated.
Beach, upper limit	Shown as a vector line on current OS Land-Line maps; historical maps generally indicate the upper limit of the beach through symbols and the abutment to a fixed feature, such as a cliff or slope base.	Observed as a sharp or smooth break in slope landward of the active beach when viewed in stereo or ortho mode and also as a major break in vegetation cover or change in sediment type.
Beach front	Not shown on OS maps.	Contact between the upper beach and lower beach or intertidal sand deposits can be detected at low water. The contact can usually be recognised as a change of sediment type, such as shingle to sand.
Barrier/ storm beach	Not generally shown on OS maps. Shingle ridges may be graphically portrayed.	Can be inferred from the limit of vegetation. Shingle ridges may be evident if vegetation is sparse. Morphology can be detected when viewed in ortho mode.
High Water Mark	Shown as a vector line on current OS Land-Line maps; historical maps from 1935-1965 indicate the HWM of Medium Tides and those pre-dating 1935 HWM of Ordinary Tides. By definition all indicate High Water of an average tide and may be regarded the same.	Cannot be detected with precision although the crest of active beach cusps can.

Measurement of feature positions

Following a detailed interpretation of the available data and landforms, feature positions were digitised to derive a series of vector lines. Accurate planimetric measurements were then made of the same landform feature vector lines observed in the various epochs of topographic maps and aerial photography, and differential offsets derived. This was conducted along pre-defined shore normal profiles representative of each landform. Feature positions were measured precisely in the GIS, using a datum at the landward end of each profile.

Coastal change analysis

Long term rates of change have been calculated by comparing the earliest and latest data sets. In the context of the historical mapping, a time span of over 140 years has been achieved by comparing the 1860 feature positions to 2002 Landline. For aerial photography, a timespan of more than 60 years has been achieved between the earliest and baseline surveys. Because the accuracy of the maps and aerial imagery is known, any changes in the relative position of features through time has been quantified in terms of the RMS errors.

Results of the analysis of historical topographic maps and georeferenced aerial imagery are presented in Table 8 and Figure 4. Analysis of orthoreferenced imagery is currently being conducted and the results were not available at the time of preparing this paper. The feature which best represents shoreline change for each landform is indicated in Table 8.

Table 8. Rates of coastal change derived from georeferenced aerial imagery

Landform	Feature	Period	Change m/yr	RMSE ±m/yr	Trend
Drigg Point and Esk estuary	Vegetation limit	1941-2002	1.25	0.06	Advance
Drigg dunes	Vegetation limit	1941-2002	0.27	0.06	Advance
Barn Scar till headland capped by sand	Cliff edge	1941-2002	-0.20	0.06	Retreat
Carl Crag bay, till cliff capped by sand	Vegetation limit	1941-2002	-0.13	0.06	Retreat
Whitriggs Scar till headland*	Cliff edge	1941-2002	0.01	0.06	Static
Seascale beach and lowland*	Vegetation limit	1941-2002	-0.13	0.06	Retreat
Calder and Ehen rivers and fan delta	Vegetation limit	1941-2002	0.25	0.06	Advance
Sellafield barrier beach, river and till cliff	Vegetation limit	1941-2002	-0.21	0.06	Retreat
Braystones beach and till cliff	Vegetation limit	1941-2002	0.04	0.06	Static
Nethertown beach and till headland	Cliff edge	1941-2002	-0.07	0.06	Minor retreat
Nethertown Station rock platform and till cliff*	Vegetation limit	1941-2002	-0.15	0.06	Retreat
Coulderton beach and till cliff	Vegetation limit	1941-2002	-0.01	0.06	Static
North Coulderton relict landslides in till*	Vegetation limit	1941-2002	0.10	0.06	Minor advance
Pow Beck river and fan delta*	Vegetation limit	1941-2002	0.22	0.06	Advance
St Bees Golf Course cliff in moraine	Cliff edge	1966-2002	-0.50	0.06	Retreat
St Bees Head rock cliff capped by till	Cliff edge	No data			

* Landform response constrained locally by coast protection measures

The results are compared with the lower projected rates of change from the SMP. The key findings of the analysis are as follows:

1. The trend and pattern of historical coastal change derived from topographic maps and aerial photographs is broadly consistent, with some notable differences;
2. The results indicate that the Drigg dunes and Drigg Point have experienced relatively high rates of accretion over the historical period, which appears to have increased since 1941. This is at odds with the SMP that indicates future recession of these landforms;

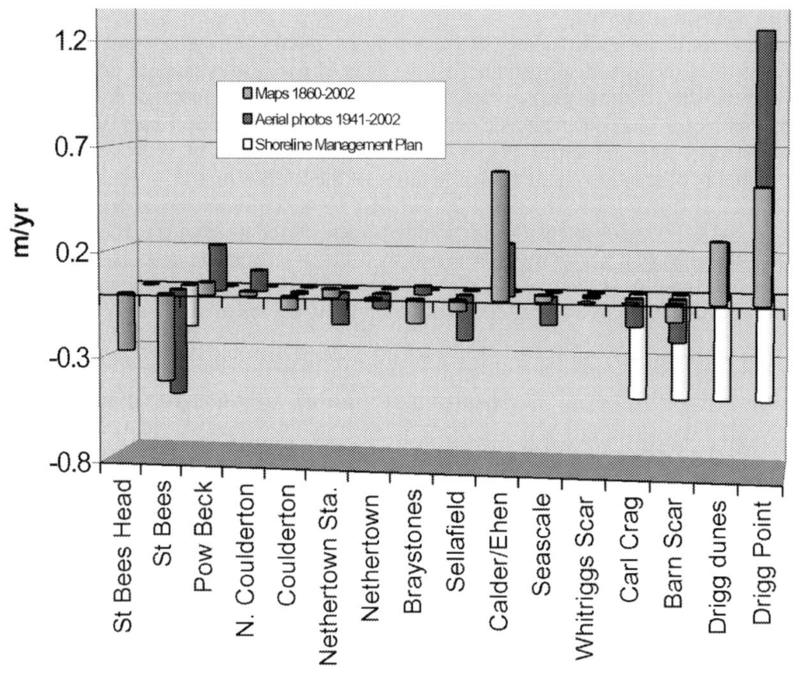

Figure 4. Rates of coastal change derived from maps and aerial imagery

3. At Barn Scar and Carl Crag there has been minor recession of the shoreline, the rate of which appears to have increased to 0.2 m/yr since 1941. The SMP indicates future recession of these landforms but at significantly higher rates than in the past;
4. The protected shoreline at Whitriggs Scar and Seascale has remained static in the long-term, although the aerial photography indicates retreat at Seascale since 1941. The Whitehaven to Furness Junction Railway is routed along the shoreline at Seascale to Pow Beck and has constrained the response of landforms since its construction in 1849 where the railway abuts the beach. Because of this, the SMP indicates there will be no change in shoreline position between Seascale and Pow Beck over the next 50 years or so;
5. The outlet of the rivers Calder and Ehen have advanced in the long-term, probably due to deposition of sand and gravels in the fan delta at the rivers outlet;
6. The barrier beach across the Ehen valley and fronting Sellafield has undergone retreat which has increased since 1941, possibly as a consequence of construction activities at the Sellafield works since the 1950s. The railway is routed across the barrier beach fronting the Ehen valley where it is at risk of breaching;
7. The landforms at Carl Crag, Seascale and Nethertown are shown by the aerial imagery to have undergone retreat since 1941 and yet the mapping data indicates no change in the long-term. Recent site observations confirm there has been erosion of these landforms;
8. The landforms at Braystones and north Coulderton indicate a reversal of the long term trend for retreat and no change from map data, respectively, to no change and advance of the shoreline since 1941 from aerial imagery;

9. The till cliffs at St Bees have undergone retreat in the long-term with annual rates of recession of up to 0.5 m/yr confirmed by the historic mapping and aerial photography;
10. The rock cliffs forming South Head St Bees have also undergone recession in the long-term with annual rates of recession of 0.27 m/yr confirmed by historical mapping; and
11. The overall pattern of coastal change indicates a coastline in retreat except at river outlets where continued deposition of fan deltas and associated landforms is apparent.

Conclusions

This case study has demonstrated the significant benefits of employing digital maps and aerial photography for the assessment of historical coastal change. These data have been integrated with a specially commissioned state-of-the-art digital aerial survey and other baseline survey data using specialist digital remote sensing methods and GIS software.

The results of this work emphasise the inaccuracies of broad scale assessments adopted by the SMP, where the errors in data and interpretation are neither quantified nor made explicit. This can lead to misleading projections of coastal change where the patterns and rates of change cannot be relied upon for decision making, particularly for high risk sites.

The digital approach provides quantitative data of known accuracy from which analysis of historical coastal change can be carried out with confidence. The analysis has been conducted within a geomorphological framework that recognises a number of key controlling landforms along the coastline. Recognition of landform features is considered fundamental to the understanding of historical and future coastal change. The framework assists the identification of the patterns and variable rates of historical coastal change taking account of the influences of coast protection and other intervention measures. Work is still in progress and it is anticipated that further improvements in accuracy and error assessment will be achieved through the use of the orthophoto models. The analysis of historical coastal change will form a key input to the next stage development of a coastal change projection model.

Acknowledgements

The authors are grateful to colleagues for their contributions to this work. The paper is published with the kind permission of BNFL Plc.

References

Bullen Consultants 1998. *St Bees Head to Earnse Point, Isle of Walney Shoreline Management Plan.* November 1998.

Graham R and Koh A 2002. *Digital aerial survey: theory and practice.* Whittley, London.

Lawrence C, Byard R and Beaven P (1993) *Terrain evaluation manual.* Transport Research Laboratory state-of-the-art review 7. HMSO, London.

Take A Strategic Approach! A solution to implementing the Habitats Directive in coastal areas

Heidi Roberts, Atkins Water, Epsom, UK
Heather Coutts, Atkins Water, Epsom, UK

Abstract

The implementation of the Habitats Directive in marine sites (and associated UK Regulations) has raised a number of issues and concerns since its introduction. The UK Government has undertaken to maintain designated habitats at a favourable conservation status, subject to natural change. The implications of this commitment are only now being realised and the lack of a clear national framework for implementation is causing potentially harmful delays to necessary coastal defence schemes in many areas.

Substantial work has been undertaken in developing strategic coastal defence policies for shoreline management plans (SMPs) and coastal defence strategy studies. This initiative has proven extremely beneficial in developing a strategic approach to shoreline management. However, this approach should be expanded to include the identification of appropriate compensation/mitigation sites in and adjacent to protected areas.

This discussion paper refers to recent experience of implementing sea defence schemes on the central south coast of England and concludes that there is an imperative need to take a strategic approach to the identification, procurement and management of suitable compensation sites. The Agency, English Nature and Government need to tackle this issue and provide a suitable framework at national and regional levels. At present SMP2 is unlikely to deliver such an environmental framework and the Coastal Habitat Management Plan (CHaMP) initiative has fallen significantly short of expectations.

Strategic habitat investigations, identifying short, medium and long term realignment sites would overcome many of the issues facing shoreline managers implementing SMPs in European sites. Such investigations should involve analysis of the environmental, social, economic, political and other issues which impede the suitability and acquisition of such sites. Investigations should also encourage the identification and procurement of sites for 'strategic acquisition' and encourage the necessary larger scale realignment of the coast which can deliver greater biodiversity benefits..

Introduction

The implementation of the European Habitats Directive (Council Directive 92/43/EEC) in marine sites has raised a number of issues and concerns since its introduction in 1992. The publication of

Coastal Management 2003, Thomas Telford, London, 2003.

'The Conservation (Natural Habitats, &c.) Regulations (1994)' (referred to as Habitat Regulations) appears to have done little to simplify the situation but has raised more issues and provided difficult answers. The rationale behind the Directive or the Regulations should not be questioned; the need to protect Europe's most important habitats is justified and necessary.

The UK government has undertaken to maintain these protected habitats in a favourable nature conservation status, subject to natural change. It is the implications of this commitment, especially within dynamic environments such as our coasts and estuaries, that are only now being realised and the lack of a clear framework to achieve this is causing potentially harmful delays to necessary coastal defence schemes in many areas.

The shoreline management plan (SMP) process has been groundbreaking as it has been the first initiative that has taken a strategic approach, both geographically and temporally, to the management of the shoreline (Jewell & Roberts, 1999). However, it has been felt that the SMP process did not go far enough and elements such as linkages to the local planning system should be strengthened. The guidance documents to help operating authorities and consultants undertake the next round of SMPs is due in spring/summer 2003, with the first two pilot plans being undertaken in parallel with the development of the Guidance. It is anticipated that the second generation of SMPs will build on the experience of the first round and are likely to have greater focus on how we plan to get from the current situation to the long term (50-100 year) policy.

However, in areas of high conservation value and protected under European legislation, local experience has shown that it is extremely difficult to implement shoreline management policies which maintain and upgrade existing coastal defences. This is because where defences are fronted by protected intertidal habitats, sea level rise is likely to result in the loss of those habitats and 'compensatory' habitat is required to maintain the integrity of the designations under the Habitat Regulations. This loss of intertidal habitat is known as coastal squeeze.

In the situation of habitat loss through sea level rise, the compensation that will be required is likely to be achieved through the managed realignment of existing defences to create additional intertidal habitat. There is currently no formal mechanism in the SMP or accompanying Coastal Defence Strategy Studies to identify such sites, and so identification of realignment/compensation sites is generally undertaken at the scheme implementation stage, usually in parallel with the production of the Environmental Impact Assessment (EIA) and Appropriate Assessment (AA). It is the difficulty in identifying and acquiring such sites in areas of high conservation value which has caused the problem and is resulting in subsequent scheme delays.

It was originally believed that the CHaMP initiative would deliver this strategic identification which could be used by practitioners as well as provide a framework for management, but as this paper will identify, although this is a good starting point, they will not address many of the problems facing managers on the ground.

The Legal Framework

'The European Union Habitats and Birds Directives are international Agreements which set out a number of actions to be taken for nature conservation (EN, 2001)'. The Habitats Directive promotes the protection of biodiversity and aims to maintain or restore habitats and species of interest to the EU to favourable conservation status. Sites designated under the Habitats Directive are termed Special Areas of Conservation (SAC). The Birds Directive 'protects all wild birds and their habitats within the EU, and there are special measures for migratory birds and those that are considered rare or vulnerable (EN, 2001)', the sites are termed Special Protection Areas (SPA). Where the sites have a marine or intertidal component, they are referred to as European Marine Sites and together the SAC/SPA network of sites across Europe is known as 'Natura 2000'

The Habitats and Birds Directives have been transposed into UK legislation through The Conservation (Natural Habitats, &c.) Regulations in 1994, referred to as the Habitats Regulations. In terms of the management of coastal defences at European Sites, the regulations specify that 'any scheme to implement the preferred flood and coastal defence policy is a 'plan or project', and must comply with the requirements of the regulations (DEFRA, 2001). The most relevant regulations in this context and which relate to the implementation of Article 6 of the Habitats Directive are:

- **Regulation 48 Assessment of implications for European site:**

Regulation 48(1) refers to the need to undertake an Appropriate Assessment (Appropriate Assessment) for any plan or project which 'a) either alone or in combination with other plans or projects would likely to have a significant effect on a European site, and b) is not directly connected with the management of that site for nature conservation (EN, 1997). An AA will assess the implications of the proposal on the conservation objectives of that site, and therefore the features of interest for which it is designated. An example of a conservation objective for a SAC estuary might be 'subject to natural change, maintain the estuary in favourable condition'.

- **Regulation 49 Conditions of overriding public interest:**

The Competent Authority can agree to a plan or project which has a negative assessment for the conservation objectives of the site if the there are no alternatives and the proposal must be undertaken for imperative reasons of overriding public interest (OPI). In cases of a negative assessment but OPI, the proposal must be referred to the Secretary of State for a decision.

- **Regulation 53 Compensatory measures**

In the case of a negative assessment and OPI, the Secretary of State can agree to a proposal if the 'necessary compensatory measures are taken to ensure the overall coherence of Natura 2000 is protected (DoE, 1994)'.

There are also a number of Regulations 60-68 referring to General Development Orders. The effect of these regulations removes normal permitted development rights from operating authorities if the activity is likely to have a significant effect and may require planning permission and appropriate assessment.

Therefore, in coastal areas fronted by (or adjacent to) European marine sites, all coastal defence work has to consider the effect it will have on the nature conservation value of the site and more specifically, the features of interest and conservation objectives for which it has been designated. If it is judged that the effect of coastal defence work could be significant on a European site, whether positive or negative, alone or in combination, then an AA will be required to supplement the

Environmental Impact Assessment (DoE, 1994).

If a coastal defence scheme has an adverse impact as measured against the nature conservation objectives, then mitigation measures must be provided to offset these impacts. 'Mitigation measures are aimed at minimising or cancelling the negative impact of a plan or project, during or after its completion (EC, 2000). These measures can include the timing of construction, limits on site access etc.

However, if the scheme is deemed to have a 'significant adverse impact', alone or in combination with other plans or projects as defined in the Habitats Regulations then the scheme cannot go ahead unless it is in the overriding public interest and there are no alternatives. If this is proved to be the case, such as the need to protect people and property, 'the Member State (Secretary of State in England, Wales and Scotland) will also seek to secure all compensatory measures necessary to ensure that the overall coherence of Natura 2000 is protected (EC, 2000)'. The compensatory measures should be independent of the project and 'provide compensation corresponding precisely to the negative effects on the species or habitats concerned (EC, 2000)'. Therefore coastal defence schemes that have a significant adverse impact on the site but have OPI *cannot* go ahead unless suitable compensatory habitat is secured.

The Effects of Coastal Squeeze

Many (of these) designated sites are adjacent to developed areas and fronted by coastal defences. In many cases the current policy is to maintain or upgrade these defences, in light of climate change and sea level rise, to protect adjacent people and property.

It has been realised for some time that encroachment into the site by upgrading or building new coast defences is likely to have a significant adverse effect and consequently compensatory habitat would be required to offset that loss and ensure the coherence of the Natura 2000 network is maintained. What has not always been realised is the need to offset the effects of coastal squeeze.

Along natural, undefended coasts, habitats and species are able to respond naturally to rising sea levels as habitats migrate landwards in response. However, along defended coasts intertidal habitat will shrink as the shoreline is constrained by hard defences and the intertidal is unable to migrate or 'roll-back' naturally. Therefore, wherever coast defences are maintained, i.e. where a 'hold the line' policy is adopted, there will be a net loss of intertidal habitat over time. In combination with other 'hold the line policies' within the designation, these are likely to have a significant adverse effect on the nature conservation objectives of the site. This is particularly the case for estuarine and harbour areas where the foreshore is wide and gently shelving.

Implications for Coast Defences

While it can be argued that sea level rise is a natural change, maintaining coast defences is not. The implications of maintaining and upgrading coast defences in estuary and harbour areas are significant.

Attempting to mitigate for the environmental impacts in most cases is unlikely to be sufficient to meet the objectives of the Directive and Regulations. Many schemes are likely to have a 'significant adverse impact, alone or in combination (EN, 1997) and require the provision of compensatory habitat under Regulation 53 of the Habitats Directive. *This compensatory habitat must be of an equivalent area of habitat similar to that which will be lost.* It is not only schemes which encroach into designated sites that will require compensatory habitat, as Government advice now advocates that merely continuing to maintain a defence may still contribute to coastal squeeze. For shoreline managers, the loss of habitat from coastal squeeze over the design life of a scheme can amount to many hectares if the scheme covers a long length of frontage.

For example, the recently published Executive Summary for the Solent CHaMP (2003) estimates a predicted loss of saltmarsh of 58-70% due to the presence of existing coast defences over the next 100 years for the full Solent CHaMP area. The CHaMP identifies that many of these areas will be replaced by mudflat but that the net effect is still 'likely to be an intertidal habitat loss of some 730-830ha, amounting to around 10% of the existing resource (Posford Haskoning, 2003)'.

Providing Compensatory Habitat

In cases where evidence of OPI has been put forward (Regulation 48), which for coastal defence schemes refers to the unacceptable risk to people and property, the scheme is referred to the Secretary of State. Through the relevant Government Office, it is the Office of the Deputy Prime Minister (ODPM), with advice from English Nature and Defra Wildlife Division, who has to agree that the scheme is of overriding public interest and make certain that sufficient compensatory measures are taken to ensure that the overall coherence of Natura 2000 is protected. This will generally involve the applicant identifying and procuring a suitable site that can be retreated in such a manner as to provide like-for-like replacement habitat to that which will be lost.

The site should be:

- suitable for the development of replacement habitat to that being lost;
- of sufficient area to compensate for the loss of habitat over the design-life of the scheme;
- adjacent to the designated site to allow the compensation area to become an integral element of Natura 2000;
- fronted by coastal defences to allow 'realignment' and the creation of new, additional habitat; and
- has little or no existing nature conservation value.

In areas of high conservation value, such as the Solent there are generally a number of designations protecting different areas of the site. For example, much of the marine and intertidal areas are designated as marine SAC and SPA/Ramsar sites. Many of the adjacent terrestrial areas are also designated freshwater SPA/Ramsar and have other SAC designations for saline lagoons and other protected habitats. This results in additional difficulty in finding suitable realignment sites.

The Solent CHaMP has identified a number of smaller sites which could be 'valuable for the creation of upper saltmarsh, especially where they adjoin existing areas of quality saltmarsh e.g. Western Yar valley, Hamble Estuary, West Southampton Water and Chichester Harbour (Posford Haskoning, 2003)'. It appears that the majority of these sites are already protected by European or other conservation legislation and therefore likely to be unsuitable without additional compensatory measures. The CHaMP also identified a number of large scale sites for habitat creation including Pennington and Keyhaven and the Eastern Yar. The Eastern Yar estuary is already designated as a cSAC for its complex of saline lagoons and SPA for its freshwater grazing marshes. Realigning this site to compensate for loss of marine SPA/cSAC would also result in the need to further compensate for the loss of lagoon and freshwater habitat. This is likely to result in a spiral of assessment of significant adverse effect and habitat compensation.

In practical terms, the site also has to be available to buy or manage at a price that can be justified by the operating authority and Defra, who in the majority of cases will be providing the grant aid. Demand for suitable sites is now increasing dramatically, especially in areas where private developers also need suitable compensation sites, and not surprisingly land prices are now far in excess of their agricultural value.

Finding sites that fulfil all of these criteria at a suitable price is proving extremely difficult in areas of significant conservation interest.

Risks associated with a non-strategic approach

As stated previously, the acquisition of sites is currently undertaken on an ad hoc basis, which is resulting in delays to implementation of essential flood defence schemes. There are a number of risks associated with these delays. For example, Flood Defence Operating Authorities have set out their flood defence policies within their areas of jurisdiction within a number of documents, including the SMP and other regional and local flood defence strategies. Delays caused by the lack of habitat for compensation mean that these programmes are being held up, and the assets that they protect are being put at ever increasing risk of flooding. There are also financial risks associated with ongoing scheme delay as the cost of individual site investigation on a scheme-by-scheme basis cannot be quantified at the project inception, and can result in spiralling costs if sites are rejected and investigations must be restarted.

Finally, there is also a risk of further damage to sites protected under the Habitats Directive as a result of ad hoc site investigation. Examples from the United States, where a 'no net loss' policy for wetlands, similar to the principles of the Habitats Directive, was formally introduced in 1993 (Zinn, 1997), have shown that simple area for area compensation does not always provide adequate compensation for the functional value of the habitat destroyed (McCrain, 1992). If the habitat lost is evaluated on the basis of its function and value, a simple area for area compensation may not necessarily replace the total value lost from the natural wetland. McCrain (1992) concluded that area-for-area compensation would result in approximately two thirds of habitat lost not being effectively mitigated by this approach. Past experience in the US has also shown that project-specific mitigation is more likely to fail, and is of less ecological benefit than larger areas of created habitat (Gilman, 1997; Crooks and Turner, 1999). Therefore an ad hoc approach to compensatory site identification may actually result in further damage to the UK's most valuable habitats.

A strategic approach is required to adequately address these risks, and to ensure that coastal assets are effectively protected without resulting in undue expenditure from the public purse while maximising benefits for nature conservation.

Existing Management Framework

A great deal of work has been undertaken for the development of strategic coastal defence policies through shoreline management plans (SMPs) and coastal defence strategy study initiatives being led by Defra and their associated requirements for grant aid. This initiative has proven extremely beneficial for the development of strategic shoreline management policies. However, there has not been sufficient focus on the process for delivering those policies and a strategic approach is required for the identification of compensation areas to aid implementation of plans and policies in areas of high conservation importance.

Shoreline Management Process

Coastal defence schemes are developed and refined through a number of plans and strategies through the Shoreline Management Plan (SMP) process. This process identifies the issues and objectives for the strategic lengths of coastline, including environmental objectives, and through technical, economic and environmental appraisal, identifies the most suitable policy for each particular frontage.

The first generation of SMPs were generally well received and were useful documents which collated a lot of information and provided the first strategic assessment of coastal defence policy (Jewell & Roberts, 1999). Although useful documents and the inclusion of environmental as well as technical and economic appraisals when determining the most appropriate policy, the first generation of SMP did little to identify the consequences of those policies and how these were going to be addressed during implementation. These plans were reviewed by the Ministry of Agriculture, Fisheries and Food (MAFF) in 2000 and amended Guidance published by Defra in 2001. This stated that 'where flood and coastal defence works are likely to have an adverse effect on the integrity of a site, they will also identify the amount of replacement habitat that is required to maintain nature conservation status, and should *ideally* indicate suitable locations where this new habitat will be created (Defra, 2001)'. This is undertaken through the production of a 'habitat loss/gain account' (Defra, 2001). The guidance also states that the SMP should identify areas that are likely to be suitable for habitat creation to help meet the national Biodiversity and Habitat Action Plan targets.

The inclusion of a loss and gain account and reference to consideration of impacts in European sites was an important first step. The account identifies the strategic losses and gains for each of the chosen coastal defence policies, namely; no active intervention, limited intervention, hold the line, advance the line, managed realignment. However, where there are no sites within the plan area that are immediately identifiable for habitat creation, identification is ad hoc.

With regard to European sites, the SMP Guidance states that the implications of the plan on such sites need to be identified. Furthermore 'consideration needs to be given as to whether or not the proposed shoreline management policies are likely to have a significant effect and therefore resulting strategies and schemes require an appropriate assessment' under the regulations (DEFRA, 2001).

When the majority of SMPs were produced, the implications of coastal squeeze in European sites was not fully recognised and therefore little or no work undertaken to identify potential realignment sites, the exception being the Habitat Migration Study undertaken as an element of the

SMP in the Humber Estuary. Therefore the current situation at Strategy levels appears to be little more than the identification of potential impacts such as coastal squeeze and reference that mitigation or compensation may be needed to meet the requirements of the Habitats Regulations during implementation.

This view is supported by the recent Managed Realignment Research undertaken by Halcrow for the Defra/Agency Policy Technical Advisory Group (Policy TAG), which identified that '16 managed realignment sites were identified that are not included in the SMPs, of which 5 (31%) are known to have been implemented and the remaining 11 are still in the planning/design process. Since five out of 13 implemented schemes and 11 out of 20 schemes in the planning/design were not included in SMPs, this does raise the question as to whether SMPs have proved highly effective in identifying sites for managed realignment (Halcrow. 2001)'. Halcrow pointed out that 'one reason for this may be that some sites in planning/design are driven by the requirements of the Habitats Regulations, the implications of which were not fully addressed in many of the first generation SMPs (Halcrow 2001)'.

Humber Approach

The Humber Estuary has experienced such issues on a large scale. As a variation to the norm, the Agency undertook a number of additional studies to support the development of the SMP in 2000. The Habitat Migration Study was one such study to supplement the SMP and was designed to address some of the key nature conservation issues relating to strategic coastal defence.

The study undertook a strategic review of areas adjacent to the estuary which had the potential for habitat creation. It also identified the broad requirements for habitat creation and identified land where new habitats could be created. In an approach similar to that undertaken by Atkins on the south coast, the Habitat Migration Study undertook a three stage screening process which identified areas of land within the study area which were completely unsuitable for realignment and should be removed from further investigation due to physical and environmental constraints; the second and third stages assessed additional criteria. The approach sought to 'identify areas with most constraints upon habitat creation rather than target any particular areas as suitable' (Binnie, Black & Veatch, 2000).

As this was a strategic assessment to support the SMP, the study did not investigate the site-specific issues. However, this report is important because it makes the link between SMP policy and addressing the constraints of the Habitats Directive.

2nd Generation of SMPs

The forthcoming guidance for the second generation of SMPs will clarify the process to achieve a number of the aspirations outlined in the DEFRA SMP Guide for Local Authorities (2001). The new guidance (currently unpublished) will recognise that many of the short term policy options are likely to remain unchanged (such as maintain the line in areas protecting coastal developments) but will focus on the process of identifying and implementing long term sustainable policies.

SMP2 identifies the two key strands for developing sustainable policies, namely

- Appreciation of the physical environment (coastal processes), and
- Understanding the issues relating to human interests and nature conservation, which determine those policy options that we wish to consider (Halcrow, unpublished)

The SMP2 Guidance will also provide new guidance on the identification and appraisal of issues and objectives which drive the SMP and promote a new transparent and auditable decision-making process upon which to appraise the various coastal defence options. The revised approach has followed the Quality of Life Capital Approach (QLC) which was originally developed by English Nature, English Heritage and the Agency. The objective of the approach is to help those preparing the SMP to understand what is important to a given area, rather than designating or protecting a certain number of 'best' areas and as such, values its uniqueness, representativeness and diversity and not just quality. The approach will have a number of benefits including greater linkage to the planning process and greater understanding of why the feature (such as a beach or harbour) is important either locally, regionally or nationally.

However, in terms of addressing the issues raised in this paper, SMP2 does not provide any additional guidance to address the specific issues encountered when implementing the Habitats and Birds Directives in coastal areas. Although to date of writing this paper the SMP2 Guidance has not been published, discussions with those involved lead us to believe that there will be no direction to identify potential realignment sites during the SMP and no reference to initiatives such as the Migration Study for the Humber Estuary.

The forthcoming Strategic Environmental Assessment Directive may be an opportunity for incorporating the necessary environmental investigations into the strategic decision-making process. Strategic plans and programmes being promoted by the Agency are already introducing SEA's prior to the Directive's introduction in 2006, most notably Catchment Flood Management Plans. Unfortunately SEA's are not being promoted as part of the SMP process and the authors believe that this is a missed opportunity for improved environmental assessment in the policy development process.

CHaMPs

It was hoped that the CHaMPS initiative would provide the strategic identification necessary and Defra envisaged that 'a CHaMP will contain a similar level of detail to a Strategy Plan for flood and coastal defence, an in time it should be possible to merge the two processes to produce a single, all encompassing strategic plan (Defra, 2001).'

However, the recent publication of the Executive Summary for the Solent CHaMP (January 2003), summarising the work and outlining the major findings, has led us to believe that, at least with this first generation, identification will be undertaken at a very high level only. The objectives of the Solent CHaMP are:

- to offer a long-term strategic view on the balance of losses and gains to habitats and species of European interest likely to result from sea level rise, and the flood and coastal defence response to it;
- to develop a response to these losses and gains by informing the strategic direction for the conservation measures that are necessary to offset predicted losses;

- identify suitable areas for new habitats that may need to be predicted; and
- make recommendations to SMPs to ensure flood and coastal defence options address the requirements of the Habitats and Birds Directives.

The CHaMP states that 'the aim has not been to identify definitive mitigation areas, but instead to identify possible sites where physical and biological conditions could be favourable for the creation of new intertidal habitat. Built up areas are unlikely to be feasible for habitat creation and all such areas have been eliminated from this appraisal. Potentially complex issues associated with land uses, property rights, local politics and economics of predominantly undeveloped lands have not been considered at this stage. It is recommended that the sites here represent a broad initial inventory that would then require subsequent detailed appraisals to filter and to test actual feasibility and define optimum boundaries for selected sites (Solent CHaMP, 2003).'

Although this is a useful initial step, it will do little to help operating authorities identify sites which meet all the necessary criteria. For example, when referring to the Solent CHaMP it appears that there are considerable suitable realignment opportunities in the area, both large and small scale, the CHaMP stating 'some 4,800 ha of land suitable for creation of intertidal habitats has been identified within the boundaries of the Plan (Solent CHaMP, 2003).' The CHaMP goes on to estimate that realignment of only 15-17% of this land would provide the compensation needed to maintain all the current coast defences in the CHaMP area. However, in practical terms every site will have issues limiting its suitability for realignment and acquiring only 1ha of these 4,800 has proved extremely difficult. Such prohibiting issues include:

- Increase flood risk to adjacent people/property
- Existing nature conservation interest
- Reduction in flood storage capacity
- Pollution or contamination of the site
- Long-term sustainability of the site (can it roll-back with sea level rise?)
- Health and safety concerns

For example, the Agency have applied to upgrade coastal defences along a frontage in Chichester Harbour and coastal squeeze losses have been predicted over the design-life of the scheme (although the defence will remain in the existing footprint), as such, a suitable compensation site is required under the direction of English Nature and Defra as required by the Habitats Regulations. On behalf of the Environment Agency, Atkins has undertaken a staged investigation to identify suitable sites in the area for realignment.

The study covered the four main south coast harbours, Portsmouth, Langstone, Chichester and Pagham. The first stage focused on the physical conditions necessary for realignment and 40 potential sites were identified. A workshop of key stakeholders was organised including relevant English Nature, Agency, Wildlife Trust, local authority and harbour management staff. The objective of the workshop to discuss each site, reject sites deemed completely unsuitable (e.g. due to adjacent development, increase flood risk etc), gain the opinions of stakeholders for each site and identify issues which may need to be addressed during further stages of investigation.

The sites were initially ranked on a range of high level generic issues, which resulted in sites with the fewest constraints having the highest ranking. A more detailed investigation was then undertaken after this initial 'filtering' process, based on technical, environmental, legal, economic and social issues. The methodology of this process is described in the paper presented by Heather Coutts, "Streamlining Site Selection for Managed Realignment" (Abstract reference 053).

The most suitable sites were identified via this process, but to date no land has been procured due to the complex social, political and, more importantly, nature conservation issues in the area. Of the initial 40 sites identified, 8 were taken forward for further investigation. However, because of issues that have been raised at each site, the acquisition of a piece of land has now been delayed for some time. The issue of compensation/land purchase was also raised as the most important constraint by stakeholders by the recent Halcrow Research, other constraints identified by Halcrow (2002) through stakeholder consultation are:

- *The Habitat Regulations*
- Potential loss of land with high property value
- Lack of support from public opinion
- Insufficient consultation
- Potential high cost of realignment
- *Potential loss of terrestrial and freshwater habitats*
- *Managed realignment is ineffective if carried out on a piecemeal basis*
- Lack of access to or information about suitable funding
- Insufficient robustness about coastal defence
- *Difficulty of recreating an environmentally diverse habitat*

This illustrates that there are a number of common issues associated with the identification of suitable realignment sites and although the Solent CHaMP will provide an important first step for the strategic identification of sites, it will not address many of the issues currently faced by many coastal managers and engineers in the region and will not provide any substantial progress towards the identification of suitable compensation sites.

This site identification has been undertaken during the scheme implementation stage and the difficulty in identifying and acquiring a site has resulted in substantial delays. However, beginning this investigation at the SMP stage and refining it in parallel with the Coastal Defence Strategy Studies would have a number of benefits including:

- Identifying issues which may impede the identification and acquisition of suitable sites
- Begin to address such issues at a strategic level
- Engage key stakeholders such as English Nature
- Encourage the identification of large sites which could be used as compensation for a number of schemes
- Encourage the identification of sites which may be suitable in the medium and long term
- Reduced delays in implementing schemes and maintenance programmes developed through the SMP process.

Discussion

It is apparent that the shoreline management process has been extremely important in developing strategic policies and overseeing the long term sustainable management of the shoreline. The

SMPs are intended to be strategic documents which outline the future direction of shoreline management for specific lengths of frontages, the policies are then examined in greater detail and proposals for implementation developed through Coastal Defence Strategy Studies. This is an extremely important process which develops the future vision of the coast and defines the process through which this can be implemented. However, the complications that have arisen through the implementation of the Habitats and Birds Directives and the associated Habitats Regulations have led to confusion on the coast which is yet to be fully resolved. Indeed the recent Halcrow research (Halcrow, 2002) has highlighted the Habitats Regulations as both a driver and a constraint to the implementation of managed realignment schemes in the UK.

Implementing new coast defences or advancing the footprint into designated sites potentially has adverse impacts on the designations and compensation is required to maintain the integrity of the Natura 2000 network. In addition, maintaining or upgrading existing coast defences in estuary and harbour areas is likely to contribute to coastal squeeze. Within European sites this will, in many cases, have a significant adverse impact, either alone or in combination with other plans and projects and necessitates the provision of compensatory habitat.

The identification and implementation of compensatory habitat is currently undertaken on an ad hoc basis during scheme implementation and important coastal defence schemes are being delayed as identification and procurement of suitable sites is difficult. In complex coastal areas such as the south coast of England, which has high levels of coastal industry, recreation and tourism but which retains significant areas of high conservation interest, further work needs to be done at a strategic level to aid implementation of SMPs and reduce future delays. The CHaMP initiative has fallen significantly short of expectations and will do little to help managers on the ground, therefore such investigations should involve more detailed investigation of the social and technical issues associated with each site.

The strategic assessments should investigate both small scale and large scale sites which could be suitable over the short, medium and long term. Operating authorities and other competent authorities should also work together with Defra and English Nature, through such investigations, to identify sites which could be used as ‘strategic land acquisition’ or ‘landbanks’. Such investigations will be a vehicle for compliance with the Habitats Regulations in protected areas and also contribute to the UK Biodiversity Action Plan (BAP) Targets.

More specifically, in areas covered by European sites, a strategic investigation of habitat creation sites should be undertaken as part of the SMP process.

Conclusions and Recommendations

- The need to maintain and upgrade coastal defences along much of our developed coastline is becoming increasingly imperative
- The first generation of SMP did not address the issues associated with implementing shoreline management policies in areas protected under the Habitats and Birds Directives. Therefore identifying and acquiring compensation sites for managed realignment is undertaken on an ad hoc basis at scheme level, which is resulting in significant delays to coastal defence schemes

- A strategic approach to the identification, procurement and management of suitable compensation sites is required in protected areas. This should be undertaken at a regional or sub-regional level depending on the requirements of the SMPs
- The CHaMP initiative has fallen significantly short of expectations and will do little to overcome the practical problems facing shoreline managers on the ground
- The Environment Agency, English Nature and central Government need to address this issue strategically and provide a suitable framework at national and regional levels, it does not appear that SMP2 will fulfil this role
- Investigations should include detailed assessment of the technical, social, political and other issues associated with individual sites
- Sites should be investigated with a view to short, medium and long term realignment which will provide the strategic framework to address issues of strategic land acquisition.

References

Atkins (2002) Managed Realignment Opportunities in Portsmouth, Langstone, Chichester and Pagham Harbours, Phase 1 Report (Atkins, Epsom)

Atkins (2002) Managed Realignment Opportunities in Portsmouth, Langstone, Chichester and Pagham Harbours, Phase 2 Report (Atkins, Epsom)

Binnie Black & Veatch (2000) Humber Estuary Shoreline Management Plan Habitat Migration Study Volume 1

Crooks, S and Turner, R.K. 1999. Integrated Coastal Management: Sustaining Estuarine Natural Resources. *Advances in Ecological Research: Estuaries* **29**: 241-289.

Department of the Environment (1999) The Town and Country Planning (Environmental Impact Assessment) (England and Wales) Regulations (HMSO, London)

Defra (2001) Shoreline Management Plans A Guide for Coastal Defence Authorities (DEFRA, London)

Defra (2003) Managed Realignment: Land Purchase, Compensation and Payment for Alternative Beneficial Land Use (Defra, London) http://www.defra.gov.uk/corprate/consult/closed.htm

Defra (2003) Guidance on the Role of Flood and Coastal Defence in Nature Conservation in England (Defra, London) http://www.defra.gov.uk/corprate/consult/closed.htm

English Nature (2001) Solent European Marine Site..English Nature's advice given under Regulation 33(2) of the Conservation (Natural Habitats &c.) Regulations 1994 (English Nature Peterborough)

English Nature (2001) Habitats Regulations Guidance Note 1 (English Nature, Peterborough)

English Nature (2001) Habitats Regulations Guidance Note 2 (English Nature, Peterborough)

English Nature (2001) Habitats Regulations Guidance Note 3 (English Nature, Peterborough)

English Nature (2001) Habitats Regulations Guidance Note 4 (English Nature, Peterborough)

European Commission (1992) Council Directive 92/43/EEC on the conservation of natural habitats and of wild fauna and flora (EEC, Luxembourg) http://www.ecnc.nl/doc/europe/legislat/habidire.html

European Commission (2000) Managing Natura 2000 Sites The provisions of Article 6 of the 'Habitats' Directive 92/43/EEC (EC, Luxembourg)

Gilman, E.L. 1997. A method to investigate wetland mitigation banking for Saipan, Commonwealth of the Northern Mariana Islands. *Ocean and Coastal Management* **34(2)**: 117-152.

Halcrow (2002) Managed Realignment Research Stage 1 Report (Halcrow Water, Swindon)

Halcrow (unpublished) SMP2 Procedural Guidance

Imperial College Consultants Ltd (2001) SEA and Integration of the Environment into Strategic Decision-Making (ICON, London)

Jewell and Roberts (1999) Managing the Coast of Central Southern England (Isle of Wight Centre for the Coastal Environment, Newport)

Posford Haskoning (2003) Living with the Sea: Solent Coastal Habitat Management Plan Executive Summary (Posford Haskoning, Peterborough)

MAFF (2000) Flood and Coastal Defence Appraisal Guidance Overview (including general guidance FCDPAG1 (MAFF, London)

MAFF (2000) Flood and Coastal Defence Appraisal Guidance Environmental Appraisal FCDPAG5 (MAFF, London)

McCrain, G.R. (1992). Habitat Evaluation Procedures (HEP) Applied to Mitigation Banking in North Carolina. *Journal of Environmental Management* **35**: 153-162.

Zinn, J. *Congressional Research Service Report for Congress. Wetland Mitigation Banking: Status and Prospects.* 12 September 1997.
http://www.cnie.org/NLE/CRS?Detail.cfm?Category=wetlands

North Norfolk – a regional approach to coastal erosion management and sustainability in practice

Dr COG Ohl, HR Wallingford Ltd, Wallingford, UK
PD Frew, North Norfolk District Council, Cromer, UK
PB Sayers, HR Wallingford Ltd
G Watson, North Norfolk District Council
PAJ Lawton, St La Haye Ltd, Plumstead, Norwich, UK
BJ Farrow, North Norfolk District Council
Dr MJA Walkden, University of Bristol, Bristol, UK
Dr JW Hall, University of Bristol

Introduction

The North Norfolk coast is characterised by soft cliffs and dunes, with discrete towns and communities fronted by coastal defences. The Shoreline Management Plan (SMP) for this region reflects this varying land use with a succession of 'Hold the Line,' 'Do Nothing,' and 'Managed Retreat' policy options, as shown in Figure 1 (Halcrow 1996). These policies consider the impacts of cliff recession on beach levels in the region, as well as the necessity of preserving unique environmental and geological conditions. However, due to the coarse nature of analysis in SMPs, such regional impacts are not subjected to exhaustive analysis.

Management plans resulting from the SMP (even if they only recommend maintenance without any fundamental change) will cause impacts on updrift and downdrift processes that were not formally considered by the SMP. Thus, as the soft cliffs along this coast contribute to the supply of beach material, a 'Do Nothing' policy along a given frontage may encourage release of sediments, providing a measure of protection for assets downdrift and yielding an economic benefit at a regional level. Similarly, a 'Hold the Line' policy at one location could potentially lead to beach starvation, accelerated erosion, and eventual outflanking of defences at another location downdrift (i.e. a regional economic disbenefit).

Addressing both the interdependency of management units and sustainability of coastal defences on the North Norfolk coast, a coastal strategy study for the frontages of Overstrand, Trimingham, Mundesley, and Bacton (approximately 18km of coastline) has been conducted (HR Wallingford 2003). The littoral processes and future shoreline evolution have been considered at the regional level, enabling the development of a more robust, long-term plan, founded on an integrated process of strategic options appraisal. In particular, regional shoreline evolution modelling and GIS-based economic analyses have allowed quantification of the effects of management options, enabling appraisal in accordance with MAFF (1999).

To arrive at an appropriate policy at both the local and regional level, the level of detail within the broader study was not uniform over its length. Thus, to understand the performance of existing defences fronting the coastal towns, more refined surveys and decision-making processes have been applied to inform options development. However, the central tools used for economic options appraisal were a regional coastline evolution model and a GIS-database.

Coastal Management 2003, Thomas Telford, London, 2003.

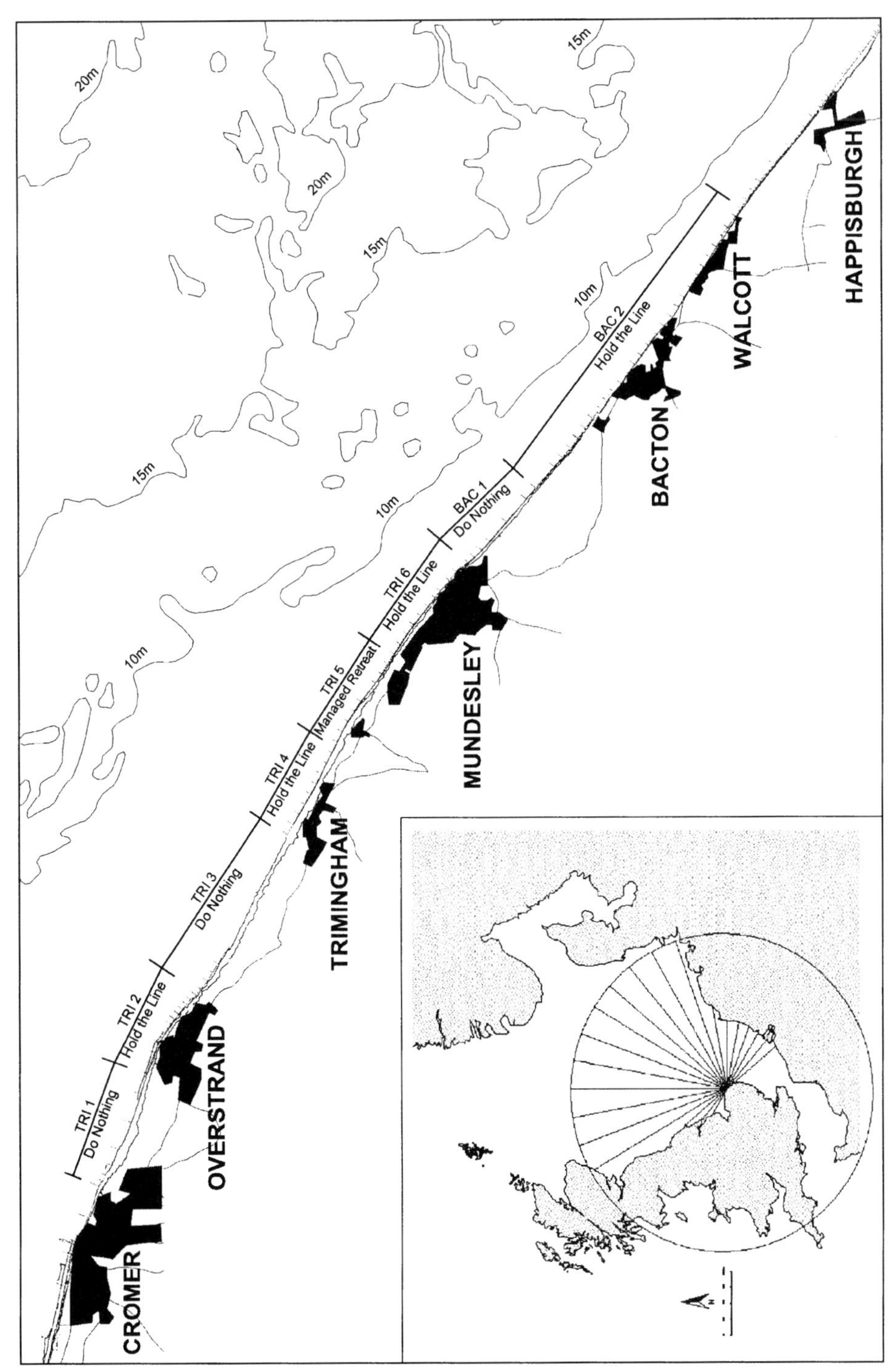

Figure 1. SMP policy options and location plan with 500km circle

The coastline evolution model incorporates results from the cliffSCAPE model (Walkden & Hall 2002) to provide probabilistic predictions of cliff recession. Future recession of the soft coastal cliffs was predicted for various management scenarios (i.e. 'Do Nothing,' modify groyne field, 'Hold the Line,' etc.). Storing geo-referenced property values and infrastructure replacement costs estimates, the GIS-database was used to assess the benefits of each option identified. Thus, at a regional level for each scenario, probabilistic predictions of clifftop position were used calculate benefit in terms of monetary risk.

Background

The North Norfolk coastline has been subject to erosion and retreat since the end of the last Ice Age, when the North Sea basin filled (again) with water. The main processes causing the coastal changes can be summarised as follows:

- Variations along the coast in the rate of beach sediment transport (longshore drift);
- Erosion of the nearshore seabed (which, north of Mundesley, is comprised of a chalk platform and, south of Mundesley, is of similar soft rock to the cliffs);
- Landwards migration of the beach profile in response to sea level rise;
- Wave attack on the cliff face at and above the high water mark;
- Cliff weathering and erosion (e.g. by winds, rainfall, freeze-thaw etc); and
- Landslides of the cliff faces due to saturation caused by groundwater flows.

Prior to the construction of coastal defences in the study area, the rate of cliff recession due to all these causes was approximately 0.65m to 0.75m/year (Cambers 1976). However, there have been substantial short-term variations in this rate along the coast in response to varying weather conditions, variations in the glacigenic cliff material, and the frequency of wave attack on the cliff base. Furthermore, dramatic, short-term recession rates have been in response to the removal of coastal defences, such as 50m retreat over a 3-year period (from 1996-1999) recorded at Happisburgh following the failure and subsequent removal of the timber palisade defences.

The construction of coastal defences, especially seawalls, has significantly altered the natural processes. While a reduction in natural cliff recession rates was achieved in some areas, this generated increased recession on the undefended, downdrift sides of the coastal defences for the following reasons:

- The coastal defences reduced the erosion of the cliffs behind them, thus reducing the supply of sediment to the beaches locally; and
- The defences (particularly groynes) tended to trap beach sand travelling along the coast, (typically from north-west to south-east).

Both of these effects reduced the amount of sand arriving on the beaches in front of the cliffs immediately east of the defences (drift starvation). Because the sediment drift on the unprotected coast was not supplied by sand from the defended frontage, the beaches (and shortly afterwards the cliffs) eroded to make up the deficit in the sediment budget. Such downdrift erosion often resulted in construction of more coastal defences further downdrift.

Cliff processes

The coastal cliffs along the frontage are developed in a variable sequence of weak, glacigenic deposits (Anglian-age and earlier Pleistocene deposits) overlying an eroded Chalk platform. Seawalls currently protect the cliffs at Overstrand, Trimingham, Mundesley, Bacton, and Walcott, while timber palisades and groynes provide protection for much of the intervening cliffline. These defences appear to have reduced the recession rate rather than prevented erosion. Whilst slope degradation behind defences generally involves relatively small and minor events, large-scale landslide events do occur.

Cliff recession model

Whilst surface form, geology, and landslide processes have provided the basis for estimating the potential recession models, the cliffline is noted for the rapid facies changes within the glacial materials. Therefore, it is difficult to be precise about what materials can be expected at a particular section and what materials are likely to be encountered behind the present cliff face.

Cliff recession in the region involves a repeated cycle of the following three stages:

1. Basal undercutting of the intact cliff toe by wave action, leading to steepening of the cliff profile and a reduction in slope stability;

2. Cliff failure, involving either small-scale shallow slides, large deep-seated landslides, or a combination of both;

3. Deposition of debris at the base of the cliff, protecting the cliff toe and (if substantial) temporarily acting as a groyne; and

4. Removal of debris from the foreshore by wave action, leading to the onset of basal undercutting (stage 1 above).

A range of landslide processes, reflecting the variable geology, has shaped the cliffline into a series of steep to near-vertical cliffs and undercliffs. In places, the cliffs have been stabilised by a variety of prophylactic measures, including drainage and retaining structures. In front of the Bacton gas terminal site (to the north-west of Bacton), the cliffs have been stabilised by a variety of landslide remedial measures, including re-grading and slope drainage.

Based on the surface form, geology, and the known or inferred landslide processes, 26 separate cliff behavioural units (CBUs) have been identified along the frontage. The condition of the CBUs was described in terms of the following categories: actively retreating, actively unstable, marginally stable, and relatively stable. Furthermore, field assessments of characteristic slope angles has provided a broad indication of the limiting angles prior to failure (i.e. approaching the end of stage 1, above) and post failure (i.e. stage 2) within each cliff unit. These angles (and associated variances) have formed part of the input data for the numerical modelling of cliff recession (see below).

While cliff recession is driven by wave attack at the cliff toe, removing landslide debris from the cliff toe and undercutting the exposed in-situ materials (Cambers 1976), shore platform lowering exerts a significant control on the rate of cliff recession. Significantly increasing the water depths in front of the structure, affecting overtopping performance, and increasing the risk of undermining, shore platform lowering can be an important consideration in the long-term performance of coastal defence structures.

Environmental considerations

Various environmental designations have been assigned to areas within the limits of the strategy study, and these require consideration during options development and appraisal. Use of GIS methods to plot the geographic limits of these designated areas has aided identification of potential environmental impacts of management options.

In addition to three County Wildlife Sites and the Norfolk Coast AONB (Area of Outstanding Natural Beauty), sections of cliff at Overstrand, Trimingham, and Mundesley have been designated as Sites of Special Scientific Interest (SSSI). Furthermore, Overstrand Cliffs have been designated a Special Area of Conservation (SAC), reflecting the unique geology of the cliffs. With particular bearing on coastal management in the region, the Joint Nature Conservation Committee (1991) describes the principles of conservation for the Sidestrand to Trimingham Cliffs SSSI site as follows:

> "The scientific interest of the Trimingham site is the fact that the cliffs exhibit slumping and landslipping. Any constructions that stop or limit this movement therefore detract from the science."

Furthermore, while English Nature (1994a and 1994b) provides various action points for the SSSIs along the frontage, those particularly relevant to the consideration of coastal processes are:

- Any proposals for coastal defence works should be carefully analysed for their impact on the geological interest of the site. Encourage soft engineering options where possible;
- Oppose all proposals which will damage, obscure, or reduce access to the geological interest of the site; and
- Ensure that the natural erosive processes continue at the site (reduced rate is acceptable).

Thus, environmental considerations are central to the development and appraisal of options for the region.

Hydrodynamics

Aside from human interventions in the defence of coastal settlements, the major forces for change in the North Norfolk coastline (its beaches and cliffs) are the combined effects of waves and tides. While the study area is exposed to waves from 300°N to 90°N (approximately), waves at the coastline are predominantly from 0°N to 70°N, due to the long fetch lengths (over 500km) for this sector (demonstrated by the circle in the inset location plan of Figure 1). To predict longshore sediment transport rates and the future evolution of the shoreline, information on nearshore wave and tidal conditions is necessary, and various numerical models have been applied in this study.

Used previously for strategic studies of sediment transport in the southern North Sea (HR Wallingford 2001b), a regional TELEMAC tidal flow model was used to appraise tidal currents close to this shoreline. Developed by LNH Paris, the finite element based model TELEMAC uses an unstructured grid to enable the detailed simulation of a particular area of interest while keeping imposed boundary conditions distant.

Offshore wave conditions were predicted using the HINDWAVE numerical model, which simulates the growth of waves under the action of winds. For the transformation of wave conditions from offshore to nearshore, the TELURAY model was used, applying the concept of following wave rays between the inshore locations and the seaward edge of the grid. To provide long site-specific sequences of wave and tidal data as input to the cliff modelling, wave and tidal conditions were generated for a 1000-year sequence (from an original 10-year sequence of measured and simulated data), retaining proper seasonality, extremes, sequencing, and dependence between parameters.

The results of this hydrodynamic modelling showed that the largest waves are likely to arrive from approximately 030°N, whereas the most frequent wave directions are from the north-west (330°N). In contrast, the largest surges tend to be associated with winds from the north-west and north. Therefore, broadly northerly sea conditions are likely to be the worst case for potential impacts at the coast (to include large erosion events and damage to coastal structures). This analysis informed a subsequent joint probability assessment of the simultaneous occurrence of large waves and high water levels, allowing for the identification of appropriate coastal defence options through informed conceptual design. Thus, at Mundesley, the 100-year significant wave height is 5.4m and the 100-year water level is 3.94mODN, while the joint probability of a 4.4m significant wave occurring height with a 3.22mODN water level equates to a 100-year event (amongst other possibilities).

Littoral sediment processes

Demonstrating the effects of environmental conditions (tides, waves, and rainfall) on sediment transport and cliff recession, the simplified flowchart in Figure 2 sets out the main littoral processes and their interrelationship. The feedback loop, whereby changes to the beach have an impact on cliff recession (which in turn produces beach sediments that alter the beach) is crucial to the understanding of shoreline evolution for this frontage. Missing from this figure is the longer-term (of the order of 100 years) impact of cliff recession on the plan-shape of the coastline, which has a direct impact on waves incident on the coast through alteration of wave transformation processes (e.g. refraction and diffraction).

The prevailing waves from the north-west create a net drift along the North Norfolk coast from Sheringham towards Great Yarmouth. Previous work at the University of East Anglia (Clayton 1977; Onyett and Simmonds 1983) emphasised that:

- Estimated longshore drift rates along this part of the Norfolk coast are very large (as high or higher than anywhere else in the UK);
- The longshore drift rate increases eastwards along the coastline from Sheringham (where the rate is very low) to the Mundesley and Happisburgh area (where the rate is maximum); and
- Further east and south, the drift rate decreases to nearly zero south of Great Yarmouth.

Fundamentally important to the coastline evolution in the study area, the second point implies that the drift rate out of the eastern end of the frontage (towards Bacton) is likely to be higher than the drift rate into the western end (i.e. from Cromer). Thus, this difference in sediment transport volume (leading to beach erosion and cliff recession) is a natural phenomenon caused by the gradual changes in orientation of the Norfolk coastline and the character of the waves generated in the North Sea. The seawalls along the frontages at the coastal towns prevent additional sediment being added to the beaches to compensate for this deficit in sediment transport volume, exacerbating the underlying erosion trend. The traditional solution to this problem has been to reduce the longshore drift rate by installing groynes.

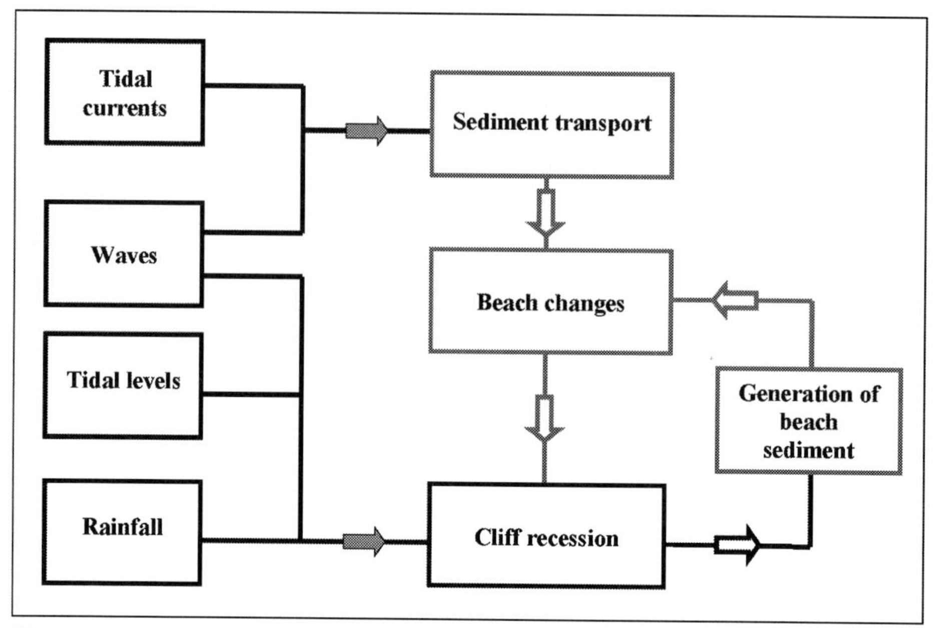

Figure 2. Simplified flowchart of littoral processes and coastal erosion

Potential longshore drift rates, beach volume changes, and sediment analyses

Based on the wave conditions discussed previously, estimates of potential longshore drift rate were made using the standard CERC formula. While the coastal cliffs are comprised of a mixture of muds, sand, and gravel, calculations have been carried out that ignore the mud and gravel components of the beach sediment, as their contribution to the overall sediment budget is negligible (McCave 1973). The resulting calculations indicated that the open-beach drift rate generally increases from west to east along the study area, thus implying the likelihood of beach erosion along the frontage.

North Norfolk District Council has surveyed the beach surface along 18 shore normal profiles (from Cromer to Walcott) in the summer and winter months from 1992 to 2000, and offshore bathymetric surveys have been carried out at five-year intervals. BDAS (Beach Data Analysis System) was used to calculate changes in beach levels over the monitoring period, thereby providing an indication of the direction and magnitude of beach movement.

In addition to numerical analyses of potential longshore drift, surface sediment samples were taken from various locations along the frontage. Previous work by McCave (1978) relied primarily on this analysis to determine longshore drift directions along the coast, whereby the median particle grain size was shown to increase downdrift (as the finer sand is winnowed away and lost offshore). In this study, the median particle diameters from the near surface samples indicated a clear increase in grain diameter from north-west to south-east (i.e. away from the drift divide at Sheringham), confirming the McCave's previous conclusions.

Cliff recession and potential sediment yields

The successful protection of lower cliffs by seawalls and groynes has reduced local sediment inputs in the defended areas. Where defences have been constructed during the period after the publication of the second edition of large scale OS maps (1905), these reductions have been estimated to be no greater than 100,000m^3/year below the natural rate of sediment input along the Anglian coastline (Vincent, McCave, & Clayton 1983).

Based on estimates of sediment yield from the cliffs (in terms of m^3 per metre cliff recession and sand, mud, and gravel percentages) (BGS 1996) and the cliff height, the potential sediment yields from the cliffs have been calculated. Following this, the potential input of sediment into the regional sediment budget from Cromer to Walcott has been estimated using the observed long-term retreat rate.

Conceptual sediment transport model

A conceptual model of sediment transport was developed using the range of data discussed above, to include:

- Input of beach-forming sediment from the retreating cliffs and shoreface;
- Changes in beach volume; and
- Longshore transport rates based on numerical modelling.

(While cross-shore sediment transport can be an important factor in littoral sediment transport processes, both the Southern North Sea Sediment Transport Study (HR Wallingford 2002b) and the Futurecoast project (Halcrow 2002) have indicated that no significant net cross-shore drift is apparent on this coastline.)

Sediment transport rates across the region boundaries were obtained from the following previous studies:

- Cromer Coastal Strategy Study (HR Wallingford 2002a), for input into the north-western boundary; and
- Ostend to Cart Gap Coastal Strategy Study (HR Wallingford 2001a), for output across the south-eastern boundary.

Figure 3 provides the results of this conceptual model, quantitatively depicting the sediment movement in the study area based upon the findings in this study. The accompanying Figure 4 compares these determined values of net drift from the conceptual model with numerically modelled potential drift discussed previously (net and potential sediment transport rates lie within the ranges reflected by the upper and lower bands). Both the increasing trend in drift rate (along the coast from north-west to south-east) and the magnitude of net drift broad agree with earlier work by the University of East Anglia (Vincent, McCave, & Clayton 1983).

Interaction with adjacent Coastal Management Units

West of the study frontage at Cromer, there is a stated policy to 'Hold the Line' (i.e. to continue to hold the line of the existing defences), and (as shown in Figure 1) this strategy is also the preferred option along the Overstrand, Trimingham, Mundesley, Bacton, and Walcott frontages.

Between Cromer and Overstrand, between Overstrand and Trimingham, and between Mundesley and Bacton, the policy is not to further intervene in the protection of the coastline (i.e. 'Do Nothing') due to the environmental importance of this stretch of coastline. Similarly, between Trimingham and Mundesley the stated policy in the SMP is of 'Managed

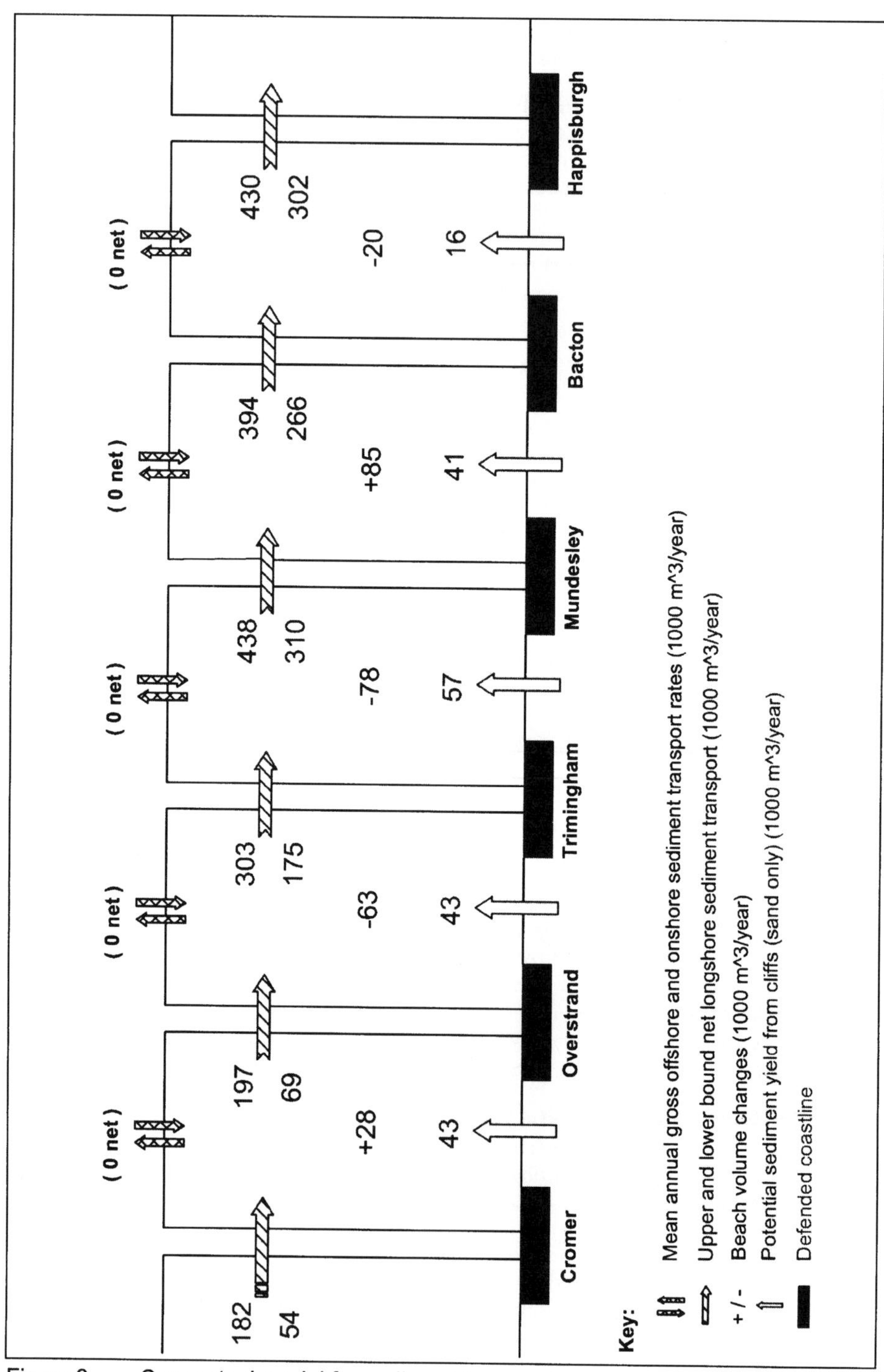

Figure 3. Conceptual model for sediment transport

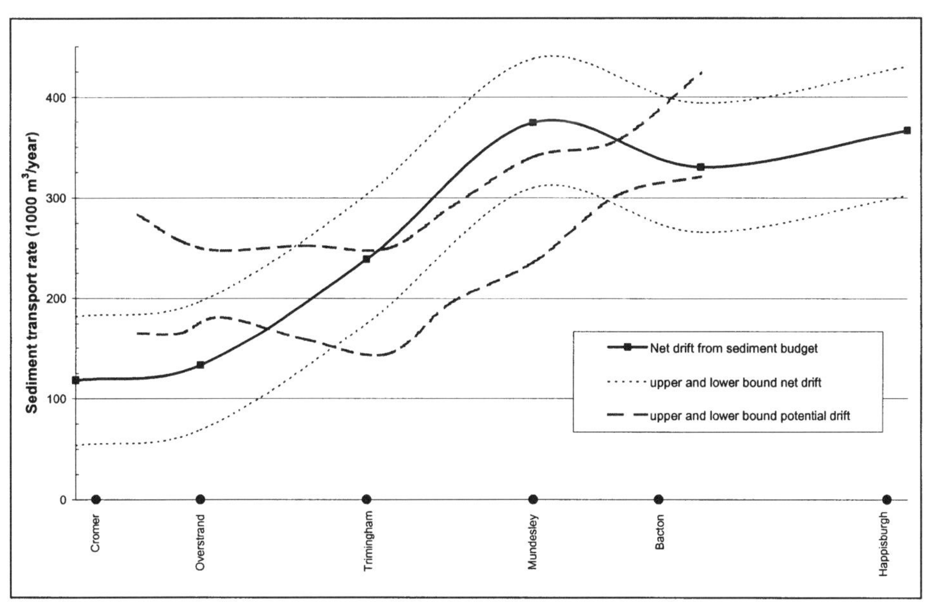

Figure 4. Comparison of potential net longshore drift and net longshore drift rate

Retreat' (i.e. setting the present coastal defences further landwards and accepting some cliff recession). In lieu of specifying 'Do Nothing' for this section, the SMP stipulates 'Managed Retreat' to ensure 'continued provision of a beach access point . . . which is important for coastal defence maintenance and in support of the local tourist accommodation.'

Along this coastline, the 'Do Nothing' and 'Managed Retreat' policies adopted between the protected sections will result in continued cliff retreat along these sections. These policies will maintain natural processes and continued sediment supply from the cliffs, meeting geological and environmental interests. In addition, these policies will generate more beach sediment, which will tend to propagate to the downdrift frontages in the study area. Conversely, the 'Hold the Line' policies, retaining the present line of defences through maintaining cliff toe protection (seawalls) and beach control structures (groynes), will tend to encourage the formation of wider beaches on the defended frontages and reduce the rate of sediment transport out of these management units.

Regional approach

While the impact on coastal processes is considered within all coastal strategy studies, the broader (and perhaps more important) strategic questions as to the sustainability of a selective 'Hold the Line' / 'Do Nothing' management policy are not always addressed. Discussing appropriate boundaries for benefit-cost analyses and potential influences outside the immediate area of a scheme, MAFF (1999) recognise this, stating that 'the assumptions made should be realistic and not simply convenient.' On frontages where coastal erosion is a principal issue, and where erosion is closely linked to littoral sediment transport, questions of sustainability must be addressed at a scale appropriate to the sediment transport regime. Thus, in such circumstances it is not appropriate to consider analysis within single management units, if the potential effects outside of the area of consideration are significant.

In the case of the North Norfolk coast, qualitative guidance with respect to sediment transport pathways has been provided by the Southern North Sea Sediment Transport Study (HR Wallingford 2002b). Due to the predominant south-easterly direction of longshore drift, the regional scale adopted for the study was set to the boundaries given in Figure 1 (i.e. TRI 1 to BAC2), and the wider Sediment Sub-Cell 3B (Sheringham to Lowestoft) was not considered in its entirety. Furthermore, the down-drift effects on more distant frontages south-east of Walcott (e.g. from Happisburgh to Winterton) were not analysed. However, given the nature of the options considered (i.e. broadly within the SMP framework) and considering the results of numerical modelling (indicating the likely extent of downdrift impacts), the analysis domain is considered 'realistic' in accordance with MAFF (1999).

Benefit / disbenefit calculations and shoreline evolution modelling

Recognising the interdependence of policy decisions in adjacent management units, the following analysis tools are desirable to support regional management of coastal erosion:

- A shoreline evolution model which can account for the downdrift impacts on adjacent frontages; and
- A geo-referenced database (e.g. GIS-based system), capable of calculating benefits (both positive and negative) for each management option.

For the calculation of scheme impacts extending beyond the boundaries of the study area, MAFF (1999) provides the following guidance:

> "For example, the value of the sediment generated from continued erosion in the project area to beach building downdrift is a valid benefit of the 'Do Nothing' option. The value of that sediment may be calculated from the cost of importing beach material to provide defences to the same standard as presently exist provided that such defences are economically worthwhile."

While the value of sediment released from a 'Do Nothing' frontage must be assessed to complete benefit-cost analyses, cost estimates of imported beach material may be both difficult and contentious in many studies. Additional difficulties exist in assessing whether or not the downdrift location receives a direct benefit from increased sediment supply (e.g. for downdrift frontages with extensive hard defences).

Alternatively, given an understanding of sediment transport processes and their effect on coastal erosion in the study area and beyond, an appropriate shoreline evolution model may be constructed. Such a model can be used to assess the effects of various options on both the defended frontage and downdrift frontages, resulting in both benefits and disbenefits (assessed as costs in accordance with MAFF guidance). For the neighbouring Happisburgh frontage (Ostend to Cart Gap), this approach was adopted previously through the BEACHPLAN sediment transport analysis program (HR Wallingford 2001a). The sediment transport calculations within BEACHPLAN allowed sediment released from receding cliffs to contribute to the overall sediment budget (thus providing protection to downdrift frontages).

In the current study, cliffSCAPE (a cliff modelling program including mobile beach and shore platform representations) has been used to model the regional coastal processes (Walkden & Hall 2002). Representing the evolution of the shore platform in response to predicted hydrodynamic conditions, sediment transport, and beach behaviour, the program predicts long-term (of the order of 100 years) plan shape evolution of the coastline. The cliff evolution model was calibrated through comparison with historical cliff recession rates during three periods: 1880–1905, 1905–1946, and 1946–1967 (Cambers 1976). While representations of littoral processes in the model were supported by the aforementioned investigations, the effects of coastal structures (e.g. concrete seawalls, timber palisades, and groynes) were based on site condition surveys and estimates of residual life.

Evaluation of cliff recession losses: probabilistic approach

Given predictions of cliff toe position from cliffSCAPE and basic estimates of geotechnical properties for the CBUs along the frontage, a stochastic model of cliff evolution was applied to provide probabilistic predictions of clifftop location over the 50-year period required for benefit-cost analyses. This approach allows the simulation of clifftop position for multiple regional management scenarios, the results of which may be compared against that of the 'Do Nothing' scenario for benefit and disbenefit calculations.

In accordance with the recommendations of Lee and Clark (2002) and applying the method of Hall, Lee, and Meadowcroft (2000), the consequences of cliff recession were evaluated in terms of the discounted risk of cliff top assets being eroded. Thus, the present value of the loss in any year is given by

$$\text{Present Value} = \text{Asset value3Probability of loss3Discount factor.} \quad (1)$$

An example section through the cliff, showing the present day and possible future cliff positions as solid and dashed lines (respectively), is presented in Figure 5. The accompanying probability density function represents the probability of a given clifftop location as a function of distance from the present day cliff toe.

Key issues and modelling scenarios

The resulting study has attempted to address the following most critical questions:

- Can the 'Hold the Line' policies be sustained in the medium to long term whilst adopting 'Do Nothing' / 'Managed Retreat' policies between?
- If yes, how is this policy best achieved?
- If no, over what time scales will this remain viable; what management actions are required to maximise the period of viability; and what is the preferred long-term approach?

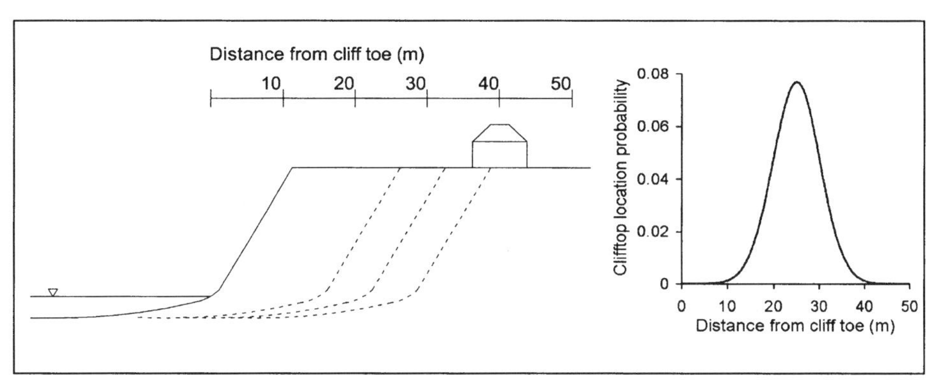

Figure 5. Example section through cliff showing probability of cliff recession (Hall, Lee, & Meadowcroft 2000)

To this end (in addition to various sensitivity analyses) the shoreline evolution modelling has focussed on predicting clifftop positions for the following scenarios:

1. Open coast – removal of all defences (seawalls and groynes) at year 0;
2. Do Nothing – groynes and seawalls removed at the end of their residual life;
3. SMP policy options – groynes and seawalls in 'Hold the Line' frontages (as shown in Figure 1) held in their present alignment and condition;
4. Revised SMP policy options – as for scenario 3 above, but with 'Do Nothing' applied to the Trimingham frontage;
5. Groyne modifications – as for scenario 3 but with enhanced groyne fields (i.e. net drift decreased by 20%) in the following scenarios (✓ indicates modified groynes):

Scenario	Cromer	Overstrand	Trimingham	Mundesley	Bacton
5A	✓	✓	✓	✓	✓
5B	✓	✓		✓	✓
5C	✓			✓	✓
5D	✓				✓
5E	✓				

While the 'Do Nothing' scenario is required to establish the baseline, informing benefit-cost calculations, the open coast scenario has been included to assess the potential recession which would occur upon mass removal or failure of the coastal defences. Scenario 3 identifies the results of the policy recommendations of the SMP, while scenario 4 is included to consider the implications of not defending the frontage at Trimingham (the smallest of the coastal towns immediately at threat on this frontage). Recognising the impacts of coastal defences on downdrift frontages, the matrix of groyne modification scenarios 5A to 5E cover the range of likely management interventions.

For each scenario, probabilistic predictions of clifftop position were made for the duration of the 100-year appraisal period. The resulting probabilistic data was then used to assess a range of potential management interventions across the frontage, through the use of a GIS-database storing asset valuations and estimates of infrastructure replacement costs. By considering the whole-life costs of each option over the 100-years (with benefits and disbenefits arising from the modelling as discussed) sustainable long-term strategic options were identified.

Conclusion

This integrated study has facilitated the development of solutions to local coastal defence problems within the context of a longer-term, sustainable regional strategy. The following principal components have provided the basis for the analysis:

- An understanding of the regional geomorphology and cliff recession processes;
- Recognition of the unique coastal cliff environment;
- Extensive numerical modelling of nearshore wave and tidal conditions;

- Multiple studies of littoral sediment transport, to include numerical modelling, sediment sample analyses, and reviews of previous research;
- Regional scale modelling of coastal processes and shoreline evolution; and
- GIS-based economic analyses to support options appraisal at a regional level.

Thus, central to this integration has been an understanding of the physical processes governing coastal erosion, the interdependency of adjacent management units (in terms of sediment transport), and GIS-based economic analysis to assess regional impact of management options. While similar strategy studies may be conducted within the confines of a single management unit, accurate impact assessments for management options will be frustrated by poor understanding of impacts outside of the analysis domain.

Acknowledgements

Funding from the Department for Environment, Food, and Rural Affairs has supported the strategy studies providing the basis for this paper.

References

BGS (1996) Sediment input from coastal cliff erosion. Technical Report 577/4/A. Environment Agency.

Cambers G (1976) Temporal scales in coastal erosion systems. Trans. Inst. Brit. Geogrs. 1, 246-256.

Clayton KM (1977) East Anglian Coastal Research Programme, Report 5 – Beach Profiles: Form and Change.

English Nature (1994a) Geological Site Documentation/Management Brief: Mundesley Cliffs SSSI, (EC), Norfolk. S Graham, Earth Science Branch, March.

English Nature (1994b) Geological Site Documentation/Management Brief: Overstrand Cliffs SSSI, (EC), Norfolk. S Graham, Earth Science Branch, March.

Halcrow (1996) Sheringham to Lowestoft Shoreline Management Plan Sediment Sub-Cell 3B. Phase 2. May.

Halcrow (2002) Futurecoast. DEFRA.

Hall JW, Lee EM, and Meadowcroft IC (2000) Risk-based benefit assessment of coastal cliff recession. Proc. ICE: Water and Maritime Engineering, Vol. 142 (September) pp.127-139.

HR Wallingford (2001a) Ostend to Cart Gap Coastal Strategy Study. Report EX 4342. August.

HR Wallingford (2001b) Southern North Sea Sediment Transport Study Phase 2 Inception Report. HR Wallingford, CEFAS/UEA, Posford Duvivier, Dr B D'Olier. Report TR 117. January.

HR Wallingford (2002a) Cromer Coastal Strategy Study. Report EX 4363.

HR Wallingford (2002b) Southern North Sea Sediment Transport Study, Phase 2: Sediment Transport Report. Report EX 4526. August.

HR Wallingford (2003) Overstrand to Walcott Strategy Study. Report EX 4692.

Joint Nature Conservation Committee (1991) Sidestrand and Trimingham Cliffs, North Norfolk, (N 11, 3 SILS). GBNCRS Site Report No. 23E. Prepared for English Nature by R Cottle. Date of issue 16/8/91.

Lee EM and Clark AR (2002) Investigation and management of soft rock cliffs. Thomas Telford, London.

MAFF (1999) Flood and coastal defence project appraisal guidance: FCDPAG3: Economic appraisal. MAFF Publication PB 4650.

McCave IN (1973) Mud in the North Sea. In: ED Goldberg, ed., North Sea Science, MIT Press, Cambridge, Mass., 1973, 75-100.

McCave IN (1978) Grain-size trends and transport along beaches: example from eastern England. Marine Geology, 28: M43-M51.

Onyett D and Simmonds A (1983) East Anglian Coastal Research Programme Final Report 8: beach transport and longshore transport.

Vincent CE, McCave IN, and Clayton KM (1983) The establishment of a sand budget for the East Anglian Coast and its implications for coastal stability. Shoreline Protection, Thomas Telford Ltd, London.

Walkden MJA and Hall JW (2002) A model of soft cliff and platform erosion. Coastal Engineering 2002, Proc. of the 28th Int. Conf., Cardiff UK, July 8-12.

The Habitats Directive: A Case of Unintended Consequences?

Robert Harvey MA, Halcrow, Burderop Park, United Kingdom
Richard Ashby-Crane BSc MSc, Halcrow, Burderop Park, United Kingdom

Abstract

The Habitats Directive was introduced in 1992 and implemented in the UK in 1994 by the Habitat Regulations. Many of the Special Areas of Conservation (SACs) and Special Protection Areas (SPAs), designated and protected under the Regulations, are located in coastal environments and it is there that many test cases on the implementation of the Directive have arisen.

This paper discusses a range of case studies and other examples, which illustrate where the Habitats Regulations appear to be delivering nature conservation benefits and where they do not. Interpretation and application of the Regulations has evolved significantly since their introduction. Is current interpretation what the Directive intended? Was the Directive designed to cater for the inherently dynamic nature of coastal habitats? What are the implications for project costs, programmes and feasibility?

The paper focuses on the interpretation of three tests that the Directive introduced into the planning system for projects that are "likely to have a significant effect" on an SAC or SPA:

- When should projects be considered "directly connected with or necessary to the management of the site", enabling them to go ahead without other tests being applied?
- What represents an "adverse effect on site integrity" and how should this be defined?
- What alternatives should be considered, how do we determine when there are "no alternatives" to a proposed project and to what extent should cost be a factor?

Case study experience suggests that the first test is being interpreted very strictly and that few projects progress because they are directly connect with or necessary for site management. Application of the remaining two tests appears to be very wide, resulting in many projects being deemed to have an adverse effect on site integrity, with a significant proportion of these subsequently being promoted on the basis that there are no alternatives. The result can be a time-consuming and expensive consent process to implement even small capital and maintenance projects, coupled with a requirement to provide compensatory habitat.

This paper argues that "directly connected with or necessary for the management of the site" could be interpreted more broadly, to include projects that will enhance the long-term sustainability of coastal habitats by working with coastal processes. On the other hand, "adverse effect on integrity" should be a strict test, intended to prevent protected sites being destroyed or irreparably harmed. It should not necessarily be applied to projects that affect only a small part of a habitat that is in any case highly dynamic. The intention of the Directive is to prevent implementation of projects that adversely affect the integrity of sites, other than in exceptional cases where there really are "no alternatives".

The authors believe that a reappraisal of the way in which the Directive is applied could address legitimate concerns amongst stakeholders, make it more effective in protecting our most valuable sites from serious damage and benefit nature conservation policy.

Introduction

The Habitats Directive was introduced in 1992 and implemented in the UK in 1994 by the Habitat Regulations. The Habitats Directive requires European Member states to establish a network of protected sites known as "Natura 2000" to protect specific habitats and species ("qualifying features"). Amongst other provisions, the Regulations introduced new consent procedures into the UK planning system for proposed "plans or projects" that are likely to affect these Special Areas of Conservation (SACs) and Special Protection Areas (SPAs), together termed "European Sites".

Many of the European sites protected under the Habitats Regulations are located on or adjacent to the coast (Figures 1 and 2). The inherently dynamic nature of coastal environments has given rise to some particular challenges in implementing the Directive. This paper outlines the most important provisions of the Directive and the Regulations, reviews guidance on their implementation from English Nature and a number of case studies and asks whether present practice reflects the intent of the Directive and how it relates to the practicalities of the coastal environment.

Although the UK government has stated that, as a matter of policy, the same considerations are to be applied to Ramsar sites, these are, strictly speaking, not covered by the Habitats Directive or the Regulations and have not been specifically considered in this paper.

What does the Directive say?

Council Directive 92/43/EEC of 21 May 1992 on the conservation of natural habitats and of wild flora and fauna contains 24 Articles and is seven pages in length. The Habitats Directive establishes a network of protected areas called "Natura 2000" across the European Union comprising Special Areas of Conservation (SACs) and Special Protection Areas (SPAs). The provisions of most direct relevance to coastal and flood defence projects are contained in Articles 6 and 7. These create an obligation on Member States to:

- Establish necessary conservation measures for Special Areas of Conservation; and
- Avoid the deterioration of habitats and disturbance of species within SACs and SPAs; and
- Agree a plan or project not connected with the management of a SAC or SPA and which is likely to have a significant effect on it, only if an appropriate assessment shows it will not adversely affect the integrity of the site; or
- Provide compensatory measures if a plan or project, that would adversely affect site integrity, is agreed for imperative reasons of over-riding public interest and in the absence of alternative solutions.

What do the Regulations say?

The transposition of the Directive into UK law proved complex as the new procedures had to be bolted on to the existing planning system and cater for situations that under existing planning legislation constitute permitted development. This complexity is illustrated by the fact that the Conservation (Natural Habitats, &c.) Regulations 1994 incorporate 108 Sections and run to 77 pages. The Regulations define a "Competent Authority" for the purposes of agreeing proposed plans or projects as "*any Minister, government department, public or statutory undertaker, public body of any description or person holding a public office*". It is, however, significant that the government conservation bodies (English Nature, Scottish Natural Heritage and the Countryside Council for Wales) are not responsible for consenting projects and their role is to advise the relevant Competent Authorities. Regulations 3, 48, 49 and 53 implement Articles 6 and 7 of the Directive and the relevant provisions are as follows:

(i) Regulation 3(4): "*Every competent authority in the exercise of any of their functions shall have regard to the requirements of the Habitats Directive so far as they may be affected by the exercise of those functions*".

(ii) Regulation 48: "*A Competent Authority, before deciding to undertake or give any consent, permission or other authorisation for a plan or project which:*

- *is likely to have a significant effect on a European site in Great Britain (either alone or in combination with other plans or projects); and*
- *is not directly connected with or necessary to the management of the site;*

shall make an appropriate assessment of the implications for the site in view of that site's conservation objectives...

In the light of the conclusions of the assessment, and subject to Regulation 49, the authority shall agree to the plan or project only after having ascertained that it will not adversely the integrity of the European site."

(iii) Regulation 49 goes on to state that a plan or project may be agreed notwithstanding a negative assessment for the site only if there are no alternatives and there are imperative reasons of over-riding public interest. In such a case, Regulation 53 requires that necessary compensatory measures are taken to ensure that the overall coherence of Natura 2000 is protected.

The route through Regulations 48 and 49 of the Habitats Regulations for projects subject to planning permission is summarised as a flowchart in Figure 3.

What happens in practice?

The (former) Department of the Environment published Planning Policy Guidance Note 9 on Nature Conservation in 1994. This includes an explanation of the procedures introduced by the Regulations, particularly relating to development proposals affecting SPAs and SACs.

English Nature has produced a set of Habitats Regulations Guidance Notes (HRGNs), endorsed by Defra and other agencies, of which the following have so far been published or are in preparation:

1. The Appropriate Assessment (Regulation 48)
2. Review of existing planning permissions and other consents
3. The determination of likely significant effect under the Regulations
4. Alone or in combination
5. Directly connected with or necessary for site management (in preparation)
6. The condition imposed on Permitted Development by the Regulations
7. Compensatory measures (in preparation)

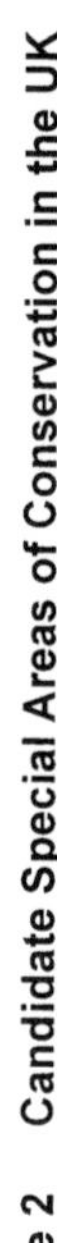

Figure 2 Candidate Special Areas of Conservation in the UK

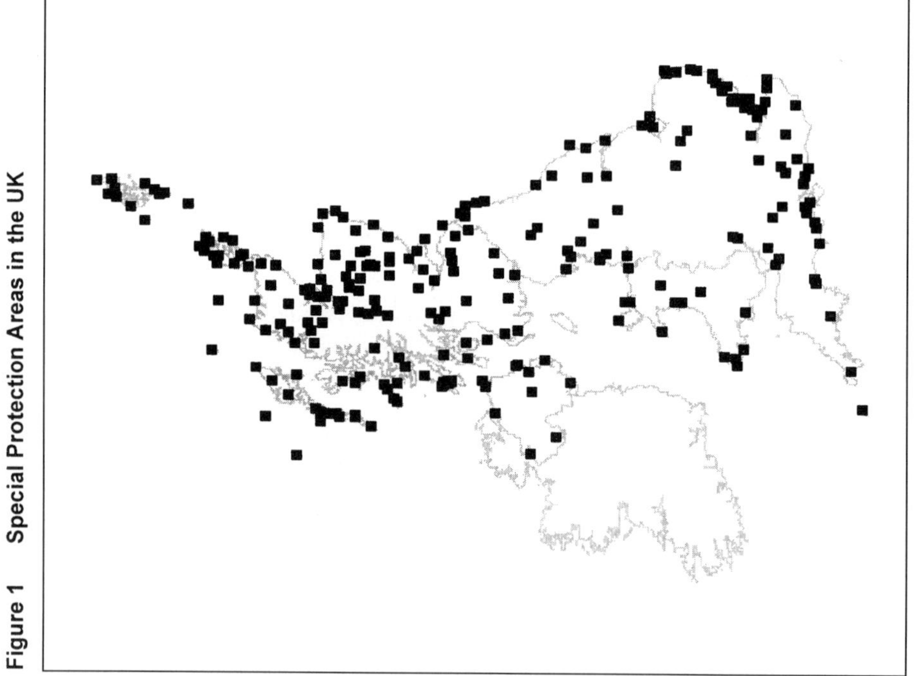

Figure 1 Special Protection Areas in the UK

Figure 3 Consideration of Development Proposals Affecting SPAs and SACs

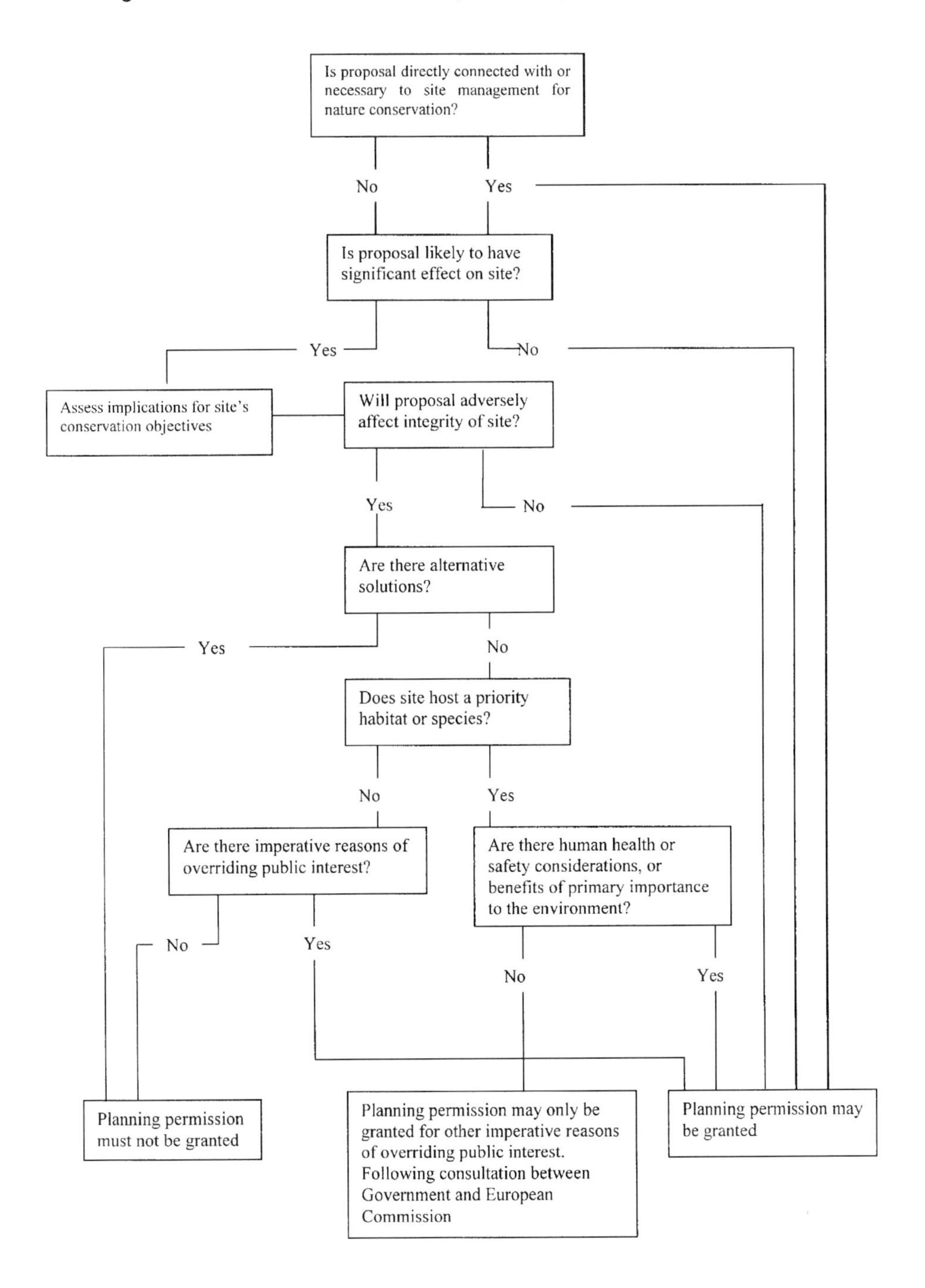

Each European site has a set of conservation objectives supported by favourable condition tables for each qualifying feature, prepared by the respective government conservation bodies. These are intended to inform a decision about whether the site is in favourable condition, but they also provide frames of reference for evaluating the effects of a plan or project on the integrity of the site.

Perhaps the most significant indication of what happens in practice is derived from experience of gaining consent for projects and the advice given by the government conservation bodies. Three case studies relating to flood and coastal defence are briefly reviewed in Table 1.

What are the difficulties with the Directive and the Regulations themselves?

(i) *Fixed site boundaries*

The Habitats Directive requires European Member states to establish sites to protect specific habitats and species within defined boundaries. There is no provision within the Directive or the Regulations to alter site boundaries in order to take account of the effects of natural processes, which on the coast can result in radical changes to the nature, extent and distribution of habitat within relatively short time scales. However, some Member States, such as Sweden and Finland, have classified sites within wide boundaries that cover extensive buffer zones as well as the interest features themselves (English Nature, 2003). This approach has not been followed in the UK, which has based its European sites on previously designated Sites of Special Scientific Interest, many of which are small and have boundaries tightly defined around the features of nature conservation interest.

(ii) *Preservation versus Conservation*

Regulations 48 and 49 create a strong presumption that habitats are to be preserved in their present location. Only where "no alternatives" and "imperative reasons of over-riding public interest" are demonstrated can provision of compensatory habitat be an acceptable alternative to *in situ* conservation. Whilst "over-riding public interest" is often straightforward to demonstrate in the case of flood defences protecting life and property, there are often a number of alternative ways of achieving this objective, particularly since the cost of alternatives is not an explicit factor in the Regulations. This can constrain the type of coastal defence solution adopted, for example because:

- Many freshwater habitats, such as grazing marsh, are protected by the Regulations, and would be transformed into intertidal habitat if Managed Realignment were implemented.

- Sustainable management of an estuary may benefit from relocating areas of salt marsh from one site to another, but this runs counter to the presumption of *in situ* preservation.

Table 1 Coastal schemes and the Habitats Regulations - Three Case Studies

Pett Frontage Sea Defences, East Sussex

The Environment Agency proposes to upgrade the sea defences at Pett in East Sussex by recycling shingle from Nook Point to recharge the beach between Winchelsea and Cliff End. Shingle at Nook Point has collected against Rye Harbour arm where it supports perennial vegetation and potentially could support annual drift line vegetation, two of the qualifying features of Dungeness cSAC. English Nature advised that the loss of 2.1 ha of vegetated or potentially vegetated shingle as a result of recycling operations may constitute an adverse effect on the integrity of the cSAC and this was accepted by Rother District Council.

Urgent Tidal Defence Works in the Humber Estuary

Between 1997 and 2002 a number of tidal defence works in the Humber Estuary were progressed as 'urgent work' in advance of the completion of the Humber Estuary SMP, two of which were:

- At Pyewipe, it was proposed to refurbish a damaged open-stone asphalt revetment and provide new toe protection to part of the length. Although mostly within the existing footprint of the defence, small sections encroached up to 1.8m into the intertidal area, leading to a loss of 0.15 ha of mudflat. It was also considered that 'holding the line' at this location also promoted intertidal habitat loss through 'coastal squeeze', estimated at 0.08 ha over 10 years. These losses, although very small in the context of the estuary, were advised by English Nature to represent a possible adverse effect on the integrity of the Humber SPA.
- At Thorngumbald, the existing tidal defence largely comprised a grassed embankment, with partial rock armouring, but of insufficient height to provide the required standard of flood protection. The proposed solution comprised landward realignment of the defence, creating 75ha of new intertidal habitat between the old and new banks. The proposals were considered by the Environment Agency and English Nature to have many long-term benefits in terms of creating new habitat of SPA quality and providing sustainable long-term defence. However, English Nature advised that they would adversely effect the integrity of the SPA on grounds of scour to existing intertidal habitat at the breach location, loss of 0.5 ha of saltmarsh on the existing bank, disturbance during construction and loss of 10.5ha of the existing mudflat over 50 years due to sea level rise.

In both cases, the schemes have progressed under imperative reasons of over-riding public interest. Despite their urgent nature it took three years to develop schemes and a further year to obtain planning permission. Compensatory intertidal habitat is being provided for both these schemes, and others in the estuary, through the Thorngumbald scheme itself, in the ratio of 3:1 for direct habitat losses and 1:1 for losses due to coastal squeeze.

Halvergate Marshes Flood Defence Scheme, Norfolk and Suffolk Broads

Improvements to the flood defences at Halvergate are required to protect freshwater habitats within Breydon Water SPA, Broadland SPA and the Broads cSAC, together with a substantial area of agricultural land. The proposed scheme comprises a mixture of flood bank strengthening (including widening to increase resistance to breach) realignment (setback) and a small amount of replacement sheet steel piling. The works will impact on the European sites either when flood banks are set back or widened and encroaching onto grazing marsh. In the case of the lengths of setback, extensive areas of new brackish reedbed will be created. In most locations bank strengthening (with replacement steel sheet piling) could provide an alternative to realignment that would avoid the effect on grazing marsh. However, historically piling has lead to the straightening and deepening of river channels, contributed to the loss of flood plain habitats and there is general agreement that its continuation is not a sustainable solution. English Nature initially advised that the works would represent a significant effect on the European sites due to the reduction in the extent of lowland wet grassland. However, it was eventually agreed that the proposed scheme, including defence realignment, could be promoted as 'necessary for site management', thus avoiding the need for appropriate assessment and adoption of an alternative.

It is questionable whether a strong presumption of maintaining habitats in their present locations is appropriate in relation to dynamic coastal habitats, particularly when they can be readily re-created as an alternative means of maintaining the area of habitat locally. Murby (2002), citing the Cley and Salthouse flood defence project in North Norfolk as an example, argues that the present presumption that habitats be preserved *in situ* not only runs counter to natural processes, but leads to schemes being brought forward that are actively damaging to nature conservation interests.

We therefore consider that there is tension between the idea of the ***preservation*** of habitats and species inherent in the idea of designating sites with fixed boundaries and the ***conservation*** of ecosystems in a dynamic environment. English Nature has addressed this by making conservation objectives for habitats and species "subject to natural change", but this gives rise to potential inconsistencies. For example:

- Is sea level rise a natural or an anthropogenic change? How can the contributions of the different factors affecting sea level be isolated?
- Is realigning defences to achieve a more sustainable estuary shape an anthropogenic change or is it facilitating natural processes? What if the defences are instead allowed to fail rather than being deliberately breached?
- Are small changes to a habitat induced by human activity to be considered an adverse effect on the integrity of a site when significantly larger changes to the same habitat induced by natural processes are accepted?

(iii) *SPAs versus SACs*

In some situations, there is a conflict not only between maintaining different qualifying features of a European site, but between maintaining the interest of different European sites. For example, a freshwater wetland forming part of an SPA may be protected from saline inundation by a sea wall, the maintenance of which is causing loss of intertidal habitats forming part of an SAC as a result of coastal squeeze (e.g. the Halvergate Marshes case study). If natural processes are allowed to take their course then the defence may be realigned (or allowed to fail), leading to the loss of the freshwater interest. On the face of it, the requirements of the Regulations to maintain favourable condition can be fulfilled for only one of the two sites. This is one of the problems that Coastal Habitat Management Plans (CHaMPs) set out to address, but its resolution within the decision making process created by the Habitats Regulations is not easy.

(iv) *"In combination" Effects*

The Habitats Regulations require that projects be subject to appropriate assessment where they are "*likely to have a significant effect on a European site in Great Britain (either alone or in combination with other plans or projects)*". Whilst consideration of cumulative effects from several projects is clearly relevant to an assessment, there is no provision or guidance that enables 'de minimus' effects to be discounted. At least in theory, any activity, however inconsequential, in an area within which other significant projects are proposed, could be seen as contributing to the 'significant effect' and therefore requiring appropriate assessment. It appears to us that a more practical approach may be to apply the "in combination" test at the time of assessing adverse effect on site integrity and that only schemes that on their own are likely to have a significant effect should progress to this stage.

What are the difficulties with the way the Regulations are being implemented?

There are several respects in which present practice in implementing the Regulations, as illustrated by our case studies, appears to run counter to the need for flexible application within an essentially dynamic environment such as the coast:

(i) *Projects directly connected with or necessary to the management of the site*

Very few projects have been progressed under the criterion that they are "directly connected with or necessary to the management of the [European] site". The significance of this clause is that if a plan or project is promoted by this route, the other steps in Regulations 48 (including the need for appropriate assessment) and 49 (e.g. demonstrating "no alternatives") do not have to be followed.

Although "*directly connected with or necessary to the management of the site*" is not defined in the Directive or the Regulations, English Nature's guidance in HRGN 1 and advice on specific projects append the words "*for nature conservation*" to the clause as worded in the Regulations. Whilst it is arguable that this addition is implied by the context (since "*the site*" refers to a site designated for nature conservation), it is equally arguable that the clause has been interpreted in too narrow a manner.

English Nature's draft HRGN 5 indicates that plans or projects will meet this test if they are proposed for:

- "*the management of those European habitat and/or species for which the site has been designated*"; or
- "*from a management plan specifically designed for the benefit of European interests* [e.g.] *a Coastal Habitat Management Plan*".

However, the guidance goes on to suggest that where there is more than one interest feature within a site, plans or projects designated to benefit one feature should be assessed against the impact on the others. This interpretation is at variance with and significantly more onerous than the words of the Directive and the Regulations, which clearly refer to works being necessary for management of a "site", not for individual "features".

In our opinion, projects that enhance sustainable management of an estuary or river and therefore the habitats within it could fall into the category of being "*directly connected with or necessary to management of the site*". Coastal defence interventions often protect a number of different assets from flooding, including both residential property and freshwater habitats, and are developed to meet a range of environmental, social and economic objectives. In some cases, interventions that protect existing habitat or work with natural coastal processes to achieve a more sustainable coastline (including realignment proposals that aim to address the long term habitat loss from 'coastal squeeze') may reasonably be classed as directly connected with or necessary for site management.

Our experience indicates that the government conservation bodies have been very reluctant to sanction the use of this provision, except in relation to plans or projects promoted by conservation organisations with the sole aim of managing or enhancing site features. However, there are also indications that in some instances this is changing, where an alternative route through the Regulations would clearly deliver a worse outcome for nature conservation:

- At Two Tree Island in Essex, deteriorating sea defences could lead to the release of waste landfill material into the intertidal mudflats of Southend and Benfleet Marshes SPA. English Nature has agreed that works to maintain and upgrade the defences are necessary for the management of the SPA, thus opening up the possibility of Defra funding for the works, even though there are no assets subject to traditional economic valuation, such as property, at risk. If this route were not agreed then works to the defences may be considered to have an adverse effect on the site through perpetuating coastal squeeze of the intertidal habitat, requiring the provision of compensatory habitat if they were to go ahead for reasons of over-riding public interest.

- At Halvergate marshes in the Norfolk Broads, the operating authority Broadland Environmental Services Ltd is seeking to realign flood defences to landward in localised areas to enhance sustainability. The scheme is within Breydon Water SPA. and will involve a change of habitat from grazing marsh to reedbed within the setback areas. It has now been agreed in principle with English Nature, RSPB and Defra that this can be taken forward on the basis that it is necessary for the conservation management of the site, although this was not the position taken two years earlier. If this were not the case, it is likely that the effect on grazing marsh may be deemed an adverse effect on integrity, leading to the adoption of an alternative of extensive on-line piling, which not affordable, does not deliver substantial environmental benefits overall and is not now viewed by stakeholders as a desirable or sustainable long-term solution.

(ii) *Adverse Effect on Integrity*

In many cases, a very wide definition of what constitutes "adverse effect on integrity" is being promoted. This appears at odds with the major effects on the extent and distribution of coastal habitats that occur as a result of natural processes. For example:

- In the case of Pett Frontage Sea Defences, the project will directly affect 2.1 ha of perennial vegetation of stony banks, which is less than 0.1% of the total resource in the Dungeness cSAC. Moreover, vegetated shingle is a dynamic and potentially re-creatable habitat, of which significant areas are gained and lost within the site by natural processes over short periods. Whilst the extraction of up to $90{,}000 m^3$ per year of shingle from Nook Point is being regarded as an adverse effect on site integrity, the Dungeness Coastal Habitat Management Plan proposes the removal of Rye Harbour arm, leading to the mobilisation of some 1.5 million m^3 of the same shingle.

- The proposed development of a major new UK port is predicted to cause accelerated accretion at the northern end of nearby mudflats within an adjacent SPA, which may result in longer exposure of existing mudflats during the tidal cycle and potentially expansion of saltmarsh at the expense of mudflat. This impact is the reverse of the process of saltmarsh conversion to mudflat that occurs in adjacent locations as a result of maintaining sea defences and associated coastal squeeze, which elsewhere is itself treated as an adverse effect on the integrity of European sites.

In both these cases, in accordance with English Nature's advice, the effect on the European sites is being treated as an adverse effect on site integrity. Indeed, in some instances, such as these two and the Humber urgent works case study referred to earlier, "adverse effect on integrity" appears to be treated almost as synonymous with "significant effect on the site". We consider this approach to be questionable. Planning Policy Guidance note 9 defines integrity as "*The coherence of the site's structure and function, across its whole area, or the habitats, complex of habitats and/or populations of species for which the site is or will be classified*". This implies that small, temporary or reversible effects on a site that do not materially affect its overall nature conservation interest in relation to its qualifying features should not be deemed an adverse effect on its integrity.

(iii) *Mitigation versus Compensation*

Under case law, re-creation of a habitat outside European site boundaries is not allowed as mitigation for the purpose of determining adverse effect on integrity. It is allowed as compensation, but the requirements of Regulation 49 (no alternatives and over-riding public interest) must first be met. This approach further reinforces the presumption of *in situ* preservation referred to above.

Alternative approaches to providing mitigation might include:

- Designating European sites in such a way that their boundaries include areas where mitigatory habitat could be created to offset losses and so avoid an adverse effect on site integrity. This would be complementary to the designation of 'buffer zones' around protected areas, as practised in many other countries.

- Allowing mitigatory habitat to be created outside the site boundary and the boundary then revised to incorporate it. However, whilst there is nothing in the Regulations that prevents mitigation from being delivered outside the site, there is presently no provision for re-drawing site boundaries. There is understandable resistance within some organisations involved in nature conservation to the idea of treating habitat as a tradable commodity in order to facilitate development proposals in the coastal zone. However, relocation of nature conservation assets seems to be a logical response to the inherently dynamic nature of the coastal environment, particularly given that many habitats such as salt marsh can readily be re-created and birds are inherently mobile organisms.

(iv) *No Alternatives*

Whilst demonstrating "imperative reasons of over-riding public interest" is relatively straightforward in relation to coastal and flood defence schemes that protect people's lives and property, the criterion that there should be "no alternatives" before a scheme may proceed gives rise to more difficulties. Given that the objective is coastal and flood defence, there are usually a number, often a large number, of different ways in which this may be secured. For example, hold the line is an alternative to managed realignment where freshwater sites would be lost and a rock revetment may be an alternative to beach recharge where intertidal mudflats would be adversely affected.

In the case of Pett sea defences, 47 alternatives were considered in order to satisfy the requirements of Regulation 49. Many of these were rejected on the grounds that they would also be liable to adversely affect the integrity of the Dungeness cSAC, though establishing the magnitude of such impacts may be very difficult unless exhaustive analysis of each option is undertaken, which would be very time-consuming and expensive.

Some alternatives may not be acceptable to other interest groups and consenting bodies such as planning authorities on grounds of adverse effects on visual or recreational amenity, and may be unlikely to gain consent. It is often argued that the Habitats Regulations over-ride all other planning considerations in determining whether an alternative should be selected, but to date we have not encountered cases where a conflict between the Habitats Regulations and other planning considerations has arisen in selecting an alternative.

The Directive and the Regulations do not place any financial or other constraints on what alternatives to a plan or project must be considered before it can go ahead on grounds of over-riding public interest. However, in relation to Pett sea defences, Defra has stated that options that would be too expensive to qualify for grant aid (i.e. not affordable) do not constitute alternatives for the purpose of Regulation 49.

In relation to coastal projects promoted by private enterprise, the criterion that grant aid should be available does not apply and, in contrast to coastal defences, it is frequently appropriate to examine alternative locations as well as different designs. In relation to the proposed port development referred to earlier, the conclusion that the project may adversely affect the integrity of an adjacent SPA led to an extensive review of alternative locations around the entire UK coastline to support the case that there are no alternatives.

(v) *Referral to Secretary of State*

If projects may adversely affect the integrity of European sites, the Regulations provide for the Secretary of State to review the application before the planning authority can grant it. In practice this has meant that applications are in effect "called in" for detailed consideration, which can take up to a year. Although the Directive makes the provision of compensation an obligation on the Member State, in practice the granting of planning permissions has been made conditional on the applicant demonstrating that compensation measures are in place.

At the time of preparation of this paper, Defra advised that six flood and coastal defence schemes had been referred to the Secretary of State seeking approval for 'reasons of overriding public interest' and 'no alternatives'. These were as follows:

- Four approved (Pyewipe, Thorngumbald, Hullbridge and Morecambe Bay);

- Two still under consideration (Pett Frontage and Hayling Island); and
- None rejected.

One scheme, Cley and Salthouse flood defence project in North Norfolk, was developed to be put forward under this clause but, on the basis of further discussion with English Nature, has subsequently been withdrawn and an alternative solution is being considered.

Although the numbers are still relatively low, there seems to be a strong presumption for approval, generally on the basis that the compensation package has been negotiated with appropriate stakeholders prior to referral.

(vi) *Delays*

The complexities that the Habitats Regulations have introduced to the planning process, including the need for extensive data collection to support an appropriate assessment and referral to the Secretary of State, are a major cause of protracted delays in coastal and flood defence schemes forward for implementation. In the case of the Pett Sea Defences, appropriate assessment studies cost £55,000 and progressing the scheme through the Habitats Regulations has added at least three years to the programme for implementing defences to increase flood protection for 390 houses. Such delays have a number of undesirable consequences:

- Risks to life and property are prolonged.
- It may be necessary to implement emergency works to manage these risks, which would otherwise have been avoided. Such works may themselves have adverse effects on a European site, requiring retrospective assessment and possibly compensation.
- Some stakeholders may become frustrated and opposed to the Habitats Regulations.

Some of these issues have been examined in a recent study "Options for maintaining features of European Importance in dynamic coastal situations" undertaken as part of English Nature's *Living with the Sea* project (English Nature, 2003). This considered issues such as whether we should aim to conserve or preserve habitats, whether site boundaries should be flexible or tightly defined, how habitats should be accounted for, how favourable conservation status relates to site integrity, involving the community in decision-making and practice in other European countries. The draft report suggests four options for maintaining features of European importance in dynamic coastal situations:

- Wider site boundaries to allow for habitat migration;
- Habitat recreation where migration in response to natural processes is not possible;
- Realignment of defences to enable landward migration of intertidal habitat; and
- The natural functions of coastal systems should be restored where possible.

Conclusions

- The designation of European sites within tightly drawn, fixed site boundaries creates a strong presumption of *in situ* habitat preservation. This makes it difficult to achieve sustainable coastal management through allowing natural habitat migration and by artificial habitat re-creation to replace what is lost as a result of intervention elsewhere. Neither the original Directive, nor the way in which it has been implemented in the UK, appear to have taken into account the essentially dynamic nature of the coastal environment and its associated habitats.

- Experience to date indicates that a strict test of what constitutes "necessary for site management" combined with a wide definition of what constitutes "adverse effect on site integrity" are being applied. This combination makes it likely that the "no alternatives" test will be triggered before schemes can be approved. In many cases, there is more than one alternative approach available to achieve flood and coastal defence objectives, and hence under these criteria an alternative approach would have to be followed, even if it represents a less sustainable approach for long-term coastal management.

- Many coastal and flood defence schemes deemed to have an adverse effect on site integrity have been progressed on the basis that there are over-riding reasons of public interest and no alternatives. This results in compensatory habitat being provided, but does not prevent the negative impact on the original site and incurs long delays and associated expense in the consent process. It is not clear that this is a cost-effective approach to delivering nature conservation, nor that it engenders wide support amongst affected communities and some other stakeholders in coastal defence schemes.

- The intention of the Habitats Directive was to prevent the most valuable wildlife sites in Europe being destroyed or irreparably harmed. We endorse this objective. Indeed, there are many examples of where impacts have been avoided or minimised and project environmental performance has been improved as a result of early consideration of how to address the requirements of the Habitats Regulations.

- However, in relation to coastal and flood defence, the Regulations are frequently being used as a means not to stop seriously damaging development, but to secure compensatory habitat for scheme impacts, even when these are relatively small. We do not consider that the intention of the Directive was to allow a large number of schemes to progress under the provisions for over-riding public interest and no alternatives; such instances should be exceptional.

Recommendations

The following suggestions are put forward to address the issues that have been identified with the objective of improving the way in which the Habitats Directive works in the coastal zone and the effectiveness with which conservation resources are targeted. Some may entail changes to the legislation itself, whereas others can be implemented through revised guidance on the application of the Regulations.

- It is considered that there should be a review of European Site boundaries to identify 'buffer zones' that are wider than the constituent SSSIs.

- European site boundaries should be made subject to change/review in response to habitat change, whether natural or by man (e.g. to enable inclusion of 'compensatory' or other habitat creation schemes once they are of sufficient nature conservation value/potential).

- Habitat creation within or adjacent to European sites should be allowed as mitigation (given the support of nature conservation agencies and appropriate risk management measures), thus enabling total habitat resource to be maintained without schemes having to progress under time consuming and expensive OPI/no alternatives/compensation route.

- 'Directly connected with or necessary for site management' should be interpreted more flexibly to allow smoother progression of projects, such as many major realignment schemes, which promote greater sustainability in the long term management of coastal and estuarine habitats. Such schemes, which will often have some negative impacts on a site, would have to demonstrate a clear nature conservation benefit overall.

- Consideration should be given to excluding 'in combination effects' from the assessment of 'likely significant effect', or at least to identifying some *de minimus* criteria, that would allow relatively innocuous projects to go ahead without the requirement for the expensive and time consuming 'appropriate assessment' process.

- Adverse effect on site integrity should be interpreted more strictly and applied only where sites are seriously threatened, with a strong presumption that such schemes do not then progress and that 'true' alternatives are identified.

References

The Conservation (Natural Habitats &c) Regulations 1994, HMSO ("Habitats Regulations")

Council Directive 92/43/EEC of 21 May 1992 on the conservation of natural habitats and of wild fauna and flora

Department of the Environment (1994) Planning Policy Guidance (PPG) 9: Nature Conservation

English Nature, February 2003, Living with the Sea: European Framework Draft Report

Environment Agency, 2000a. Urgent Works Paull to Kilnsea & Whitton to Pyewipe: Environmental Statement Urgent Works 1, Thorngumbald Clough to Little Humber.

Environment Agency, 2000b. Planning for the Rising Tides: The Humber Estuary Shoreline Management Plan.

Environment Agency, 2000c. Urgent Works Paull to Kilnsea & Whitton to Pyewipe: Environmental Statement Urgent Works 15 to 17, SCM Jetty to East of Oldfleet Drain.

MAFF (2000) Flood and Coastal Defence Project Appraisal 5: Environmental Appraisal

Murby, P, (2002) *Why do we pay so much to protect wildlife from nature?* Coastal Futures Conference, London, January 2002

Towards spatial planning in the marine environment

Dr Duncan Huggett, Senior Policy Officer, RSPB, Sandy, UK
Mark Southgate MRTPI, Head of Planning and Regional Policy, RSPB, Sandy, UK
Dr Sharon Thompson, Marine Policy Officer, RSPB, Sandy, UK

Abstract
The marine environment around Europe is in a severely degraded state and its capacity to recover fully is doubtful. The UK Government, the European Union and regional seas conventions such as the North Sea Conference have all concluded that many of the problems facing the marine environment are a consequence of the current planning and management framework. In particular, root causes include a lack of strategic planning context within which individual consenting authorities can make decisions, poor co-ordination between sectoral interests and a lack of forward planning. A common conclusion emerging is that a spatial planning framework for the marine environment is urgently needed.

Marine spatial planning should not be seen as something complex or radically different from terrestrial spatial planning. Key principles apply both on land and sea. However, the land based planning approach should not be blindly reproduced as it has its own inherent weaknesses that need to be addressed. In addition, there are critical differences between land and sea environments that must be taken into account.

Large-scale spatial planning and development has received in-depth consideration at a European level. Aspects of *Guiding principles for sustainable spatial development* and the *European Spatial Development Perspective* may be helpful in developing marine spatial planning. However, the developing approach to European integrated coastal zone management probably provides the most useful indication on the way forward. Key principles that need to underpin marine spatial planning include integration and better co-ordination, long-term and forward-looking planning, participation and inclusiveness, and plan-led system based on an assessment of environmental capacity.

Introduction
Marine conservation is a major challenge in northern European waters, and for the North Sea in particular. There has been a long history of intense use of the area, and with eight countries, bordering the North Sea, co-ordination and complimentary action on marine conservation is legally and politically extremely difficult. The environment itself is heavily altered and degraded in many areas, and therefore action needs to be focused on recovery as well as preventing further damage. While it may be too late to work with "pristine" habitats in most of the North Sea, there is still a tremendous responsibility to pursue marine conservation of this environment.

Much effort is currently being put into the development of Marine Protected Areas (MPAs) of one kind or another, to deliver varying degrees of conservation of marine species and

Coastal Management 2003, Thomas Telford, London, 2003.

habitats in the North Sea. The European Commission is committed to the establishment of Natura 2000 throughout European marine waters. The Convention for the Protection of the Marine Environment of the North-East Atlantic (OSPAR) has adopted a recommendation (2003/3) on a network of marine protected areas (the purpose of which is to establish by 2010 an ecologically coherent network of well managed marine protected areas in the OSPAR maritime area, see figure 1).

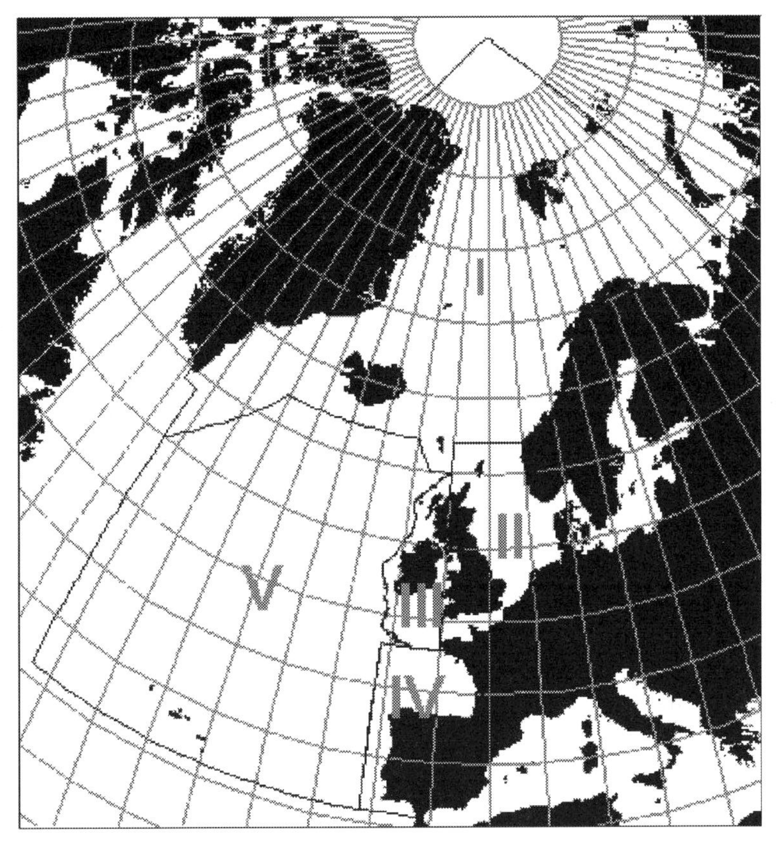

Figure 1: The area of the Convention for the Protection of the Marine Environment of the North-East Atlantic (the OSPAR maritime area) showing the OSPAR regions (map courtesy of the OSPAR Commission)

However, MPAs need to be set in the context of environmentally based planning and management of the whole of the North Sea if they are to be successful. The challenge is to develop an approach, based on marine spatial planning and management, that will support MPAs and help ensure that they do not become just islands of wildlife within an otherwise barren sea. Marine spatial planning is also the logical tool for promoting sustainable development in the marine environment based on concepts such as the ecosystem approach and the precautionary principle.

Where we are coming from: the political impetus

International level

The fifth International Conference on the Protection of the North Sea was held in Bergen, Norway, 20-21 March 2002. Ministers agreed that in order to prevent and resolve the potential problems created by conflicts between conservation and restoration of the marine environment and the different uses of the North Sea, the strengthening of cooperation on the spatial planning process of the North Sea States in relation to the marine environment was needed (Bergen Declaration, 2002).

Consequently, Ministers invited OSPAR within the framework of its biodiversity strategy to improve arrangements for the exchange of information and national experiences in the spatial planning processes of the North Sea States. Furthermore, OSPAR was invited to investigate the possibilities for further international cooperation in planning and managing marine activities through spatial planning of the North Sea States taking into account cumulative and transboundary effects.

In response, at the meeting of the OSPAR Commission in June 2002, Contracting Parties agreed that in order to meet the request, appropriate elements should be included in the work programme for the Biodiversity Committee (BDC), including marine spatial planning (OSPAR 2002). Consequently, the OSPAR Secretariat presented proposals to the 2003 meeting of the BDC. BDC concluded that if any progress was to be made with the requests from the 5th North Sea Conference, it was desirable to bring together a range of those involved from the different North Sea states to exchange their views on the problems involved and possible ways of addressing such problems (OSPAR 2003a). BDC also noted that an exchange of this kind was likely to be more successful if some preliminary understanding of what arrangements were in place in the various North Sea states could be achieved.

On consideration of the OSPAR Secretariat paper and subsequent discussions, the BDC of OSPAR agreed that the Secretariat would, in consultation with contracting parties and observers, develop a report that described the current spatial planning systems operated by the North Sea states. After the report was completed, a small workshop would be organised (possibly with the assistance of the Netherlands) to discuss how progress could be made (OSPAR 2003b).

At the same time OSPAR was considering how best to begin developing marine spatial planning in the North Sea, the European Commission published its marine thematic strategy (European Commission 2002a). The Commission undertook to use the key tools of marine spatial planning, combined with strategic environmental assessment to integrate nature protection measures and the various sectoral activities impacting on the marine environment. Furthermore, a stakeholder conference organized during the Danish presidency of the EU to discuss the thematic marine strategy concluded that the need for spatial planning should be considered to establish a good basis for a more integrated approach of the marine area (Danish Government 2002).

National level

The Review of Marine Nature Conservation

In September 1999, the UK Government initiated a review of marine nature conservation (the RMNC). Whilst the work of the RMNC is continuing, Defra (then DETR) have published an interim report (DETR 2001). A recurring theme of the interim report was the need for a delivery framework for marine nature conservation. The report noted that below low water mark there is no equivalent to the Town and Country Planning system of development control that brings together much of the regulation over a wide range of land use activities in a common framework. In addition, there is no one regulator who carries the breadth of duties and powers possessed by local authorities on land.

It concluded that the lack of an overall framework for marine nature conservation had led to a number of significant problems including variation in the interpretation and duplication of powers and duties in the marine environment. The solution was seen to be an integrating framework for national marine conservation measures within UK waters. It was suggested that the current reactive approach, with lack of any effective, cohesive supporting legislative or policy base, could be developed into a strategic, coordinated, proactive approach that provides practical mechanisms to manage marine biodiversity.

The interim report did not propose what such a framework would look like. It suggested that at the broadest level, there was some support for a marine equivalent to the Town and Country Planning system that used existing consenting mechanisms. It acknowledged that such an approach might require a new agency with wide-ranging powers to plan and manage marine areas. At a local level, one approach might be to extend planning control to cover internal waters.

The RMNC had commissioned a review of the extent of local authority jurisdiction in the marine environment (Tyldesley *et al.* 2000). The report painted a very complex picture concerning local authority jurisdiction in the marine environment, one that has led to considerable misunderstanding and confusion. The report recommended that in the interests of legal certainty and consistency, should Government wish to introduce new powers or duties in respect of marine nature conservation below the low water mark that involved local authorities, then it would be necessary to introduce a statutory provision to extend local authority jurisdiction.

Whilst the RMNC interim report arrived at no firm conclusions about the way forward, it was clear on one point: without the basic architecture to provide the context, detailed marine nature conservation proposals would flounder. Therefore, to explore what practical approaches might be appropriate, the RMNC established the Irish Sea Pilot Project (ISPP). An important feature of this work, especially in the context of developing marine spatial planning, is that it carries forward and tests ideas relating to a 'seascapes' or marine landscape approach to planning and management of marine areas (Day & Roff 2000, Laffoley *et al.* 2000a, Laffoley *et al.* 200b).

Of critical importance to the development of marine spatial planning will be the refinement of the 'marine landscapes' classification of ecological units, which together with the definition of the 'regional sea' could form the basic planning units. This work is not due to be completed until early 2004.

The Marine Stewardship Report

In a parallel process to the work of the RMNC, Government established a process for the development and implementation of a strategy for the conservation and sustainable development of the marine environment. The first element of this – a 'Marine Stewardship Report' – was published in 2002 (Defra 2002a). In this strategy, Government makes a clear commitment to explore the role of spatial planning for the marine environment. Clearly, an important contribution to this will be the experience and findings of the ongoing ISPP. However, the ISPP has a relatively restricted scope, focusing as it does on marine nature conservation interests in particular.

The need to make progress on marine spatial planning is further elaborated in Government's consultation paper on the Marine Stewardship Report (Defra 2002b). Government identifies the incorporation of mechanisms for decision-making (e.g. the control element of spatial planning) as a fundamental element of an operational framework needed for the delivery of an ecosystem-based approach to marine management. However, there is some way to go before such an approach has been elaborated to the extent it can be operationalised.

Fundamental problems with management of the North Sea

As identified by the 5th North Sea Conference, the European Union and the RMNC interim report, many of the problems facing the North Sea marine environment are a consequence of the current marine 'planning and management' system. In particular, there is a lack of a strategic planning context for individual consenting authorities and decisions in the marine environment, poor co-ordination between different consenting authorities and regimes and no forward planning for the sea.

It can be concluded that the coastal and marine environment of the North Sea, indeed of the UK maritime area in general, urgently needs an effective strategic spatial planning and management framework. This would introduce a plan-led approach to marine use and development, and rationality to sectoral decision-making systems. However, it would not necessarily replace them. Whilst important in their own right, current approaches to developing Strategic Environmental Assessments (e.g. for oil and gas exploration and offshore wind energy production) do not constitute marine spatial planning not least because they are sectoral in their approach.

What differentiates marine planning from terrestrial planning?

There is a misconception that marine spatial planning is something radically different to land-based spatial planning. In principle, the differences between the needs of planning for the marine environment and for the terrestrial environment are *not* significant. There is a long history of terrestrial land-use planning experience, which can be translated to the marine environment.

That is not to say that the approach to terrestrial spatial planning can simply be transferred to the marine environment. The Royal Commission on Environmental Pollution's report identified a number of important shortcomings of the current land-based system (Royal Commission on Environmental Pollution 2002). These include:

- The need for a comprehensive, mutually consistent and unambiguous policies that place the protection and enhancement of the environment as the foundation for sustainable development;

- The introduction of a clear statutory purpose for the planning system. The purpose of planning in protecting and enhancing the environment must be made explicit in a way that recognises other purposes;

- Integrated spatial strategies should rationalise the current plethora of plans and cover all aspects of Government policies;

- A statutory requirement for sub-regional and local plans to comply with the Regional Spatial Strategy;

- Integrated spatial strategies must cover all forms of land use, not just built development; and

- The planning framework must have the confidence of the public and include full participation of all stakeholders.

In addition, there are a number of key differences between land and sea environments that must be taken account of when developing a spatial planning framework for marine areas. First, marine planning must have regard to the three-dimensional nature of the marine environment. Sea use and development can occur on the seabed, the sea surface, and sometimes points in between. Second, marine planning must recognise the dynamic nature of the marine environment. The constantly changing marine environment demands an adaptive approach to forward planning and management. Third, marine planning must take account of the relative lack of information and understanding about the marine environment and the effects of development upon it. This demands a more precautionary approach to planning than we are perhaps used to on land. Finally, marine planning will be both more complex and the monitoring and enforcing of development consents more costly.

All of the countries around the North Sea have land use planning systems of some kind but extension of the concept to the marine environment has only been tried in a few cases and only within territorial waters. For example, programmes such as those undertaken by Norway and Sweden, where the planning system extends to areas of sea, are national in their scope and therefore limited in their application. Often, where forward plans do exist, they are scattered amongst different agencies – there is no strategic overview, integrated plan, or hierarchy of plans, as found in many land-based systems.

Examples of large scale spatial planning development

The development of spatial planning in the UK (and elsewhere in Europe) has traditionally been small scale and at a local level. Such approaches do not lend themselves to direct application to potentially vast areas of seemingly homogeneous marine environment. However, there are a number of examples where principles for developing large (even massive) scale spatial planning have been or are being developed.

Guiding principles for Sustainable Spatial Development

At the 12th Council of Europe meeting of European Conference of Ministers responsible for Regional Planning, a series of series of guiding principles for sustainable spatial development of the European Continent were agreed (CEMAT 2000). The purpose of the guiding principles was to define measures of spatial development policy by which people in all member states of the Council of Europe could achieve an acceptable standard of living. They aimed at bringing the economic and social requirements to be met by individual territories into harmony with their ecological and cultural functions, and therefore contribute to long-term, large-scale and balanced spatial development.

The guidelines acknowledged that subsidiarity and reciprocity were a prerequisite and that the principles needed to apply to the various political and societal bodies working at various levels inside and outside governments and administrations. Vertical and horizontal co-operation were regarded as being of particular importance. The guidelines identified that one of the most significant tasks of spatial planning policy was to provide private investors, in accordance with the objectives of planning policy, with forward-looking development perspectives and planning security. In all, ten guiding principles were agreed:

- Promoting territorial cohesion through a more balanced social and economic development of regions and improved competitiveness;
- Encouraging development generated by urban functions and improving the relationship between town and countryside;
- Promoting more balanced accessibility;
- Developing access to information and knowledge;
- Reducing environmental damage;
- Enhancing and protecting natural resources and the natural heritage;
- Enhancing the cultural heritage as a factor for development;
- Developing energy resources while maintaining safety;
- Encouraging high quality, sustainable tourism; and
- Limitation of the impacts of natural disasters.

The guiding principles then went on to develop more detailed spatial development measures that reflected different European regions. These regions were broadly defined in terms of their landscape characteristics: cultural landscapes; urban areas; rural areas; mountains; coastal and island regions; 'Euro corridors'; flood plains and water meadows; redundant industrial and military sites; border regions.

The Council of Europe's guiding principles on sustainable spatial development provide some useful pointers for the development of marine spatial planning, although much of it is predominantly relevant only to terrestrial spatial planning. However, the overall purpose and aims, and the regionalisation of approaches to spatial planning could be useful in the marine context.

A European Spatial Development Perspective

In May 1999, an Informal Council of Ministers responsible for spatial planning, agreed the European Spatial Development Perspective (ESDP) (European Commission 1999). This identified the aim of spatial development policies as working towards a balanced and sustainable development of the territory of the EU. It proposed that balanced spatial development was a fundamental part of the definition of sustainable development as envisaged by the UN Brundtland Report (1987). In pursuing this aim, Ministers agreed that three goals must be achieved equally: economic and social cohesion; the conservation and management of natural resources and cultural heritage; and more balanced competitiveness of the EU territory.

Overall, the ESDP identifies 60 spatial development policy options. These relate to three strategic areas of policy: polycentric spatial development and a new urban-rural relationship; parity of access to infrastructure and knowledge; and wise management of the natural and cultural heritage. Key to the application of these policy options is close co-operation amongst the authorities responsible for sectoral policies and spatial development at each respective level, and between actors at the Community level and the trans-national, regional and local levels (ie horizontal and vertical co-operation).

The ESDP proposes that the 60 spatial development policy options can be grouped into seven key themes. These are:

- Promotion of the networking of urban regions;
- Better accessibility to regions;
- Development of 'Euro corridors' (such as Trans-European Networks);
- Strengthening of the cities and regions at the external borders of the EU;
- Conservation and development of biodiversity in the EU regions;
- Development of the European cultural heritage; and
- The need for integrated coastal zone management.

The application of the ESDP to the marine territory of the European Union is likely to be difficult, although horizontal and vertical integration are clearly essential in the marine environment too. Most of the policy options identified are only applicable to terrestrial situations and it is very focussed on achieving equity between European regions, urban areas and transport connection between them. Ironically, it is an issue on which the ESDP has least to comment on – integrated coastal zone management (ICZM) – that perhaps provides the most useful indication of the way forward for spatial planning in the marine environment.

Implementation of Integrated Coastal Zone Management in Europe

In 1996, the European Commission established a demonstration programme on ICZM. The programme was designed around a series of 35 demonstration projects and 6 thematic studies. The aim was to gain consensus regarding the measures necessary in order to stimulate ICZM in Europe through providing technical information about sustainable coastal zone management and by stimulating a broad debate among the various actors involved in the planning, management or use of European coastal zones.

The programme concluded with the publication of a Communication (European Commission 2000a) and a Proposal (European Commission 2000b) regarding the implementation of

ICZM in Europe. Council and Parliament finally adopted the Proposal as a Recommendation on 30 May 2002 (European Commission 2002b).

Central to the recommended approach to ICZM in Europe are six key themes:

- Taking a strategic, ecosystem based approach to coastal management which protects the coastal environment and recognises risks (especially posed by climate change), whilst ensuring sustainable economic development and social cohesion (especially of remote and peripheral communities);

- Taking stock of existing coastal management frameworks at all levels and for all sectors;

- The adoption of broad principles that should underpin the approach to coastal zone management (such as taking a long-term perspective, adaptive and inclusive management and working with natural processes);

- The development of national coastal zone strategies which identify the role of different administrations and instruments for implementing the principles of ICZM;

- The encouragement of co-operation between member states; and

- The instigation of a programme of monitoring and review.

The ICZM recommendation probably provides the best European level definition of what large-scale spatial planning in the marine environment might entail. Notably, the recommendation refers to the development of a strategy, or strategies, which may be specific to the coastal zone or be part of a geographically broader strategy for promoting integrated management of a larger area. As such, ICZM strategies could be seen as sub-regional marine spatial plans.

In response to the EU Recommendation, the UK Government has undertaken a 'stock-take' of the UK coastal zone. The project is currently assessing the environmental, social and economic characteristics and natural resources of the UK coastline. It is identifying the different laws, agencies and other stakeholders that influence the planning and management of activities on the coast and analysing how these bodies integrate with each other. The stock-take should publish its conclusions in spring 2004, indicating what issues will need to be resolved by the development of national strategies for ICZM during 2004/05.

It is notable that there is also a growth in large scale, national, spatial strategies in the terrestrial environment with, for instance, the publication of the National Spatial Strategy for Ireland 2002 – 2020 (Irish Government 2002), the Regional Development Strategy for Northern Ireland 2025 (Department for Regional Development 2001), and the development of national spatial strategies for Scotland and Wales.

Conclusions

What are the potential advantages of marine spatial planning

The development of marine spatial planning will require a fundamental shift in approach in the way the marine environment is managed. This will inevitably require legislation and

Parliamentary time. A critical question that must be addressed is "why should we bother"? The answer is simple – because it will be worth all the effort and expense in the long-term. An effective marine spatial planning framework will deliver tangible benefits. It should:

- Provide a context or framework within which decisions concerning marine use and development can be made by existing sectoral regulatory authorities. In other words, we can build on what we have got rather than have to start afresh;
- Assist in the integration of policies and decisions both vertically and horizontally;
- Provide a long term view – a forward-looking vision of marine use which should allow developers in particular to plan and invest with increased certainty;
- Identify sites and areas important for sectoral interests (e.g. nature conservation) so helping to avoid conflicts with other interests;
- Allow full 'stakeholder' participation right from the point of plan development, through to making decisions within the context of the plan.
- Facilitate an 'ecosystem-based' approach to planning where planning units reflect 'seascapes' and marine processes.

These and other advantages are surely worth the investment and effort now!

What should marine spatial planning consist of?

It is clear that there is growing political momentum building for the introduction of some kind of large scale spatial planning for marine areas both within UK waters as well as those of European Union and the OSPAR area. Spatial planning and management for the marine environment should not be unduly complex or difficult. However, it is critical that the basis of any marine spatial planning framework is founded on sound principles. None of this is 'rocket science' and has already been promoted by NGOs in the UK both at a UK and OSPAR level (Wildlife & Countryside Link 2002, BirdLife 2003). These key principles include: -

- Sustainable planning – marine planning should have a statutory purpose to promote sustainable development;
- Provide integrated ('joined-up') thinking – bringing together various sectors and issues in the one plan, integrating them so that conflicts are avoided or at least minimised, and synergies optimised;
- Long term and forward looking – vision and direction are currently lacking in marine planning and management which tends to deal with current developments and trends or, at best, the very near future;
- Plan-led – it should be the plan that establishes whether a development is acceptable or not in principle;

- Provide spatial context – plans should cover large geographical areas, include all of the sea to extreme high water mark, and zone areas for use and development plus those to be avoided (in practice, the sea surface may often be capable of multiple use, so zoning does not necessarily equate with single use zones);

- Map both constraints and opportunities for development – a marine spatial plan should help developers to target areas where uses and development are acceptable and desirable, and conversely avoid areas where they could result in problems, such as adverse environmental impacts, or conflicts with other incompatible developments or uses. This would help reduce the risks users and developers are exposed to;

- Be subject to clear national policy statements (for development in coastal and marine waters) that set out principles for subsequent marine plans;

- Promote resource efficiency – planning and management should promote resource efficient development, ie 'making more with less';

- Promote participation and conflict resolution – marine planning and management should be participative and strive for consensus. 'Balance' and 'compromise' should be avoided – win-win solutions that genuinely contribute towards sustainable development should be sought;

- Be based on proper assessment of the environmental capacity of the marine environment to accommodate development. This may require enhanced environmental information;

- Be subject to Strategic Environmental Assessment (SEA); and

- Restore as well as protect and conserve the marine environment, wherever possible. A vital feature of this must be the establishment of a coherent network of marine protected areas (including strictly protected and no-take areas) throughout the marine environment.

What might a marine spatial plan look like?

Marine spatial plans are likely to be large scale. There may be merit in adopting a regional seas and coastal planning approach, especially where connections could be made to the terrestrial regions and regional planning. However, there should be no presumption that regions for the sea would match terrestrial regions. Rather, every effort should be made to define regions of the sea as functional units.

To be truly effective, a marine plan needs to be the responsibility of, and prepared by, a single authority or plan-maker. Where no such authority currently exists, lead responsibility for preparing a regional plan would need to be clearly established although plan development would have to involve all sectoral interests/authorities. However, the actions flowing from the plan (implementation) would be the responsibility of all the sectoral interests, not just the plan-making authority.

The relationship between marine planning and Integrated Coastal Zone Management (ICZM)
In some respects, ICZM could be seen as a response to a lack of effective marine planning as well as a lack of integration between marine and terrestrial environments. If an effective marine planning and management system is established, it may encompass much of the rational for ICZM so long as it makes strong links to, and interacts effectively with, the land-based planning system. Indeed, the objectives, principles and components for developing true integration in the coastal zone identified by the Recommendation, should in fact underpin the development of a marine spatial planning framework for the whole of the North Sea.

References

Bergen Declaration 2002 Ministerial declaration of the fifth international conference on the protection of the North Sea, 20-21 March 2002, Bergen, Norway.

BirdLife 2003 Developing a Framework for Marine Spatial Planning In the North Sea: Principles. Paper submitted to OSPAR, 23-27 June 2003, Bremen (ref. OSPAR 03/4/12).

CEMAT 2000 Guiding principles for sustainable spatial development of the European Continent. Adopted at the 12th session of the European Conference of Ministers responsible for regional planning (CEMAT) on 7-8 September 2000 in Hanover.

Danish Government 2002 Presidency conclusions of the meeting *Towards a strategy to protect and conserve the marine environment,* 4-6 December 2002, Køge, Denmark.

Day J & Roff J 2000 Planning for representative Marine Protected Areas – a framework for Canada's Oceans. Report prepared for WWF Canada, Toronto, April 2000.

Defra 2002a Safeguarding our seas: a strategy for the conservation and sustainable development of our marine environment. Publ. Defra, Scottish Executive & Welsh Assembly Government, May 2002.

Defra 2002b Seas of Change. The Government's consultation paper to help deliver our vision for the marine environment. Publ. Defra, November 2002.

Department for Regional Development, Northern Ireland Office 2001. Shaping our Future – Regional Development Strategy for Northern Ireland 2025, publ Corporate Document Services, September 2001

DETR 2001 Review of Marine Nature Conservation – interim report. Publ. 18 may 2001.

European Commission 1999 European Spatial Development Perspective (ESDP). Towards balanced and sustainable development of the territory of the European Union. Agreed at the Informal Council of Ministers responsible for spatial planning, in Potsdam, May 1999.

European Commission 2000a Communication from the Commission to the Council and the European Parliament on Integrated Coastal Zone Management: a Strategy for Europe (COM/2000/547), adopted 27 September, 2000

European Commission 2000b Commission proposal for a European Parliament and Council Recommendation concerning the implementation of Integrated Coastal Zone Management in Europe (COM/2000/545), adopted 8 September, 2000

European Commission 2002a *Towards a strategy to protect and conserve the marine environment* (COM(2002) 539 final, Brussels, 02.10.2002

European Commission 2002b Recommendation of the European Parliament and of the Council of 30 May 2002 concerning the implementation of Integrated Coastal Zone Management (2002/413/EC).

Irish Government 2002 National Spatial Strategy for Ireland 2002 – 2020: people, places and potential, Stationery Office, Dublin

Laffoley, D.d'A, Connor DW, Tasker ML & Bines T 2000a Nationally important seascapes, habitats and species. A recommended approach to their identification, conservation and protection. Prepared for the DETR Working Group on the Review of marine Nature Conservation by English Nature and the Joint Nature Conservation Committee. Peterborough, English Nature Research Report 392.

Laffoley, D.d'A, Baxter J, Bines T, Bradley M, Connor DW, Hill M Tasker ML & Vincent M 2000b An implementation framework for the conservation, protection and management of nationally important marine wildlife in the UK. Prepared by the statutory nature conservation agencies, Environment Heritage Services (Northern Ireland) and JNCC for the DETR Working Group on the Review of marine Nature Conservation. Peterborough, English Nature Research Report 394.

OSPAR 2002 Summary Record of the meeting of the OSPAR Commission (OSPAR 02/21/1-E), Amsterdam, 24-28 June 2002.

OSPAR 2003a Spatial Planning. Paper presented by the Secretariat to a meeting of the Biodiversity Committee (BDC), 20-24 January 2003, Dublin.

OSPAR 2003b Summary Record of the meeting of the Biodiversity Committee (BDC) (BDC 03/10/1-E), Dublin, 20-24 January 2003.

Tyldesley D & Associates 2000 Extent of local authority jurisdiction in the marine environment. Report to DETR, in association with Browne Jacobson Solicitors, December 2000 (doc. Ref. 1273rpt).

Royal Commission on Environmental Pollution 2002 Environmental Planning. 23rd Report, CM 5459, March 2002.

RSPB 2003 Areas of RMNC work where legislation may be needed. Paper LSG 1/5/5 submitted to the Review of Marine Nature Conservation legislation sub-group meeting of 28/02/03.

United Nations 1987 World Commission on Environment and Development: Our Common Future. The 'Brundtland' Report, New York, 1987.

Wildlife & Countryside Link 2002 Marine spatial planning in the UK. A Joint Wildlife, Countryside and Environment Links discussion paper, November 2002.

Managing recreational activities – a guide for maritime local authorities in England

Jonathan McCue, Atkins, Warrington, UK.
Terry Oakes, Terry Oakes Associates, Lowestoft, UK.
Tim Badman, Dorset County Council, Dorchester, UK.

Introduction

Estuaries and coasts, with their unique features, provide great diversity as well as opportunities for recreation and tourism. Commercial organisations, recreational groups and individuals all use the coast for enjoyment through participation in a wide variety of activities. The range of interests/activities often requires some form of management to reduce potential conflicts between disparate user groups and/or the natural environment. Within England, Maritime Local Authorities (MLAs), in conjunction with Harbour Authorities (HAs) and the Maritime and Coastguard Agency (MCA), are best placed to facilitate appropriate management of coastal related activities, as they understand and live with the local issues and pressures that arise.

Atkins was commissioned by Defra to produce a Guide for Maritime Local Authorities (MLA's) to assist current MLA managers on the implementation of coastal byelaws and voluntary initiatives to aid the management of coastal recreation. The Guide has been produced by Defra in answer to a number of recommendations put forward by a Government Inter-Departmental Working Party which reviewed local authority byelaw powers for the coast in 1994 (report published in 1998) and recommended that local authorities would benefit from such a Guide.

This paper summarises the issues surrounding coastal recreational activities, the approach taken to gather information relevant for such a Guide, outlines the content of the Guide and provides a description of the preferred stepped approach for MLA's to help advise on appropriate management approaches for coastal recreational activities.

Managing Coastal Recreational Issues

For the majority of cases, at least some of the coastal activities taking place within MLA jurisdiction will require some form of management. The reasons may be related to health & safety, nature conservation, nuisance, abuse, allocation of resources or a need to maximise capacity. Personal Water Craft (PWC) and 'new' activities such as kite surfing seem particularly difficult to regulate, often because of a lack of formal user groups or organisations which can disseminate good practice or codes of conduct. Other activities such as barbeques and dog walking on beaches can also be difficult to manage because users do not need slipways or specific access points which can be monitored.

The growth in the number, intensity and variety of recreational activities at the coast is leading to an increased risk to people and the environment in some areas and places a responsibility on local authorities to manage the often conflicting activities appropriately. The challenge for MLAs is therefore to manage the combination of users and pressures in a practical and cost-effective manner whilst at the same time enabling multiple use of coastal amenities.

The two main aims of recreational management should be:
- Public safety and protecting amenity.
- Preventing environmental damage.

Management should focus on:
- Being proactive.
- Allowing multiple use of amenities.
- Ensuring protection of people, amenities and the environment.
- Minimising conflicts between activities and the environment;
- Minimising conflicts between user groups.

Project Approach

The methodology approach used for producing the Guide combined the following key techniques:
- Literature review of published/unpublished work.
- Wider consultation with external parties.
- Workshop Event.
- Preparation of Issue/Activity Sheets.
- Completion of Case Studies.

A questionnaire was used as an important mechanism for consulting with key organisations. The project area was restricted to England though copies were forwarded to consultees in Scotland and Wales upon requests and responses included in the questionnaire replies. The use of the questionnaire allowed consultation to be targeted at a wide range of national sectors, such as all of the maritime local authorities and sea fisheries boards. The exercise helped to raise the issue of coastal byelaws to the wider community, many of whom were pleased to see this initiative was taking place.

The Workshop event provided an opportunity for all of the consultation group members to become further involved in the development of the Guide. It also allowed the elucidation of details of specific cases and examples of management of coastal activities from around the country, which could not be communicated in detail through the questionnaire.

A case study approach was adopted to provide the opportunity for discussions with stakeholders on the criteria associated with successful byelaws, the weaknesses in current arrangements, the issues raised at the workshop and any developments flowing from previous studies. This approach also allowed the team to explore which of the constraints identified in the questionnaire responses really present obstacles in practice. From this work, the information in order to prepare the Guide could be undertaken.

Issues of Importance

The following provides a brief summary of the main findings of the project. These were then developed further to establish clear advice on the appropriate management of coastal recreational activities within the Guide.

Voluntary Approaches

Managing conflicts requires an approach specific to the problems being experienced. If action is required, then control can be introduced around statutory or voluntary frameworks, or through a combination of instruments. A statutory approach to recreation management at the coast will give a clear basis for enforcement. However, evidence required to support a prosecution can be difficult and costly to collect. As byelaws can be promoted only to deal with local circumstances and issues, they are by their nature, reactive.

Voluntary and self-regulatory approaches to the management of coastal activities should be explored in the first instance. In this way, voluntary codes of practice can be produced and recreational users can be encouraged to behave responsibly through education and information provision. Voluntary arrangements which address local needs and compliment other local initiatives have a good chance of securing local support which in turn increases the likelihood of success of the scheme.

The Guide sets out some practical information concerning some of the statutory and non-statutory tools that are available to MLAs when developing such management strategies. It presents a number of different tools which can be used, such as Zoning, establishing "No–Take" zones, vessel registration, developing Codes of Practice/Conduct (promoted through the development of activity specific guidelines), Self Regulatory Groups and Partnerships (Recreation users forming their own user groups related to a particular activity) and Voluntary Wardening (members of staff employed by MLAs and HAs to patrol recreation areas on a voluntary basis).

Encouraging voluntary arrangements will remain a challenge. The lack of enforcement options appears to be one of the major stumbling blocks with voluntary initiatives; often it is the people who have caused the problems in the first place who fail to adhere to such initiatives. Managing recreation through the positive provision of facilities, voluntary co-operation and 'education rather than legislation' is the accepted ideal.

Byelaws

Where voluntary schemes are unlikely to succeed or have already failed, there is no alternative to a regulatory approach to the management of conflicts. Byelaws are legislative tools that have been used to manage a wide range of coastal activities. Byelaws supplement national law and can be tailored to meet local situations. They give statutory powers to authorities to enable a local issue or danger to be dealt with at the local level.

MLAs, HAs, English Nature, National Trust and other organisations have powers under different Acts of Parliament to make byelaws. The Guide outlines in more detail the powers available to introduce byelaws and the appropriate use of existing model byelaw text.

The making of a byelaw requires the promoting authority to provide evidence of the local issue or problem, for example, complaints of drunkenness on the beach. The promotion of byelaws should be supported by evidence of need, e.g. the results of crime and disorder surveys. Byelaws are useful as an adjunct to an overall strategy and to deal with local issues and problems, however, it can be difficult to achieve and maintain local support for byelaws and enforcement can be difficult without appropriate funds. Their use should be avoided where problems are minor or occasional. The Guide provides clear advice on how to develop a specific byelaw using the current framework in place.

Consultation

MLAs need to consult widely about the implications of proposed schemes, whether they are to be implemented through voluntary arrangements or enforced through byelaws. Consultation should be undertaken with local groups and organisations, parish councils and residents to ensure that their opinions are taken into account and that any operational matters (such as appropriate signing) can be considered. MLAs should ensure that adequate local consultation has taken place before introducing a new management approach.

Experience shows that one of the best ways to gain acceptance for schemes is to involve stakeholders in the decision making process. Involving stakeholders is a way of informing and explaining the need for control. Authorities taking a proactive approach towards consultation are seen to understand the needs of users.

Selecting the best consultation techniques is fundamental. For instance, meetings with several different user groups are usually less successful than meeting single user groups because they each tends to have views and requirements which conflict with those of other groups. Where pleasure craft are involved, the MCA-run local District Marine Safety Committees are a useful source of contact.

The Guide outlines a variety of appropriate consultation techniques available for MLA's.

Enforcement

The sea is a more difficult place to police than the land. Enforcement action is likely to be more difficult and expensive, with enforcers more thinly spread. If regulatory controls or volunteer schemes are not supported by ALL end users it is likely that enforcement measures will be needed. Such measures can take the form of policing of the site by relevant authorities, however, this is often a very expensive method of control both financially and in terms of time required to monitor and enforce regulatory measures. Resources for enforcement are scarce and specialist advice is not cheap.

It is clear that penalties are modest for byelaw offences and generally these are seen as a "slap on the wrist", although in some cases this can be effective. Offences such as dangerous navigation are treated more seriously. For example, the MCA has eight new offshore boats based around the UK which are used to carry out inspections at sea to help enforce Codes of Practice.

Enforcement can be initiated through peer pressure if there are active user groups in the area on an ad-hoc basis or through a more structured approach geared towards monitoring and regulating activities on weekdays, at weekends and during Bank Holidays.

There have been calls for breaches of byelaws such as navigation offences to be treated more seriously. Fines are often considered to be quite low in coastal situations but they have to be put in context with the other cases coming before the Courts. For example, the fines imposed for a fishery or dog fouling offences may reflect that fact that they do not appear to be as serious as a burglary or fraud.

The Guide outlines details on the national penalty scale for offences as set out in Section 37(2) of the Criminal Justice Act 1982. In summary, this provides a standard scale, giving maximum fines (up to £5000) for an adult on conviction for summary offences. Spot fines are often around £25 for first offences.

Dissemination of Information

Successful management schemes and initiatives are dependent on good public information. Good publicity can help solve problems by raising awareness of issues, rules and regulations. This in turn improves understanding and generates cooperation thereby reducing the difficulty of enforcement. This occurs through the following process:

Communication/Education==>awareness==>consultation==>enforcement

A large quantity of information is available to the public on a variety of recreational pursuits in the coastal zone. The Guide provides a significant list of appropriate printed material produced at both a national and local level, emphasising the requirements for inclusion in printed leaflets.

Signage

For many activities which use designated access points, information signs provide one of the best ways to disseminate information about local guidance and rules. Signs are also important to inform users visiting from outside the local area. Signs should provide simple, necessary information and wherever possible, not intrude on the local landscape character. The project determined that signage and public information needs to be providing a consistent message at a national level. The Guide produces a useful good practice checklist which can be followed for such guidance.

Buoyage

Many MLAs are looking at creating zones to separate recreational activities which would otherwise be in conflict with one another. Marker buoys are the most usual method of identifying zones. Laying buoys requires consent from under the Coast Protection Act 1949. Consent may also be needed from Trinity House (for navigation purposes) and from Crown Estates (likely landowner) in the form of a FEPA licence. The General Lighthouse Authorities will advise on the technical requirements for buoyage, but will not meet installation and operating costs (which can be substantial) unless the reason for buoyage is navigational safety. MLAs are recommended to seek advice from neighbouring or other MLAs who have introduced zoning schemes or other organisations such as the Dorset Coast Forum and MCA.

Overview of the Guide

The Guide has been produced to assist manage coastal recreational activities. It offers practical guidance to help develop appropriate forms of management of the coast as a public asset. It identifies many coastal recreation activities, as a broad spectrum of activities currently take place at the coast. They range from traditional recreational pastimes such as dog walking and beach games to the more modern activities of kite surfing and parakarting. They can be land based (barbeques, bait digging or sand yachting), water based (sailing) or air based (hang gliding and model aircraft), formal or informal, non-commercial or commercial.

The Guide focuses on coastal recreational activities that are likely to require a management view/input from a MLA. Reference to equivalent land based activities is made wherever possible. It reviews current practice for voluntary and statutory recreational management and from this, identifies good practice examples. The Guide is designed to help clarify the current management framework, provide ideas and advice on managing recreational activities in coastal areas.

Whilst focusing specifically on helping MLA's manage recreational activities, it is hoped that the Guide will be of use for HAs and other regulatory bodies responsible for promoting, managing and participating in recreational activities at the coast.

Encouraging MLAs to be proactive is a key message in this Guide. Examples are presented which outline some or all of the following key good practice principles:

- Act with full understanding of the issues.
- Act when needed and ensure action is proportionate to the issue.
- Work positively with users using effective means of consultation and communication.
- Think carefully about scheme design in partnership with users and ALL regulators.
- Use a variety of measures.

The following sections of this paper outline a summary of the key sections presented in the Guide.

Responsibilities for Management

The Guide clearly identifies responsibilities for management. A summary is provided here.

There is no overall authority responsible for managing the coastal zone and administration is divided between a large number of central and local agencies. In general, local authority jurisdiction coincides with the authority's administrative boundary. It is clear, and has been consistently agreed in the past, that coastal local authorities have administrative control and jurisdiction over areas down to low water mark, in the whole of England and Wales[2]. This is the case for open coasts but different situations apply in estuaries and harbours. Coordination should be encouraged wherever possible, as such bodies are likely to have valuable knowledge and experience in the management of coastal activities.

Maritime Local Authorities (MLA's)

MLAs have a wide ranging role at the coast. Their responsibilities include planning and development control, coastal protection; tourism and leisure, harbour management (in some cases) and environmental health and others. In respect to coastal recreational activities, most MLAs own land; directly operate facilities; manage amenities and attractions; and licence and permit activities. Therefore, in the broadest sense, the role of the MLA is to be "landowner" - to manage, regulate and help facilitate activities at the coast with a duty of care for the general best interests of the local community. In exercising this role, a MLA is able to use voluntary arrangements or can rely on legal powers where voluntary arrangements do not work.
MLAs are also seen to be facilitators in the management processes for other issues which are not under their direct control through the promotion of integrated coastal zone management (ICZM). Many MLAs prefer to encourage voluntary arrangements to manage recreational activities at the coast. In situations where voluntary techniques are unlikely to succeed or have already failed, MLAs can instigate a regulatory approach. However, their functions and ability to act are limited by the powers available to them and regulation is predominantly through the use of byelaws.

Byelaws give legal support to action on the ground and offer a clear basis for enforcement. However, obtaining a byelaw can be a long and expensive process. If the MLA can prove the need for a byelaw, Central Government has introduced procedures to reduce the time for approval. In promoting byelaws the MLA has the role and responsibility for enforcing the provisions of the byelaw.

Other key organizations include the following:

The Maritime and Coastguard Agency

The Maritime and Coastguard Agency (MCA) has an overarching role in promoting marine safety. This is carried out through its role as the UK's maritime emergency service. The MCA has a role in confirming certain byelaws and enforcing them where the MCA acts on behalf of the Secretary of State (SoS) for the Merchant Shipping Act (1995). The MCA has two byelaw model texts for navigation which apply to all vessels and enable byelaws for bathing, speed limits, licensing and other navigational issues.

Harbour Authorities (HA)

HAs have management responsibilities for port and coastal waters within their jurisdiction. They have a wide range of management powers through national and local enactments

Central Government

The ODPM overseas Local and Regional Government and planning policy. It has responsibility for confirming many of the relevant MLA byelaws. Other departments include DfT (shipping) and Defra (landscape and nature conservation, flood management, environmental and water quality).

Designing a Management Scheme

The Guide identifies a range of management approaches that may be required to effectively manage an area. This could be because of the variety of activities taking place and their associated management issues (alone or in combination) or due to the levels of impact associated with each activity. Different management methods can work effectively together, including the combination of regulatory and non-regulatory techniques. This is particularly the case for techniques such as voluntary zoning, education, awareness raising and regulatory byelaws. Often education can be used to support other practical methods of management.

When used together, the different techniques can effectively support one another and provide added value. The Guide identifies some key principles of good design :

- Think broadly when setting objectives.
- Bring all interests together.
- Act with a full understanding of the issues (see Step 1 text).
- Work positively with users.
- Review whether action is needed.
- Use structured techniques (eg: risk assessments).
- Identify partners.
- Use a variety of measures.
- Include environmental issues.
- Undertake a risk assessment.
- Review the effectiveness of scheme and modify accordingly.

Stepped Procedure

The Guide provides an approach to assist MLAs when developing management strategies. It is not definitive and allows for adaptation as appropriate. The stepped approach is advocated as it helps to identify the tools necessary to manage activities. The appropriate tools may depend on the nature and combination of the pressure/activity(s) and the resources available.

Figure 1 identifies proposed courses of action and guidance on the factors which should be considered at each stage. The basic steps of the process are summarized below:

STEP 1 Understanding the Issues

To understand the nature of the issues, information and evidence should be gathered to identify the conflicts, causes and the impacts being generated. Consultation with key local groups or individuals will help in understanding the nature and impacts of the problems and also help identify potential solutions. Analysis of associated risks is important, i.e. health and safety, environment etc. Key questions could include:

- What are the motivations of the users in question?
- What is the nature of the activity and what are the levels of participation?
- What is the trend for this activity in the area?
- Does it require specific access points and facilities?
- Who participates (when, how and who?) and are they affiliated to a local club?
- What is the impact (noise/visual/health and safety/environmental etc)?

- Is the activity weather or tide dependent?
- Who is being affected and how?
- When/ how does the problem occur?
- What are the risks?
- Can the activity be replicated elsewhere? (ie: relocated to other areas).
- Is there a cumulative impact with other activities?

STEP 2 Review experience and options

The examples in the Guide are intended to assist other MLAs facing similar pressures. The MLAs in the examples may be able to offer further advice and guidance in the light of their continuing experience. Liaison with adjacent MLAs is also important to ensure that the activity in question is not simply pushed along the coast. MLAs should collaborate with their neighbours where this helps to manage an activity. National representative groups (e.g. RYA) and national organisations (Sport England, MCA) can also provide details of relevant initiatives in other areas. Key questions may be:

- What other locations have experienced these issues and why?
- Where have other organisations tried to resolve the issue?
- What method of regulation was used (voluntary, statutory, combination)?
- What method of enforcement was used (cost of equipment and man-power, process for prosecution if applicable)?
- What level of consultation was undertaken?
- How was the scheme publicised?
- Were there any problems?
- How much did it cost?
- Was the scheme successful and why?
- Is the scheme being monitored?
- What elements of scheme development and implementation would be done differently if the process were to be repeated?

The review should identify existing tools available to the MLA. Discussions with the relevant legal department will identify where the MLA has existing byelaw powers and other regulatory measures which could be applied or adapted. Often, however, byelaws were made in the early 20th Century. These tend to have been modified over time and sometimes MLAs find it is difficult to ascertain what is still in force. A thorough review of all existing powers available to the MLA is recommended at this stage.

STEP 3 Identify appropriate forms of management

A review of resources available for implementation and enforcement is required. Following the statutory approach will be inappropriate if there are no resources available for enforcement. Further discussion is needed with stakeholders to identify opportunities for potential collaboration (e.g. self regulation by club members). The preference is for voluntary initiatives relying on user cooperation to provide a self-regulating and self–financing scheme. Such action may need to be supported by relevant statutory codes of practice. This step requires the options to be ranked according to their suitability and likely effectiveness. Key questions could include:

- Is the scheme practical?

- Will it have local support?
- What will be the cost of implementation and enforcement?
- Is the scheme cost effective?
- Will the scheme be adaptable to change?
- Will it fully address the issues?

STEP 4 Develop a management scheme

The management scheme or strategy should be drafted in consultation with key organisations/individuals. If there is the opportunity for partnership working with other groups (such as local club members helping enforcement) they should be involved throughout the development process. The scheme needs to be practical, meeting all the MLA statutory responsibilities (check for potential conflicts during development). Most importantly, the scheme needs to have local support and be enforceable. The strategy should include as a minimum:

- Background to the scheme, its need and the benefits it will provide to partners.
- The objectives and targets which are the means by which to monitor success.
- Partnership arrangements.
- The relevant rules and codes (statutory, non-statutory).
- The enforcement procedures (who is responsible, how will they be carried out?).
- How information will be disseminated (signing, zoning etc.).
- How the scheme will be monitored and reviewed.

The scheme will also need to include sufficient monitoring to determine if it is enforceable, if it is successful, and to identify any adaptations or modifications. Key points and actions include:

- The development of a structured monitoring programme to measure how successfully (or otherwise) the scheme is addressing the original problems and issues. These achievements should be measured in such a way that they can be used in Step 7
- Involve all interested parties in the monitoring work. A collaborative approach adds credibility and reduces costs.
- Seek suggestions from users and affected parties on the most appropriate location.
- Frequency of monitoring.

STEP 5 Implementation

Although many of the relevant groups/individuals should be aware of the scheme through the consultation process, it is advisable to publicise it widely to ensure that there is sufficient public awareness, especially if participants are from outside the area. Publicity could take the form of local TV, papers and radio, press releases, newsletters and leaflets (distributed to local and visiting clubs, visitor venues, holiday guides etc, meetings, signs at access points and websites.

Monitoring and enforcement needs to be undertaken at the most appropriate time – not just during office hours. This means having sufficient resources available during the most active leisure periods (i.e. the busiest, and probably the most expensive, times), during late afternoon and evenings; during weekends, Bank Holidays and school holidays; and for all seasons. MLAs need to ensure the availability of sufficient resources for long-term implementation and monitoring.

As a guide, Steps 1 to 3 may take up to 2 to 3 months to complete depending on location, issue and form of consultation. Step 4 is likely to take at least 3 months for a voluntary scheme and longer for the development and confirmation of byelaws. The ODPM can confirm byelaws following model text in 10 working days (following a month consultation period) if there are no objections but significant additional time is likely to be required for the MLA to produce the text if a model byelaw is not being followed.

STEP 6 Monitoring

Once the scheme is in operation it is essential to monitor its success in dealing with the issues and resolving the problems identified in Step 1. Sufficient resources (i.e. personnel and funds) will be required to implement a structured programme of monitoring which should examine the continuing degree of conflict and identify outstanding problems.

STEP 7 Review

Information gained from monitoring the scheme should be used to re-examine the problems and issues and to assess the success in achieving its objectives. The key issues are to agree the following points with stakeholders:

- The programme for the review of the scheme.
- The objectives for the review.
- The format of the review.
- Who will be involved in the review.
- When the review will be undertaken.

Conclusion

Encouraging MLAs to be proactive is a key message of the Guide. From the work undertaken to produce the Guide, the key messages when managing coastal activities, whether following a statutory or voluntary approach, are:

- Act with full understanding of issues.
- Act when needed and ensure action is proportionate to the issue.
- Work positively with users using effective means of consultation and communication.
- Think carefully about scheme design in partnership with all users and regulators.
- Use the right mix of measures for the situation being addressed.

Copies of the Final Guide will be available from Defra towards the end of 2003.

Acknowledgements

The authors would like to thank Sue Toland of Defra in allowing this paper to be prepared. The authors would like to thank the Project Steering Panel (Jonathan Calderbank of Sport England, Alex Steele of MCA and James Weeden of DfT for their assistance.

The Guide has been produced through detailed consultation with all Maritime Local Authorities and key relevant organisations. The project team would like to thank all those who have contributed valuable information to assist in the writing of this Final Report.

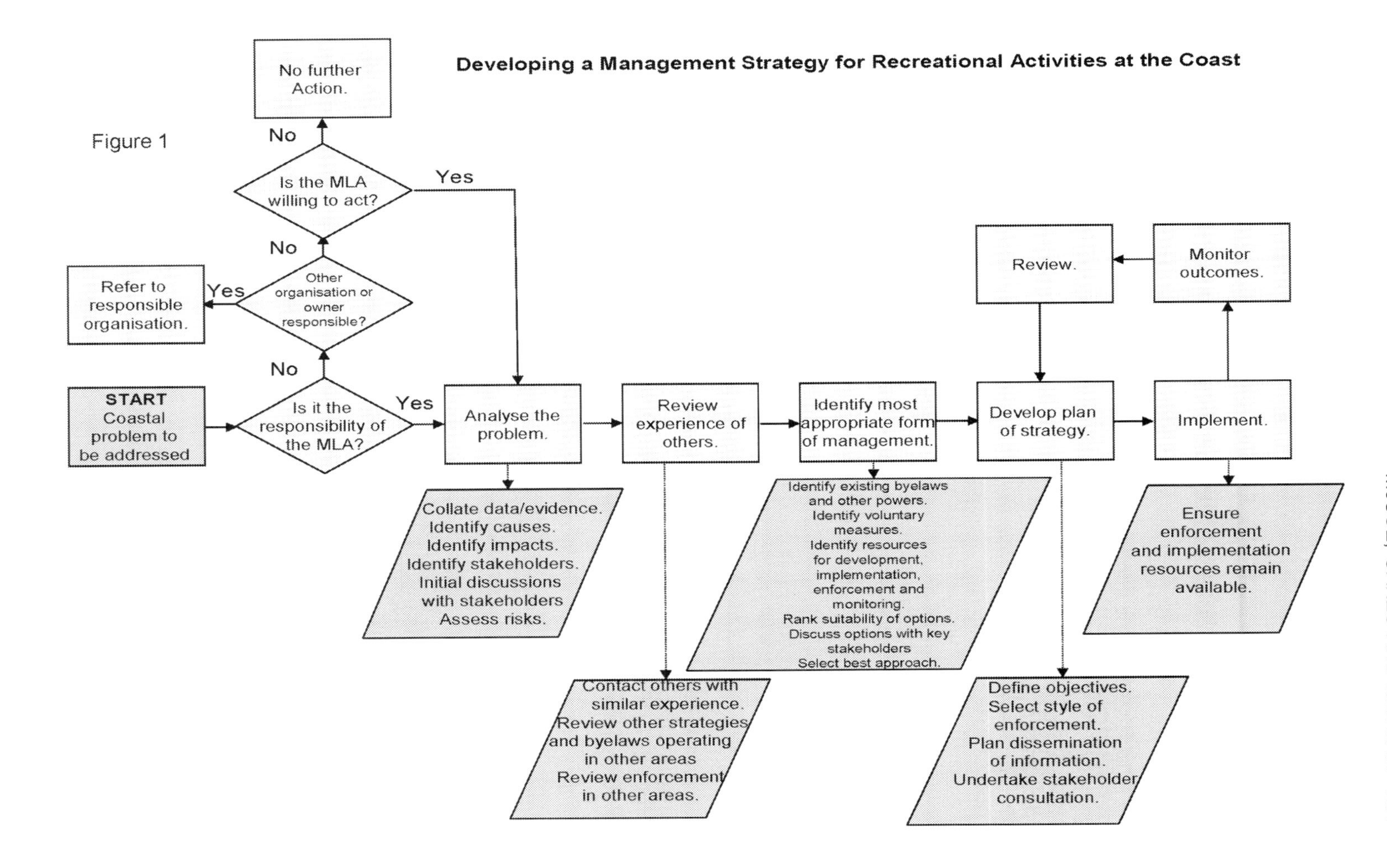
Developing a Management Strategy for Recreational Activities at the Coast
START Coastal problem to be addressed
Is it the responsibility of the MLA?
Yes
No
Other organisation or owner responsible?
Yes
Refer to responsible organisation.
No
Is the MLA willing to act?
Yes
No
No further Action.
Analyse the problem.
Collate data/evidence. Identify causes. Identify impacts. Identify stakeholders. Initial discussions with stakeholders Assess risks.
Review experience of others.
Contact others with similar experience. Review other strategies and byelaws operating in other areas Review enforcement in other areas.
Identify most appropriate form of management.
Identify existing byelaws and other powers. Identify voluntary measures. Identify resources for development, implementation, enforcement and monitoring. Rank suitability of options. Discuss options with key stakeholders Select best approach.
Develop plan of strategy.
Define objectives. Select style of enforcement. Plan dissemination of information. Undertake stakeholder consultation.
Implement.
Ensure enforcement and implementation resources remain available.
Monitor outcomes.
Review.

Figure 1

Streamlining Site Selection For Managed Realignment

Heather Coutts, Atkins, Epsom, United Kingdom
Heidi Roberts, Atkins, Epsom, United Kingdom

Abstract

Managed realignment, the deliberate breaching of flood defences to allow the formation of new intertidal habitat, has received increasing attention in the UK over recent years. Managed realignment has become a necessary shoreline management technique for two reasons; firstly as a means of providing economic and sustainable flood defences, and secondly because European law (in the form of the Habitats Directive) and the Biodiversity Convention require the UK to maintain the health and extent of a number of intertidal habitat types in the country and ensure that equivalent compensatory habitat is provided to offset loss or damage to these habitats designated under the Directive.

However, the task of finding land that is suitable for creating intertidal habitats such as salt marsh is not an easy one. Salt marshes have a number of physical and biological requirements that must be fulfilled before they will form, and these must be taken into consideration when looking for available sites. Furthermore, much of the land found in the coastal zone already has an existing nature conservation value, which can create conflicts of conservation interest when creating new habitats or carrying out shoreline management activities. Finally, private landowners are often unwilling to sell their land, which has resulted in a general shortage of suitable land available for managed realignment in the coastal zone.

There is currently no standardised procedure for identifying sites for managed realignment, which means that sites are identified on an ad hoc basis. Coupled with the numerous issues that must be considered when finding suitable sites, the time required for compensatory site identification can cause extensive delays to essential flood defence schemes located in environmentally sensitive areas.

A means of streamlining the site identification process with a view to reducing these delays has been developed by creating an analytical tool, in the form of a matrix, to assess site suitability. This paper discusses this tool, which has proved successful in bringing together the key elements involved in site selection, ranging across technical, environmental, sustainability, statutory and legal, economic, social and recreational issues. Each site is scored for suitability in all of these categories, which allows direct comparisons to be made between different sites, so they can then be ranked and the most suitable one chosen.

The matrix was developed and then tested in site identification for a coastal defence scheme in central southern England. It was concluded that the matrix is a successful means of identifying

Coastal Management 2003, Thomas Telford, London, 2003.

sites and eliminating those which are unsuitable for realignment at an early stage. The process also allows for wider geographical coverage and therefore facilitates a more strategic approach to site selection. The matrix has the potential to be adapted to other processes such as strategic planning for shoreline management.

Introduction

Project aims

Although a great deal of published literature is already available regarding managed realignment and requirements for salt marsh development, there is currently no defined methodology to identify sites that may be suitable for realignment. The aim of this project was to bring together the existing information on managed realignment and scientific research regarding salt marsh development in order to create a tool, in the form of a phased matrix, which can be used by those working in the field to help streamline the process of site identification. This also encompasses the economic, social, legal and policy issues that arise alongside the technical aspects of habitat creation, and that need to be resolved by coastal managers. Finally, the matrix was tested in the field and used to identify a potential managed realignment site as part of an Environment Agency coastal defence scheme on the south coast of England.

The need for managed realignment

Much of the coastline around the UK, especially in the south eastern region, was originally dominated by salt marsh. However, many of these areas were reclaimed for agriculture and other anthropogenic uses, mostly over the past 400 years. The end result of this coastal reclamation was that the region has been left with a large proportion of low-lying coastal agricultural land (Hazelden and Boorman, 2001). In addition, the south east suffered devastating floods in 1953, and in response extensive sea defences were built around much of the coast to protect these low-lying areas at risk from flooding (Emmerson *et al*, 1997). Coastal development continued, and in 2001 it was estimated that some 44% of the English and Welsh coasts were protected by some form of coastal defence (Oldershaw, 2001).

Under natural conditions of relative sea level rise, and with no coastal defences, the marshes would simply migrate inland and adjust their position naturally. However, when the marsh is backed by a hard defence such as a sea wall, this landward movement is prevented, and the marsh simply erodes or is submerged. This produces the phenomenon of 'coastal squeeze', a term applied to the ongoing erosion and loss of coastal habitats in the UK (French, 1999; MAFF, no publication date).

Many coastal defences are reaching the end of their design life, and are in need of repair. However, it is becoming increasingly costly and uneconomic to maintain defences protecting low-lying agricultural land with a relatively low economic value, and as a result the option of allowing these areas to revert back to salt marsh via the deliberate breaching of sea defences is being more frequently investigated. This option is called 'managed realignment' or 'set-back', and is preferable to abandonment, which can result in excessive scour and erosion (Hazelden and Boorman, 2001). The realignment option not only has beneficial cost implications, but can also confer additional flood defence benefits. The flood defence function of salt marshes fronting a sea wall are now well recognised, and studies have been completed in the UK to investigate the potential of wave attenuation over marshes (Möller *et al*, 1996; 2001). A study of wave attenuation over a 180m wide salt marsh in Norfolk compared to a 197m wide sand flat found that the salt marsh dissipated total wave energy by an average of 82%, while the unvegetated sand flat only dissipated an average of 29% of total wave energy. Significant wave heights were also reduced more by the salt marsh than the sand flat (Möller *et al*, 1996; 2001).

The need to undertake managed realignment in the UK has been further brought about by a number of key legislative instruments. The introduction of EU Directive 92/43/EEC on the conservation of natural habitats and of wild flora and fauna (or The Habitats Directive), together with the 1994 domestic Statutory Instrument that enforces it in the UK; SI 1994 No. 2716, The Conservation (Natural Habitats &c) Regulations 1994 (or The Habitats Regulations 1994) created lists (Annexes) of European species and habitats deemed to be under the most threat of deterioration and decline, and introduced requirements for Member States to create designated Special Areas of Conservation (SACs) in order to maintain these habitats/species at a "favourable conservation status" (DoE, 1994). Any plan or project that is not directly connected with the management of the site but is likely to have a 'significant effect' on it must be subject to an environmental assessment of the effects on the particular features of interest for which it was designated, known as an 'appropriate assessment'. If a negative assessment results, the plan may only go ahead for "considerations of overriding public interest" (DoE, 1994), and the Secretary of State must ensure that appropriate compensatory measures are taken "to ensure that the overall coherence of the Natura 2000 Network is protected" (DoE, 1994). EU Directive 79/409/EEC on the conservation of wild birds (or The Birds Directive) came into force in Europe in 1979, and introduced requirements for Member States to designate and conserve areas of important bird habitat, called Special Protection Areas or SPAs. Measures to offset damage to SPAs are similar to those in place for SACs, with requirements for appropriate assessment and mitigation and compensation measures for damage or destruction to the site. The network of European sites designated under the Habitats and Birds Directives are collectively known as the Natura 2000 Network.

Salt marshes are ecologically highly productive, and also provide habitat and other resources for a wide variety of flora and fauna. Tidal marshes are important invertebrate habitats; a salt marsh can typically support 12 000 *Hydrobia ulvae* snails per m^2 on its surface, and 2000 invertebrates per m^2 below the surface (Dixon *et al,* 1998). It is believed that 'outwelling' of organic materials and nutrients from salt marshes to adjacent habitats occurs, providing a food supply for a range of commercially important fish and shellfish species (Boorman, 1999; Emmerson *et al*, 1997).

Salt marshes provide sheltered nursery areas and spawning sites for numerous species of fish, act as high tide refuges for wading birds feeding on adjacent mudflats, breeding grounds for various species of waders, gulls and terns, and as feeding grounds for passerine birds and large flocks of ducks and geese (Boorman, 1999; UK Biodiversity Group, 1999; Dixon *et al,* 1998). This ecological significance is recognised in the inclusion of many areas of salt marsh within SAC and SPA designations.

In June 1992, the Convention of Biological Diversity was signed by 159 governments at the Rio Earth Summit (UK Biodiversity Group, 2001). It was the first treaty to provide a legal framework for biodiversity conservation, and called for the creation and enforcement of national strategies and action plans to conserve, protect and enhance biological diversity (UK Biodiversity Group, 2001). Following the Convention, the UK Biodiversity Steering Group was created, and in 1995 published a report containing costed action plans to conserve 116 species and 14 habitats together with recommendations for future biodiversity action plans (UK Biodiversity Group, 2001). The Habitat Action Plan for coastal salt marsh states that there should be no further net loss of salt marsh, with the creation of 100 ha of salt marsh per year to offset estimated national losses (UK Biodiversity Group, 1999). In addition, a further 60 ha of salt marsh per year should be created throughout the life of the plan to replace that lost between 1992 and 1998 (UK Biodiversity Group, 1999).

This combination of coastal defence, legislative and conservation requirements has resulted in the need for managed realignment in the UK. A number of managed realignment sites have already been created in the UK, most notably Tollesbury, Orplands, Northey Island, and most recently, Abbots Hall in Essex. The potential for managed realignment sites to eventually become valuable wildlife habitat with a significant conservation value also makes this defence option an attractive one. However, there are numerous factors that will determine the success of a habitat creation project, and the creation of a salt marsh with high conservation value should by no means be considered automatic (Brooke, 1992; Burd, 1995).

Background: Physical and Technical Requirements for Salt Marsh Formation

A review of existing published literature regarding the physico-chemical requirements and identification of conditions required for the establishment of salt marsh plants was undertaken. A consideration of a number of current issues relating to economics and current flood defence policy that also have a bearing on the suitability of a site for managed realignment, but are unrelated to its physical suitability was included, but the main emphasis was the physical suitability of the site. These factors were included as criteria in the matrix and used to rank the physical suitability of the proposed sites for habitat creation.

Site elevation, tidal inundation and surface slope

Site elevation is one of the most crucial aspects of salt marsh development. Salt marsh vegetation can only begin to colonise a mudflat when it has accreted to a certain height, and is exposed to the air for a long enough part of the tidal cycle each day (Pethick, 1984). This is because the germination of salt marsh seeds is primarily a function of the degree of tidal inundation (Emmerson *et al*, 1997). Excessive tidal inundation and waterlogging also increase stress on salt marsh plants by increasing turbidity and reducing photosynthesis (Burd, 1995).

Boorman (1999) states that in order for halophytic vegetation to establish, the substrate must only be covered by the tide for approximately six hours each day. Burd (1995) notes that the elevation of most of the previous successful managed retreat sites (those that had retained 60% or more of the vegetated marsh surface) was generally greater than 2.34 m OD, resulting in less than 300 tidal inundations each year. However, due to differences in tidal inundation at different geographic locations, surface elevation is not always indicative of the number of inundations per year. Burd (1995) therefore recommends that number of inundations rather than surface elevation should be considered during scheme design, with sites experiencing less than 450 inundations per year being the most suitable for salt marsh creation.

Site elevation is often a restricting factor in managed retreat schemes, particularly where reclaimed land is being returned to marsh. Reclaimed land may have been isolated from the sea (and all associated depositional sedimentation processes) for several centuries, and during this time the former marsh soils will have undergone compaction and dewatering, increasing their density and decreasing their porosity, while soluble salts have been leached out by rainfall and, in the case of agricultural land, fertilisers added (French, 1999; Hazelden and Boorman, 2001; Burd, 1995). Meanwhile, other undefended marshlands outside of the area will continue to accrete, which can also produce a height deficit between the proposed site and surrounding area (Burd, 1995).

If the selected site naturally slopes to the rear this may have a number of advantages in realignment. Firstly, if the rear of the site slopes into adjacent high ground, it may remove the need to provide new coastal defences behind the newly created intertidal area. Not only does this have economic benefits for the scheme in the reduction of costs, but will enable the site to 'roll back' and evolve as sea level rises, without further coastal squeeze occurring. A natural slope will also permit the formation of transitional plant communities (from saline to terrestrial) on the new marsh (Burd, 1995). These communities have become rare in the UK because many salt marshes are currently backed by hard defences which create a distinct divide, instead of a gradual transition, between terrestrial and saline habitats, and may increase the nature conservation value of the site (Burd, 1995).

Wave exposure

Salt marshes require a degree of protection from wave action before they can form, and are generally found in more sheltered areas, such as in estuaries and harbours, and behind spits and barrier islands (Pethick, 1984). Pethick (1984) also states that the formation of mudflats and salt marshes is facilitated by the settling velocity of fine-grained sediments (silts and clays) in water, and these sediments will only begin to settle out of the water column at water speeds of less than 0.00024 cm s^{-1}, which occur at slack high water in sheltered areas such as those mentioned above. Although there are exceptions where salt marshes are present on open coasts, managed realignment on open coastline is less likely to be appropriate for the creation of salt marsh habitats. It is recommended that a wind fetch distance of less than 2000m is acceptable before some kind of protection is required, although this may be no more than a simple breakwater (Brooke *et al*, 1999).

Tidal hydraulics/Tidal prism

During the site selection process for managed retreat, it is important to consider the impact of the retreat on adjacent coastal processes. Patterns of erosion and deposition at any point within an estuary will depend on a number of processes, but are primarily related to the tidal prism upstream of that point. The tidal prism can be defined as the volume of water exchanged over a single tidal period, usually within bounding sea walls, and affects the current velocity of tidal flows through the estuary (Burd, 1995). When managed retreat is carried out in the upper reaches of an estuary it may result in an increase in erosion in the lower reaches. Although there is a potential increase in flood/storm water storage in the upper estuary, because there is a greater overall volume of water entering and leaving the estuary on each tidal cycle, there is also the potential for increased downstream erosion to occur (Boorman and Hazelden, 1995). Therefore site selection for managed retreat should be a strategic process which identifies and mitigates any potential effects that may further destabilise downstream sea defences or cause additional habitat loss (Burd, 1995). These effects will have to be determined via hydraulic modelling of the estuary, as the effects will be highly site specific and will be affected by a number of factors such as the size of the site and its location within the estuary.

Accretion rates and suspended sediment concentration (SSC)

Sediment deposition on salt marshes can only occur when the marsh surface is flooded, and requires the availability of suspended sediment and the opportunity for it to be transported across the marsh (Reed, 1995). Brooke *et al* (1999) state that there should be sufficient suspended sediment in the water column to allow an accretion rate of 3-10 mm per year. SSC is likely to be one of the factors that will contribute to the development of salt marsh, but its role in facilitating accretion on a managed realignment site is not entirely clear from existing literature. In the absence of this information, accretion rates on surrounding marshes can be measured, for example through the use of sediment traps, to ensure that a rate of 3-10 mm per year will be achieved. It has also been reported that an accretion rate of more than 150 mm per year may smother new vegetation, particularly pioneer species, which may be an upper limit for site selection (Brooke *et al*, 1999).

Sediment characteristics and sediment quality

Burd (1995) indicates that loose loam to clay soils are best for the establishment of salt marsh plants, but other less suitable substrates such as sand and peat will also support plant growth after sedimentation and mixing with clay particles. Burd (1995) also states that provided there is sufficient suspended sediment to ensure that accretion occurs, the original substrate of the site should not be a limiting factor affecting the success of the scheme.

Pollutants and other heavy metals show a tendency to bind to fine grained sediments such as silts and clays, which effectively can mean that salt marsh creation may remove pollutants from the surrounding environment. However, the presence of pollutants (heavy metals, organochlorines and herbicides and insecticides from previous agricultural use) may have an adverse effect on salt marsh plants and invertebrates, possibly contributing to their decline (Leggett *et al*, 1995). Although one study found that salt marsh plants in more heavily contaminated areas exhibit indicators of physiological stress, this focussed on individual plant stress indices and did not indicate whether this stress affected the ecosystem as a whole (Padinha *et al*, 2000). Williams *et al* (1994) reviewed the accumulation and cycling of metals within salt marshes and concluded that levels of heavy metals in most estuaries were probably not high enough to adversely affect salt marsh plants. There is still a gap in existing published scientific research regarding the effects of pollutants on salt marsh ecosystems, and so potential effects of pollutants have not be considered in detail for the purposes of this paper.

Site history and proximity to existing marsh areas

A review of marsh restoration projects in the San Francisco Bay area (San Francisco Bay Conservation and Development Commission, 1988; cited in Brooke, 1992) reported that one of the key factors in determining the success of the restoration projects was the proximity of the site to an existing natural 'seed source' of salt marsh flora and fauna. If adjacent salt marshes are present, halophyte seeds will be carried onto the site with suspended sediments during tidal fluxes (Emmerson *et al*, 1997), thus facilitating the colonisation of the site by salt marsh plants. Experiments at the Tollesbury managed realignment site to assist natural colonisation found that planting plugs of soil and vegetation cut from existing salt marsh was the most effective means of artificially establishing plant cover on the site, but natural colonisation was still the most important process in vegetation establishment (MAFF, no publication date). This further suggests that the potential success of a managed realignment scheme may be increased by selecting a site close to extant salt marshes. Extending an existing site to create a larger area of habitat may also benefit wildlife more than a series of smaller, fragmented sites (Atkinson *et al*, 2001).

Reclaimed areas of former marshland are also likely to be among the best sites for realignment. Burd (1995) indicates that the presence of marshland on the site prior to its reclamation demonstrates that conditions at that point in the estuary are suitable for salt marsh development. If no marsh was present prior to the land being reclaimed or defended, it may indicate that either the physical conditions were unsuitable for salt marsh development, or the estuary had not reached a stage in its evolutionary history where salt marshes would naturally form (Burd, 1995). Site history should therefore be investigated as part of the site selection process.

Creek Network

Salt marsh creeks are a fundamental feature of natural salt marshes. They are the means by which flood tides are distributed across the marsh and subsequently drain from the marsh surface, thus producing the required hydraulic and sediment regime (Burd, 1995). Marsh channels also dissipate tidal energy and aid vegetation establishment through drainage and sediment dewatering (Reed *et al*, 1999). Large primary creeks distribute water from the estuary to the marsh, while smaller secondary creeks distribute water across the marsh surface and ensure sediment deposition across the whole surface (Burd, 1995).

If there is evidence of a relict creek network already on the site, these former creeks can then be re-excavated and reinstated on the newly created marsh. If a relict system is not visible on the site, two options remain. A new drainage system can develop through natural erosion and deposition, but this is a long-term process and may result in poor water circulation and delayed vegetation establishment (Burd, 1995). Secondly, an artificial drainage network can be dug, but this requires a complex analysis of equilibrium geometry, which is controlled by factors such as tidal prism, tidal regime, sediment characteristics and vegetation type (Burd, 1995), and it is therefore a great advantage if a site with relict drainage still visible on it can be selected. Poor design of a new creek network can lead to erosion rather than accretion in parts of a managed retreat site (Reed *et al*, 1999).

Problems associated with identifying realignment sites

There is a growing market for coastal land, and landowners can sell their land directly to those who wish to buy it specifically for managed retreat. This will normally fall under the needs of Coastal Defence Authorities such as the Environment Agency of England and Wales, Coast Protection Authorities or private developers who need to provide compensatory habitat to offset habitat losses or environmental damage from their developments or activities.

If a coastal defence authority were to purchase land for managed realignment, due to limits of available funding it is likely that the prices they would be able to pay for land purchase would follow the average agricultural grade land values. However, some landowners are unwilling to sell their property to coastal defence authorities for a number of reasons. Firstly, their land will be undefended because the authority have previously carried out a cost benefit analysis and taken an active decision not to provide defences. This can have the effect of creating ill-feeling towards the authority on the part of the landowner, and possibly cause reluctance to sell their land to the same organisation that would not defend it. Secondly, anecdotal evidence suggests that private developers have comparatively more financial resources available for land purchase than public bodies, and are therefore able to offer landowners considerably higher sums of money for their land, although no figures are available for direct comparison. Faced with having undefended land that is at risk from coastal flooding and sea level rise, it is understandable that landowners may prefer to sell it to the highest bidder.

As the vast majority of flood defence schemes are undertaken by public bodies, they are funded from the public purse, and there is therefore a limit to what a public body can justifiably spend to complete a flood defence scheme. As competition between public bodies and private developers to buy coastal land increases, and landowners become more aware of its potential value, there is a chance that land prices will be forced continually upwards in coming years. This may eventually lead to a situation whereby public bodies can no longer justify paying the prices demanded in the land market for compensation land, with the possible stalling of some flood defence schemes which require compensatory intertidal habitats to be created under the Habitats Regulations 1994 (Tim Kermode, *pers comm.*, 2002). While there is a clear need for managed realignment in the UK, the problems associated with finding sites means that an ad hoc and undefined approach is not sufficient to overcome these issues, and a more streamlined approach is required. The methodology developed for this project takes into account a wide range of issues and incorporates some of the socio-economic problems currently associated with managed realignment.

Methodology

The development of the matrix and site selection was undertaken in a two-stage process. The first stage was a broad site identification exercise, which used a Geographical Information System (GIS) to identify areas of low-lying coastal land. The second stage was a refinement process in site selection using the matrices to evaluate the suitability of each site to realignment and reject those that were unsuitable for habitat creation.

Stage 1: Geographical Site identification

Land elevation mapping

This stage of site selection used a GIS program to identify broad potential sites. Electronic Ordnance Survey map tiles (Scale 1:10000) were obtained and the following tidal levels digitised onto the maps:

- Mean Low Water: 0m OD
- Mean High Water Neaps: 1.5m OD
- Mean High Water Springs: 2.0m OD
- Highest Astronomical Tide: 2.5m OD
- Predicted 1:200 year flood level: 3.5m OD

Although published literature on salt marsh development indicates that there are numerous factors to consider aside from simple site elevation, as a general rule, salt marsh vegetation begins to develop at approximately MHWN, and is found up to about the level of MHWS. Below MHWN mud flats are formed. On this basis, all areas between MLW and MHWN were digitised on the maps, indicating potential formation of mud flat, and areas between MHWN and MHWS digitised separately to indicate potential formation of salt marsh. By digitising the sites in this way, the amount of potential habitat could be easily and accurately calculated.

At this stage, sites were identified solely on the basis of land elevation and tidal height data. The purpose of the mapping was to identify all areas that were potentially suitable for retreat/habitat creation, regardless of the issues of land ownership, existing land use or economic assets (e.g. houses or other buildings). The study area was the central southern coast of England, covering two counties, Hampshire and Sussex, and including four natural harbours: Portsmouth, Langstone, Chichester and Pagham Harbours.

The mapping exercise also collated data on existing conservation designations in the vicinity of each of the harbours. GIS data from English Nature's website were used in conjunction with the potential sites identified from the elevation mapping to produce electronic maps showing the location of both national (Sites of Special Scientific Interest) and European (SPA and cSAC) sites within the study area. These maps were used to show which sites were already subject to designations. All the sites were then ranked in two phases using the criteria identified for the matrices.

Stage 2: Site Evaluation Using Matrices

Primary site assessment matrix and stakeholder workshop

The primary site assessment matrix can be described as a kind of 'scoping exercise'. It is designed to deal principally with non-technical information about each potential site, and the scope of the matrix is intended to be as broad as possible in order to cover as many potential issues as possible. The aim of the matrix was to eliminate those sites that are wholly unsuitable for realignment, to determine which sites are most suitable for further investigation and the identification of any site specific factors (both positive and negative) that could affect the potential of the site for realignment. To this end, the topics included in the primary assessment matrix were based around very general concepts, such as whether the site had an intrinsic nature conservation value or whether there were any assets adjacent to the site that would be at risk from flooding if it were realigned. The matrix was designed to allow the assessment of multiple sites in a single table, and therefore facilitate a rapid assessment of site suitability.

As the scope of the matrix was very broad, the scoring system was limited to giving each site a positive, negative or neutral score for each item, with further space to record additional comments where necessary. The rationale behind each score was whether or not the item would contribute to the success of the scheme or made the site more suitable for realignment (+), prove to be a barrier to realignment or make it less feasible (–) or finally, have no discernible effect on the suitability or success of the site in realignment (0). At the end of the exercise, the total number of each type of score were added together to give an overall score for each site, with negative scores being subtracted from positives or vice versa. As there were a total of 13 categories on which each site was scored, the highest possible score for each site was +13, and the lowest possible score –13. Once the sites had ranked, the highest scoring areas can be taken forward for more detailed investigation using the secondary matrix. The primary assessment matrix is shown as Table 1.

Because of the number of stakeholders involved in the ownership, regulation and management of the coastal zone, a workshop was held to assist with site assessment. All regional stakeholders, excluding private landowners, were invited to the workshop to discuss the viability of each site for realignment. Those invited included representatives from coastal defence authorities, statutory and non-statutory nature conservation organisations and other stakeholders. The purpose of the workshop was to allow discussions between all stakeholders and to identify as many potential issues relevant to each site as possible, by establishing the amount of environmental, socio-economic and historical information available from those who work in the local area. This was especially necessary because the scope of site identification stretched across two administrative boundaries, Hampshire and West Sussex, and thus required information from stakeholders in both counties. During the workshop, the sites identified were discussed and ranked for suitability for habitat creation using the primary assessment matrix.

Table 1: Primary Assessment Matrix

CRITERIA	Site 1	Site 2
Environment		
What is the intrinsic nature conservation value of the site (- internationally designated, 0 local/national designations, + no designations)		
Are there any habitats/species of interest? (even if not designated) (- numerous protected species/habitats, 0 a few protected species/habitats, + no protected species/habitats)		
Would the site be able to evolve with climate change (i.e. does it slope to the rear?), will it be sustainable? (- only mudflat created, 0 saltmarsh & some room for transition, + site elevated to rear)		
Will the current use of the site pose problems for any potential retreat scheme? (- existing public amenity, 0 agriculture, + no specific use)		
Flood Defence/Economics		
Would realigning the site increase flood risk to neighbouring properties/services? (- numerous properties/services could be at risk, 0 a few properties/services could be at risk, + no properties/services at risk)		
Would rear defences be needed? (- extensive defences probably needed, 0 moderate/small scale defences may be required, + no defences thought to be required)		
Planning		
Would realignment meet/conflict with the shoreline management policy? (- improve existing defences, 0 maintain the line, + do nothing/retreat)		
Would realignment conflict with local planning policy? (- site designated for development, 0 agricultural/amenity use, + no specific planning policies)		
Are there any services running through the site which could be affected by realignment? (- numerous services would need re-routing, 0 minimal impact, + no services at site)		
Would there be any loss of amenity value to the site? (- significant loss, 0 little impact on amenity value, + positive amenity benefit)		
Other		
Who owns the land? (- unknown, 0 private landowner, + statutory authority or Wildlife Trust)		
Would they be interested in realigning the site? (- known hostility, 0 unknown, + could be persuaded)		
What is the local feeling about realignment in this area? (- hostile, 0 unknown/unconcerned, + welcome)		
FINAL SCORE		
TAKE FURTHER? Y/N		

Secondary site assessment matrix

The rationale behind the structure of the secondary site assessment was to take forward a selection of potential sites for more detailed investigation, it having been determined during the scoping exercise that there were no immediately obvious overriding factors that would make them completely unviable. At this stage there were still too many sites and too many unknown factors to make it logistically justifiable to carry out a complete site investigation on each site (for example, using hydraulic modelling or trying to determine local accretion rates), and so the focus of the secondary site investigation is primarily to collect baseline (existing) data and use it in conjunction with the known technical requirements for salt marsh creation in order to make a direct comparison of suitability between each site.

The information to be included in the secondary site assessment matrix was based around the earlier literature review. At this point, further investigation of the environmental and policy issues that would make each site either environmentally or politically unacceptable was necessary, and also an initial investigation of land ownership issues. Technical issues that could be assessed without extensive detailed investigation were also included. The desired end result of this exercise was to identify two or three of the most appropriate sites that were worthy of a final detailed technical investigation and potentially, eventual realignment.

The scope of the secondary matrix was broken down into several sections, covering:

- Statutory and legal issues,
- Environmental issues (including landscape),
- Technical issues,
- The potential sustainability of the site,
- Flood defence and economic issues, and;
- Social and recreational issues.

It was decided at this stage that a wider ranging scoring system should be employed in order to ensure issues that could have a major impact on the viability of a site were given greater weight, for example, if the site was already subject to an international designation and the statutory nature conservation authority were very unlikely to grant permission for its use. For this phase, five possible scores were assigned to each of the 28 categories from –2 to +2, giving a maximum/minimum possible score of +/– 54 respectively. Due to the amount of detail that needed to be included within the matrix, one matrix was completed for each site investigated. The secondary matrix is shown as Table 2.

Table 2: Secondary Assessment Matrix

CRITERIA	SCORE
Statutory and Legal Issues	
Level of environmental protection (-2 international SPA/SAC/Ramsar, -1 SSSI other national, 0 local, +1 partial designation, +2 no designations)	
What will be the impact on the current site designation? (-2 major adverse impact, -1 minor adverse impact, 0 no effect, +1 minor beneficial impact, +2 major beneficial impact)	
What is English Nature's view of the project likely to be? (-2 likely to oppose without negotiation, -1 opposition negotiable with mitigation/compensation, 0 unknown, +1 agreement with minor conditions such as working methods, +2 no objections)	
Will appropriate assessment be required under the Conservation (Natural Habitats &c) Regulations 1994? (-2 yes: for majority or whole site area, -1 yes: for less than half site area, 0 not determined, +1 negotiation possible due to low conservation value, +2 highly unlikely)	
Will compensation be required under the Conservation (Natural Habitats &c) Regulations 1994? (-2 yes: for majority or whole site area, -1 yes: for less than half site area, 0 not determined, +1 negotiation possible due to low conservation value, +2 highly unlikely)	
Will the site create suitable compensation habitat? (related to site elevation) (-2 significant engineering works required to create habitat, -1 mudflat only without engineering, 0 majority mudflat with some saltmarsh, +1 the required combination of mudflat/salt marsh habitat, +2 required habitat plus transition to upper salt marsh habitats)	
Land ownership - is the landowner amenable to sale? (-2 no, -1 with conditions resulting in extra costs, 0 not approached, +1 will sell exact amount required only, +2 is prepared to sell more than required for this scheme)	
Could the site be 'landbanked' for other schemes? (-2 site too small to provide initial compensation for current scheme and has to be considered in combination with another site, -1 no, site too small to bank, 0 under discussion, +1 small additional area available, +2 large area can be banked as part of a regional strategy)	
Environmental Issues	
Existing conservation value (-2 Annex I listed European species, -1 BAP/species protected under the CRoW Act 2000, 0 unrecorded, +1 locally rare or low conservation species only, +2 no species of interest)	
Potential impact on features of interest/protected species (-2 loss of numerous species with high costs for alternative habitats or translocation, -1 some expenditure for on-site mitigation required, 0 not determined due to gaps in data, +1 some impact but mitigation not required, +2 no impacts predicted)	

Will the site improve the conservation value of the area? (-2 will result in a loss/change of more than one existing good quality habitats/species, -1 will result in loss of one desirable habitat/species without much biodiversity gain, 0 scheme not likely to have an effect, +1 biodiversity or habitat quality could be slightly improved, +2 biodiversity/habitat quality could be significantly improved)	
Is there a history of pollution or contamination on the site? (-2 known history and high remediation costs, -1 possibility of minor contamination but not a barrier to retreat, 0 site investigation required, +1 some agricultural chemicals in use, can be reduced prior to breaching, +2 no history of contamination)	
Potential landscape impact/impact on visual amenity (-2 permanent major detrimental impact (e.g. from extensive new flood defences), -1 permanent minor adverse effect with mitigation available, 0 temporary detrimental impacts during transition of habitat, +1 minor improvement on existing situation, +2 major improvement on existing situation (e.g. removal of unsightly structures prior to breaching))	
Potential impact on archaeology/cultural heritage (-2 high designation (e.g. SAM) will have to be defended from flooding, -1 archaeological sites of lesser importance at risk, with associated mitigation costs, 0 unknown archaeological interest, +1 minor archaeological sites at risk or low risk with minor mitigation costs, +2 no archaeological interests at risk)	
Technical	
Proximity of extant salt marshes as a seed supply for new site (-2 no extant salt marshes, -1 salt marshes removed from site locality, 0 not assessed, +1 limited salt marshes nearby, +2 extensive salt marshes immediately adjacent to site)	
Site history (-2 always terrestrial and evidence of former development, -1 always terrestrial and used for agriculture, 0 possibly former wetland, +1 former salt marsh reclaimed >100 yrs ago, +2 recent land reclaim)	
Evidence of relict creeks (-2 no patterns visible, -1 fluvial/agricultural drainage channel(s) present, 0 information not available, +1 limited creek network visible, +2 extensive former creek network visible)	
Sustainability	
Will the site/scheme help to return the coast/estuary to a more 'natural' profile? (-2 site was never part of intertidal, -1 negligible effect because scheme is isolated, 0 information incomplete, +1 a few other realignments planned in combination, +2 scheme is part of a strategic plan for the estuary)	
Can the site be maintained (will it roll-back) with increased sea levels? (-2 site will be completely constrained by existing topography, -1 over half of site will be constrained, 0 still under discussion, +1 site partially constrained, +2 site rises naturally to the rear, will evolve with sea level rise)	
Flood Defence Issues and Economics	
Would this increase flood risk to economic assets? (-2 extensive development at risk, -1 few properties and high grade agricultural land at risk, 0 very few assets at risk, +1 low grade agricultural land/public amenity areas only, +2 standard of flood defence improved)	
Will there be a requirement for new defences? (-2 high cost defences required to 1:200yr standard, -1 lower cost defences required to 1:100yr standard for more than half of site, 0 low cost defences required to 1:50yr standard for more than half of site, +1 1:100/1:50yr defences required for less than half of site, +2 no new defences required)	
Will these new defences constrain the realigned site? (-2 whole site will be enclosed, -1 over half of site constrained, 0 not determined, +1 less than half of site will need defences, +2 no defences required)	
Will there be an impact on flood storage capacity? (-2 reduced flood storage will entail extensive mitigation such as pumping, -1 reduced flood storage but no/few properties at risk and little mitigation required, 0 modelling required but no adverse impact envisaged, +1 no change, +2 increase in flood storage volume)	
What will be the cost of land purchase? (-2 direct competition from other sources and likely to be outpriced by competitors, -1 land owner prepared to sell for well above market value, 0 under discussion, +1 may sell for slightly above agricultural market value, +2 in line with agricultural market prices)	
Would the site be eligible for payments through agri-environment schemes? (-2 no possibility of funding resulting is loss of income, -1 further investigation, no obvious schemes available, 0 no funding required, +1 eligible for a limited amount of funding, +2 funding available from a number of schemes)	
Social/Recreational	
would there be political acceptance of the option? (-2 against local policy, -1 local political opposition, 0 no view, +1 local support, +2 in line with local policy and has local support)	
Potential impact on public access (-2 extensive permanent reduction in access, -1 minor permanent restrictions, 0 temporary restrictions during implementation only with no net improvement, +1 no change at any time, +2 access improved when fully implemented)	
Impacts on rights of way (-2 permanent closures required, -1 permanent re-routing of Right of Way necessary, 0 temporary closure/diversions only, +1 no impact on rights of way, +2 Rights of Way improved through scheme)	
Will there be an impact on health and safety? (-2 permanent detrimental impact with high expenditure for mitigation required, -1 minor permanent adverse impact with some mitigation, 0 temporary impact associated with construction, +1 no impacts envisaged, +2 improvement in health and safety)	
FINAL SCORE	
FINAL RANK	

Results

The Stage 2 mapping identified a total of 40 separate sites within the study area, covering an area of approximately 770 hectares. Of the 40 sites that were identified, 15 were rejected outright during the workshop with no investigation at all because one or more factors known to the stakeholders made the site completely unsuitable for retreat. A further 8 sites were rejected after completing the primary assessment matrix as being unsuitable due to several negative factors making them less desirable. 14 sites scored sufficiently well to be brought forward as potential areas for further investigation. 3 sites were earmarked as possible for long term investigation, although they were unsuitable for immediate retreat. Because the purpose of the investigation was to identify a suitable realignment site for a specific coastal defence scheme, only sites that were immediately suitable for managed realignment were taken forward. Following the completion of this initial stage of investigation, eight sites that were potentially suitable for short term retreat were taken forward for further investigation.

One factor that caused the rejection of some sites was anomalies in the Ordnance Survey data. In some cases, sites were identified as being of a suitable elevation when changes to the topography over previous years (usually by anthropogenic means) meant that the site was actually too high to be considered for realignment.

During the workshop, it was also noted that some sites may not be suitable for realignment because of contaminated land issues; specifically the possibility of pesticide (DDT) contamination in one area from a former factory. Contaminated land has not been widely referred to in the published literature and could be a major factor that would affect the acceptability of the site from an environmental and planning perspective. Land contamination issues were subsequently added to the secondary matrix, as they could become a potential barrier to realigning a site.

At the end of the secondary assessment exercise, three of the eight sites clearly emerged as being the most suitable for realignment. The scores for the top three sites were +15, +12 and +3 (with a possible maximum score of +54). The remaining 5 sites all had negative scores, ranging from -5 to -14. These three sites have formed a starting point from which the Environment Agency can open up discussions with landowners in relation to land purchase.

Discussion

There are a number of advantages and disadvantages to using the matrices in site selection. Firstly, the use of the matrix allows a comparatively rapid, simultaneous assessment of a relatively large number of sites. Secondly, all of the major generic issues that occur when identifying suitable sites for managed realignment to be listed together, which helps to ensure that they are not overlooked in the early stages of the assessment. The use of a fixed scoring system also allows direct comparisons between sites to be made. However, the scope of the matrix is limited in that there will always be a requirement for further, detailed work such as hydraulic modelling, in order to assess the potential impact of the retreat on the lower parts of the estuary. However, it is not justifiable to spend a large amount of time and money on modelling numerous sites, so these studies must be left until there are only a few areas to choose from. If these studies reveal that the retreat of the first-choice site will have a major adverse impact on the rest of the estuary, the site may be deemed to be unviable, and another site found. Although this is only likely to happen with much larger sites than those identified in this project, it may still be beneficial to identify a number of sites prior to carrying out detailed studies as a safety measure.

One particular difficulty in the design of the matrix is that it is very hard to categorise each issue into a simple scoring system, as the environmental and physical conditions at each site will vary considerably. Therefore the scoring system should not be viewed as completely definitive, only as a guide. Every identified site will not conform exactly to a single standard, and so some flexibility in the interpretation of the scoring system needs to be used when ranking sites.

Similarly, it has been easy to focus heavily on the importance of international designations when designing the matrix, but it is difficult to incorporate un-designated existing conservation value. For example, a walkover ecological survey carried out for some of the highest-ranking sites revealed that some of these supported saline-freshwater transitional communities, which are increasingly rare in the UK, but were not designated. It may not be environmentally beneficial to retreat into these habitats, but it is difficult to try and convey this in comparison to the importance that an international designation may have to the viability of a site.

The need to address numerous aspects of conservation legislation (existing designations, the need for appropriate assessment and compensation) led to these factors giving extra weight to the final score in the assessment. Sites without an existing designation therefore tended to score relatively well within the matrix. However, certain technical aspects of retreating the site were also identified as being of overriding importance, specifically, the need and cost of providing new flood defences, and the subsequent constraint that this would place on the future evolution of the new site. However, as this only constituted a single category, it did not lend enough weight to reduce the overall score, even though it is a potentially overriding factor to reject the site, and the final score became artificially high. Therefore, although the matrix can be used to give an indication of site viability, a certain amount of external judgement and interpretation also has to be used in final site selection.

The problem of inaccuracies in Ordnance survey data highlights the need to consult at an early stage. The results of the mapping exercise did not always match up to the local experience of coastal practitioners at the workshop, which meant that some unsuitable sites were identified as being suitable for realignment. If these errors had not been identified early on, unnecessary time and financial resources could have been spent following up unsuitable sites.

A number of gaps in current published literature were identified when reviewing the technical aspects of creating new salt marshes. Specifically, these were:

- Do pollutants affect the development and survival of salt marshes? If so, how, and can anything be done to reverse the trend?
- What part, if any, does suspended sediment concentration play in facilitating accretion and salt marsh development?

It may be useful for future research efforts regarding salt marsh creation to attempt to answer these questions, in order to assist in the future recreation of intertidal habitats.

Conclusions and Recommendations

Following the analysis of sites suitable for managed realignment, a number of specific recommendations relating to both the use of the matrix and UK policy have been identified. It is recommended that this approach is used to help identify managed realignment sites wherever possible. The matrix should be applied as a means of strategically assessing the opportunities for managed realignment, and for identifying areas where habitat creation can be realistically carried out. This would serve to speed up the site identification process and would also be of use to other local strategies, such as flood defence strategy studies; firstly by identifying areas where 'do nothing' shoreline policies could be implemented with maximum nature conservation or biodiversity benefits; and secondly by identifying areas where habitat creation can be used to compensate for unavoidable damage to internationally designated sites. Finally, identifying sites and assessing whether they are likely to be suitable for habitat creation before opening up discussions about land purchase helps to manage the expectations of landowners who are approached to sell their land, which could help to maintain good relations between coastal defence authorities and the public.

Some further minor recommendations are also made:

- One of the benefits of scoring sites during a workshop is that it takes place in a group situation, which may help to remove bias from the matrix. Scoring can sometimes be biased towards a desired outcome, and by ranking sites in a group environment there may be less chance of this happening. While this may not be practical for every scheme, it is recommended that this is done whenever possible.

- A register of site investigations should be set up in the future, both as a source of information and also in order to prevent duplication of previous work and subsequent extra expenditure of resources in re-assessing sites for each new scheme. A site register also may be useful if, for example, a site is rejected for a project-specific rather than technical reason, such as being too small for the purposes of one particular scheme. It could then be recorded on the site register and possibly used for a different scheme. Although one of the functions of the new Coastal Habitat Management Plans (CHaMPs) is to record habitat losses and gains within the Plan areas, the release of the Solent CHaMP suggests that the recording process will be related only to geomorphological processes and not to created habitats (Bray and Cottle, 2003), and therefore the CHaMP process would be unsuitable for this purpose. The most suitable place for this register to be held would be the appropriate government department responsible for coastal defence.

References

Atkinson, P.W.; Crooks, S.; Grant, A. and Rehfisch, M.M. 2001. *The success of creation and restoration schemes in producing intertidal habitats for birds. English Nature Research Reports No. 425.* Peterborough: English Nature.

Boorman, L.A. 1999. Salt Marshes – present functioning and future change. *Mangroves and Salt Marshes* **3**: 227-241.

Boorman, L.A. and Hazelden, J. 1995. Saltmarsh Creation and Management for Coastal Defence. In: Healy and Doody (eds) *Directions in European Coastal Management.* Cardigan: Samara Publishing.

Bray, M. and Cottle, R. 2003. *Solent Coastal Habitat Management Plan: Final Report.* Peterborough: Posford Haskoning.

Brooke, J.S. 1992. Coastal defence: The Retreat Option. *Journal of the Institute of Water and Environmental Management* **6**: 151-157.

Brooke, J.S.; Landin, M.; Meakins, N. and Adnitt, C. 1999. *The Restoration of Vegetation on Saltmarshes: Environment Agency Research and Development Technical Report W208.* Bristol: Environment Agency.

Burd, F. 1995. *Managed Retreat: a practical guide.* Peterborough: English Nature.

Department of the Environment, (DoE) 1994. *The Conservation (Natural habitats &c) Regulations 1994.* London: HMSO.

Dixon, A.M.; Leggett, D.J. and Weight, R.C. 1998. Habitat Creation Opportunities for Landward Coastal Re-alignment: Essex Case Studies. *Journal of the Chartered Institution of Water and Environmental Management* **12**: 107-112.

Emmerson, R.H.C.; Manatunge, J.M.A.; Macleod, C.L. and Lester, J.N. 1997. Tidal Exchanges Between Orplands Managed Retreat Site and the Blackwater Estuary, Essex. *Journal of the Chartered Institution of Water and Environmental Management* **11**: 363-372.

French, P.W. 1999. Managed retreat: a natural analogue from the Medway estuary, UK. *Ocean and Coastal Management* **42**: 49-62.

Hazelden, J. and Boorman, L.A. 2001. Soils and 'managed retreat' in South East England. *Soil Use and Management* **17**: 150-154.

Leggett, D; Bubb, J.M. and Lester, J.N. 1995. The Role of Pollutants and Sedimentary Processes in Flood Defense – A Case Study – Salt Marshes of the Essex Coast, UK. *Environmental Technology* **16(5)**: 457-466.

Ministry of Agriculture, Fisheries and Food (MAFF), no publication date. *Tollesbury Managed Realignment Experimental Site: Summary Report on Research and Other Activities, 1994-96.* London: MAFF.

Möller, I.; Spencer, T. and French, J.R. 1996. Wind Wave Attenuation over Saltmarsh Surfaces: Preliminary Results from Norfolk, England. *Journal of Coastal Research* **12(4)**: 1009-1016.

Möller, I.; Spencer, T.; French, J.R.; Leggett, D.J. and Dixon, M. 2001. The Sea-Defence Value of Salt Marshes: Field Evidence From North Norfolk. *Journal of the Chartered Institution of Water and Environmental Management* **15**: 109-117

Oldershaw, C. *The Earth in Our Hands – how geoscientists serve and protect the public. Factsheet 7: Coastal Erosion.* London: The Geological Society of London, August 2001. Factsheets available from: http://www.geolsoc.org.uk Site accessed 17/04/02.

Padinha, C.; Santos, R. and Brown, M.T. 2000. Evaluating environmental contamination in Ria Formosa (Portugal) using stress indexes of *Spartina maritima*. *Marine Environmental Research* **49(1)**: 67-78.

Pethick, J. 1984. *An Introduction to Coastal Geomorphology*. London: Edward Arnold.

Reed, D.J. 1995. The response of coastal marshes to sea level rise: Survival or submergence? *Earth Surface Processes and Landforms* **20**: 39-48.

Reed, D.J.; Spencer, T.; Murray, A.; French, J.R. and Leonard, L. 1999. Marsh surface sediment deposition and the role of tidal creeks: Implications for created and managed coastal marshes. *Journal of Coastal Conservation* **5**: 81-90.

UK Biodiversity Group, Site updated 2001. *UK Biodiversity - Rio and beyond*. http://www.ukbap.org.uk/aboutBAP.htm Site accessed 14/06/02.

UK Biodiversity Group 1999. *UK Biodiversity Group Tranche 2 Action Plans - Volume V: Maritime species and habitats*. Peterborough: English Nature.

Williams, T.P.; Bubb, J.M. and Lester, J.N. 1994. Metal accumulation within salt-marsh environments – A review. *Marine Pollution Bulletin* **28(5)**: 277-290.

Coastal development and tourism: resolving conflicts of approach- two case studies

Clon Ulrick, Associate, Arup, London, UK

Introduction

There are many conflicting pressures on new developments in coastal and estuarine areas.

For:

- Recreational and tourism opportunities which are unique to the coastline
- Regeneration: An urgent need to 'reinvent' tired coastal towns with declining infrastructure and social deprivation
- The special 'sense of place' that a waterside location provides

Against:

- Planning principles, including a presumption against coastal development that is not required to be on the coast.
- The complexity and risks for marine consents and licences
- Environmental quality and sensitivity- the Birds and Habitats Directives, and the precautionary principle are all inhibiting to coastal development
- Major engineering problems of construction in the sea

This paper demonstrates how these conflicting pressures can affect project delivery, through a case study of two striking new public buildings built into the sea. These are both projects with which the author's team (the maritime engineering team of Arup) has been involved:

- **The Turner Centre, Margate**
- **The National Maritime Museum Cornwall, Falmouth**

This paper represents a maritime engineer's perspective of the broader issues that have affected these projects and which could affect similar structures in the sea.

Acknowledgements

The author acknowledges with thanks the images of the Turner Centre by Snohetta architects; images of the Maritime Museum by Long and Kentish and Peter Durrent/arcblue; and images of Turner's paintings from Tate Britain.

Description of projects

The Turner Centre, Margate

Client:	Kent County Council
Architect:	Snohetta AS and Spence Associates
Project Managers & cost consultants:	Davis Langdon & Everest
Structural Engineers:	Jane Wernick Associates
Service Engineers	BDSP
Maritime Engineers:	Arup
Capital cost (excluding fit out)	£12m (estimated)

Figure 1. The Turner Centre, Margate (Architect's impression)

The Turner Centre (see Figure 1) will be a gallery principally for contemporary art that reflects the spirit of JMW Turner and will also occasionally exhibit some Turner paintings. There are also workshop areas and ancillary facilities such as a bookshop, café/restaurant, and offices. The centre comprises two separate buildings connected by a bridge. The gallery building is on the inter-tidal foreshore immediately outside Margate Harbour. It is a 30m high organically shaped building with gallery space on three levels and storage & plant space below. The other building is located on Margate pier. This pier building is by contrast a double storey rectilinear structure. The bridge is on two levels for delivery of art at the lower level and visitor access at the upper level. The project has received planning permission and a Section 106 Agreement has been signed. The centre is planned to open in 2007.

The National Maritime Museum Cornwall

Client:	South West of England Regional Development Agency
Architects:	Long and Kentish
Project Managers:	Arup
Structural, Service and Maritime Engineers:	Arup
Cost Consultants	Davis Langdon & Everest
Capital cost (including fit-out)	£28m

Figure 2. National Maritime Museum Cornwall

The Maritime Museum (see Figure 2) comprises exhibition space including historic vessels, interactive nautical activities, workshops, a library and an underwater viewing gallery. In addition to the gallery and museum there is a separate block of offices and shops, the buildings together gathered around a public events square. The museum was constructed partially on existing reclaimed ground but extends out beyond the previous sea wall into Falmouth Harbour. It was completed in 2002 and fully opened to the public in spring 2003.

Architectural Concepts

Snohetta and Spence's concept for the Turner Centre gallery was that the artworks should be exhibited in a largely formless space, without distracting frames of reference, in the style of many of Turner's most famous paintings. See Figure 3. The external shape of the gallery building suggests, perhaps, an upturned ship or a seashell.

Figure 3. JMW Turner: Storm with Wreckage

Long and Kentish's concept for the Maritime Museum was to use the basic forms and materials of traditional Cornish buildings (particularly granite, slate and oak), but in a forward-looking design. The main gallery in particular, with boats suspended in mid air is a very large and striking space (Figure 4) and the Tidal Gallery provides a submarine experience. It has won critical acclaim from the Royal Fine Art Commission, among others, and has been awarded the Royal Town Planning Institute South West Awards Overall Winner 2002.

Figure 4 The Maritime Museum: boat gallery

Architects for both buildings were appointed through design competitions. This procurement route indicates both the importance and uniqueness of both projects and the developers' commitment to high quality and adventurous designs.

Both buildings are consciously landmark buildings designed to attract attention and transform perceptions of the area. The Turner Centre was placed in the sea specifically so that it would be highly visible and would stand out from the clutter of Margate's urban skyline. The Maritime Museum has a tower that projects vertically as a beacon and viewing platform and it also projects out into the harbour.

The buildings differ fundamentally in their relationship with the sea. The Turner Centre gallery is entirely inward looking. There are no windows in the gallery building walls (except, possibly, a few small portholes). There is only a single skylight (see Figure 5). By contrast, the Maritime Museum includes semi-open areas facing seawards and a watered area within the footprint of the building, creating a feeling that the sea surrounds and inundates the building (Figure 6)

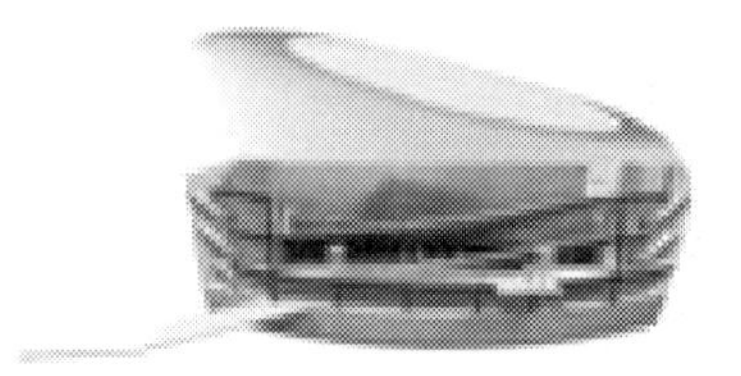

Figure 5. Turner Centre: cut away section through gallery building

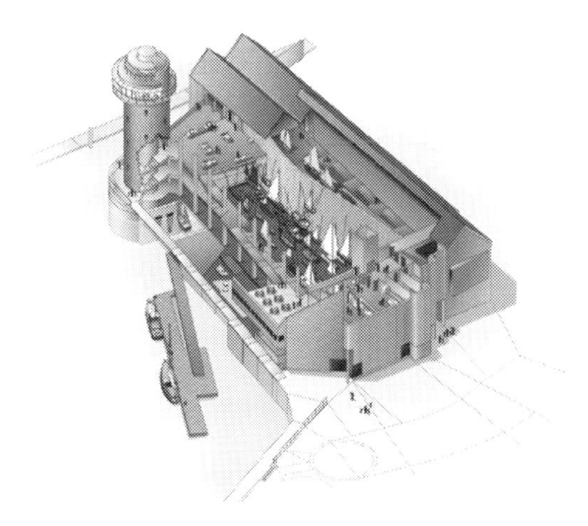

Figure 6. The Maritime Museum cut away section

Location

Why Margate?

Turner attended school in Margate and as an adult lived for many years in lodgings adjacent to Margate harbour overlooking the sea. He was attracted to Margate not only by his landlady, who was also his lover, but also by the quality of light across the water here. He sketched and painted Margate frequently (see Figure 7). Margate still has the quality of light, lovely sandy beaches and an interesting old town, but has yet to capitalise on its connection with Turner.

Figure 7. Engraving of JMW Turner's painting; 'Margate'

Why Falmouth?

Falmouth has long nautical heritage and what is claimed to be the third largest natural harbour in the world. It is a popular boating and tourist destination. Separate collections of historic boats from the existing Cornwall Maritime Museum in Falmouth and the National Maritime Museum Greenwich were brought together in the new building on the available prime waterfront site.

Why in the sea?

The relationship of both projects to the sea is indicated in Figures 8 and 9. The Turner Centre has no functional requirement to be in the sea. It is there to provide a dramatic setting to the gallery and a highly prominent landmark for Margate. Conversely, the Maritime Museum has excellent thematic reasons to be at the waters edge. Furthermore, the underwater gallery and the desire for visitors to feel a sense of interaction with the sea arguably provide a good justification for a limited encroachment into the harbour.

Figure 8. Turner Centre gallery location in the sea outside Margate pier

Figure 9. Maritime Museum location in Falmouth Harbour

Strategic Socio-Economic Issues

The background to both projects is the decline and evolution of seaside towns in the UK: The English Tourist Council has identified problems for seaside towns as including:

- changing social trends such as increasing holidays abroad rather than in the UK;
- lack of investment in the public infrastructure and accommodation;
- a poor image;
- marketing weaknesses;
- access problems.

Neither Margate nor Falmouth is immune to these problems, and both towns have surprisingly high levels of urban deprivation. For example, in 1996 the Isle of Thanet had the second highest unemployment rate in England and is currently ranked 60th in the government's Indices of Multiple Deprivation. Both projects were driven by a desire to break out of this bleak mould. Both projects are intended to be iconic structures that will change the image of the towns and be catalysts for urban regeneration. Both projects are intended to be major visitor destinations and to extend the tourist season into the winter months.

Planning issues

The planning policy context is multi-layer, from European Community Directives, through Planning Policy Guidance to Local Plans. There is a significant level of contradiction within and between the various policies.

Of particular relevance to both projects is the government's Planning Policy Guidance for coastal planning (PPG 20). PPG 20 generally adopts a precautionary principle restricting coastal development. Key guidance militating against such development include:

- Conservation: "Local authorities should recognise the need to protect remaining areas of conservation value within the developed coast" (para. 2.6)
- Development: "Development plan policies should normally not provide for development which does not require a coastal location" (para.2.9).
- Risks: "new development should not generally be permitted in areas which would need expensive engineering works, either to protect developments on land subject to erosion by the sea or to defend land which might be inundated by the sea (para.2.13). This restriction is reinforced in PPG 25: Development and Flood Risk.
- "Before major developments are permitted it will be essential to demonstrate that a coastal location is required" (para. 3.10)

However, there is also guidance promoting these types of developments:

- "Policies and proposals may include action....to regenerate rundown coastal towns and ports" (para. 2.22)
- Regeneration plans "should seek to improve the attraction of such areas as resorts and to regenerate harbour/port areas...Tourism developments are likely to play an important role in such regeneration initiatives" (para. 2.24)

The Turner Centre in particular was potentially vulnerable to the view that the structure did not need to be in the sea, because the competition had originally specified that the Turner Centre should be located on an available site located just north of the harbour and including the pier itself. However, when the planners considered both projects the strong regeneration potential and the quality of both projects were seen as overriding considerations and both obtained planning permission.

Marine Consents & Licences

For the Turner Centre, it is intended to seek both a Harbour Revision Order and Transport and Works Act order. The Harbour Revision Order will tidy up the powers of the local authority and do not relate specifically to the Turner Centre project. It was felt that Section 3 of the Transport and Works Act was applicable because of the potential impact of the structure on navigation. A problem with Section 3 of the Act is that it applies to specified forms of development, such as reclamation and wind farms, but there is no category for buildings. (Presumably, this type of structure was not envisaged when the legislation was drafted). The Act has already been recently changed by Parliament to include the category of 'viewing platforms' specifically for the Spinnaker Tower in Plymouth. At the time of writing, there appears to be a real possibility that it may be necessary to change the law again to allow the Turner Centre to be constructed.

A particular risk associated with Harbour Revision Orders and Transport and Works Act orders is that a single objector can severely delay the project and even possibly force a decision to go to Public Inquiry. Considerable pro-active consultation is planned, over and above that already carried out for planning, in order to minimise this risk.

By contrast, neither a Harbour Revision Order nor Transport and Works Act Order were obtained for the Maritime Museum. The development relied on the FEPA licence and Coast Protection Act consent.

Environmental issues

The beach on which the Turner Centre gallery is sited is designated as a Site of Special Scientific Interest (SSSI), a Ramsar site, a Special Protection Area (SPA) under the Birds Directive and candidate Special Area of Conservation (cSAC) under the Habitats Directive. This status is principally because of turnstones over-wintering on the beach and micro-organisms on the chalk reef nearby (see Figure 10).

Figure 10. Chalk reef adjacent to the proposed Turner Centre

To compensate for the permanent loss of inter-tidal beach lost to the turnstones, a purpose built turnstone platform is to be constructed inside Margate Harbour. This platform will be screened to reduce disturbance to the birds (a significant problem at the moment is that the birds have few places to roost where they will not be disturbed). A planning condition is that a bird warden will monitor the impact of construction activities on the birds.

To avoid damage to the chalk reef during construction, vessels and anchors will keep clear of the reef and construction methods must ensure that the sand is not disturbed into suspension in the water column, which might then smother the micro-organisms. Fortunately, the net littoral drift is from the reef towards the Turner Centre.

An environmental impact assessment was required for the Turner Centre. The Environmental Statement concluded that the impact on the Turnstones and chalk reef would not be significant if the proposed mitigation measures were adopted.

Falmouth Harbour was also a candidate SAC. The main environmental concern for the Maritime Museum was that the seabed is contaminated with Tributyl Tin (TBT) arising from ship building and repair in the harbour over the years. Construction in the harbour raised significant risks of disturbing TBT into the water column. Fortunately, the extended footprint of the building was mainly over some low-level sea defence revetments that had been constructed in the recent past. These revetments effectively contained any contaminants in the sea floor. The new sea wall was constructed by piling through the revetments, which were left in place and eventually covered over (see Figure 11). A further consideration was that the Environment Agency would not accept the contamination risk of wet concrete being poured in the inter-tidal zone and consequently the sea wall was constructed in driven steel piles.

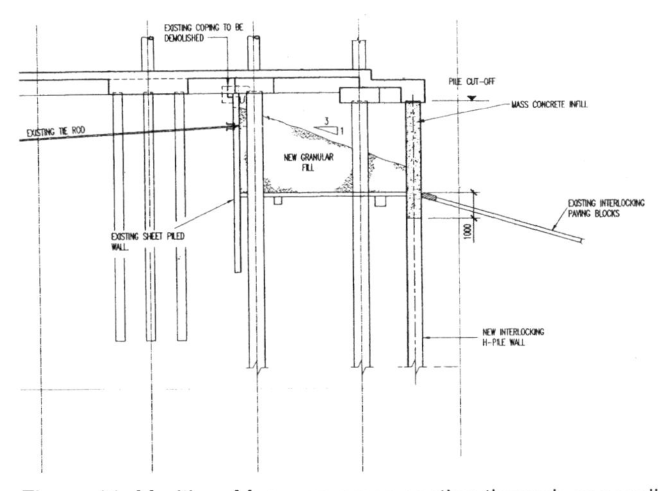

Figure 11. Maritime Museum: cross section through sea wall

Maritime- related engineering and construction issues

Sea Conditions

The Turner Centre is very much more at risk from sea conditions compared to the Maritime Museum. While the Maritime Museum is in a natural harbour, the Turner Centre gallery building is in open sea stretching northeast to Norway. Historically, severe storms have damaged the pier and several buildings on the sea front at Margate (see Figure 12).

Figure 12. Margate pier in a storm, 2001

HR Wallingford carried out a desk study and computational modelling of sea conditions at both sites. The wave loads on the Turner Centre are substantial and it is intended to construct a three dimensional physical hydraulic model to confirm these loads. The model will also confirm the areas where scour and deposition may occur. This will hopefully reassure consultees that the gallery building will not lead to deposition of sand on the chalk reef, and will also confirm the extent of toe protection to the pier required to protect against any increased scour. The Maritime Museum was intended to be open to the sea, and it was essential to confirm through modelling that this concept was feasible. The modelling also confirmed that reflected wave energy from the new vertical sea wall would not increase wave heights unacceptably for boats moored in the marina just in front of the building.

Overtopping

At both sites, overtopping was a major issue. Architects on both schemes had to limit the extent or location of glazing facing seawards, or provide storm shutters. Overtopping can also restrict safe access. While the Turner Centre gallery is an enclosed shell that can safely be occupied during the fiercest storms, it will not be safe to walk down the pier in severe conditions unless a substantial canopy shelter is provided. This would have a significant impact on the listed pier and the listed structure at the root of the pier. At the time of writing, the strategy for dealing with overtopping at the Turner Centre has not been resolved.

Disruption to construction

At the Turner Centre, with the site facing NE, the probability of significant waves disrupting floating construction plant is almost as high during summer months as in winter. It is probable that extensive temporary platforms will be required to reduce the reliance on floating plant. At the Maritime Museum this was much less of an issue. The water here is much more sheltered, and most of the marine works were carried out by cranes operating from the adjacent land.

Ship Impact

Any building in the sea is at risk from accidental ship impact from a large vessel. This has a low probability of occurrence but a high consequence. The basic answer for both buildings is to provide a robust structure that can sustain localised damage in an extreme event without catastrophic consequences. At the Turner Centre, the main risk is from coastal vessels that anchor offshore, rather than the small vessels that use the tidal harbour. A numerical risk assessment will quantify this risk so that the design loads can be quantified. The Maritime

Museum is also close to a working port with coastal vessels, although screened by the marina that surrounds it. A solid sheetpile wall with backfill was much more robust against ship impact that a suspended deck on tubular piles and this reinforced the preference for the selected design.

Interference to Navigation

Interference to navigation is a particular concern at construction stage. There are a small number of fishermen, a yacht club and the RNLI at Margate. It is particularly important that the two slipways are unobstructed to allow RNLI access to the sea at any time. For the Maritime Museum the main concerns were to avoid obstructing traffic to the marina pontoons adjacent to the museum and to avoid damage to the moored yachts. This required ongoing dialogue with vessel users throughout construction to minimise and deal with complaints, and appears to have been relatively successful.

Design life

The design life for the Turner Centre is 60 years and conventional dense concrete with high cement content, low water/cement ratio and generous cover will ensure that the design life is achieved.

The design life for the National Maritime Museum is 100 years. This was a major challenge for the steel piles for the new sea wall supporting the building. There was also the risk of future Accelerated Low Water Corrosion attacking the wall and undermining the building. An innovative technique was selected using H piles infilled with concrete (see Figures 13 and 14). At the end of the design life the exposed outer flanges of the piles will have completely corroded away leaving the web and inner flange protected by the concrete. These alone are suitable for the design loads. Further details of this design are described in a separate paper (Ref. 1).

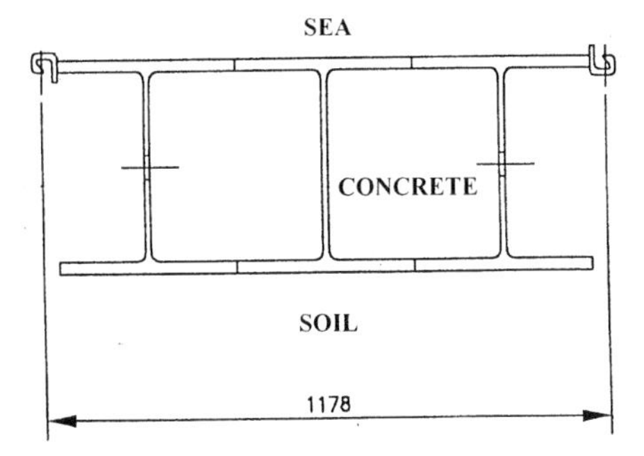

Figure 13. Plan on piles for Maritime Museum low maintenance sea wall

Figure 14. Maritime Museum low maintenance piled sea wall (elevation)

The Maritime Museum is clad in timber and it is intended that the Turner Centre gallery will be similarly clad in timber. Timber has a long tradition of use in seaside environments and the weathering of the timber will add to the visual interest. However, the design life for the timber cladding will clearly not match the structural frame.

Risk in the marine environment

There can be significant differences between actual risk and perceptions of risk, and this is particularly evident for structures in the sea.

Ship impact did not appear to be perceived as a significant risk by third parties, although it was nevertheless necessary to consider this for design purposes. There is a danger that this perception could suddenly change after a major incident.

Conversely, the perception that the building could be washed away was a major issue for the Turner Centre. This was understandable given the exposed site, the previous storm damage to buildings in the area and the fact that the value of an exhibition of Turner paintings in the Gallery would easily exceed the value of the building. This real concern has been dealt with in two very different ways.

- Firstly, the design criteria. Initial recommendations were that wave criteria should be a one in 50 year event for serviceability limit state and one in 1,000 year event for ultimate limit state, similar to the criteria that might be used for a breakwater. However a one in 1,000 year event has approximately a six percent probability of exceedence over a 60-year design life. Is this an acceptable risk, given the special circumstances of a building such as this? It is under consideration to increase the design criteria to up to one in 100 and one in 10,000 year events respectively for particular failure scenarios.
- Secondly, Snohetta dealt very effectively with the *perception* of risk by confronting this concern head on. An image of the building in stormy seas (see Figure 15) was presented in public consultation. The audience was reminded of the story of JMW Turner strapping himself to the mast of a ship in a storm so that he could paint the event from his actual experience. This image helped to convert a fear into a sense of excitement.

Figure 15. Architect's image of the Turner Centre in a storm

Perhaps the major maritime-related risk to the Turner Centre is the Transport and Works Act order. The possible need to change the Act and possible objections to the application or an Inquiry are all significant risks to programme.

Other important risks for both projects relate to obtaining funding, the business case, planning consent and other construction- related risks that are very challenging but outside the scope of this paper.

Conclusions

In the author's personal view:

- Both projects rely on their location in the sea to be truly landmark structures that can transform the perception of their respective urban environments.
- Both projects capitalise on their location in the sea in very different ways. There are common themes in this appraisal of the two projects, yet each project has unique issues arising from being in the sea that require bespoke solutions.
- There are many planning rules for coastal developments that are somewhat contradictory. The presumption in PPG 20 against any development that is not required to be on the coast is clearly sensible. However, both of these projects demonstrate that a high quality scheme can 'break the rules' with respect to planning on the coast.
- Similarly, these projects have shown that through careful design and mitigation measures, it is possible to build in environmentally sensitive coastal areas without significant impacts.
- 'Caveat Emptor'- the maritime location of these projects add additional risks. The maritime-related risks for the Turner Centre are significantly higher than for the Maritime Museum. However- these risks are manageable, given:
  - that the fundamental concept for the project is so strong that there is a will at all levels to make the project work
  - an experienced design team who can identify and mitigate at an early stage the particular risks arising from construction in the sea
  - an emphasis on risk management throughout the development of the project
- Both projects are examples of how innovative buildings in the sea can provide striking changes to the urban landscape and significant benefits to their respective coastal communities.

References

1. IJ Gunn and GWR Haigh "A low maintenance seawall for National Maritime Museum Cornwall" Proceedings of Institution of Civil Engineers conference on 'Breakwaters, coastal structures and coastlines' September 2001.

Stable bay theory and integrated coastal development: A case study

Dominic Reeve, School of Civil Engineering, University of Nottingham, Nottingham, UK.
Helen Bovey, School of Civil Engineering, University of Nottingham, Nottingham, UK.
Greg Guthrie, Coastal & Rivers Division, Posford Haskoning, Peterborough, UK
Chris Budzynski, Berwick-upon-Tweed Borough Council, Berwick-upon-Tweed, UK

Introduction

Beaches are the most natural form of coastal defence (Hsu and Evans, 1989) and so it makes sense to include them as an integral part of coastal development and management. As will be described in more detail in the following section, bays form a crenulated shape when they have reached static equilibrium, and this can be exploited by constructing artificial headlands, between which a crenulated shape bay will form. Headland control, which is based on the parabolic bay theory, has so far not been very popular as a means of protecting the coast. One reason is the uncertainty of the bay stability when there is no single predominant wave direction. This paper investigates the application of the parabolic bay theory to a site in the UK, and the uncertainty associated with variations in the key parameters that are required by the theory.

Stable bay concept

Littoral drift or longshore transport occurs as waves carry sediment along a coastline. If one looks at natural coastlines, it can be seen that bays are orientated so that they are transverse to the direction of approach of the persistent swell waves (Silvester and Hsu 1997). Observing the orientation of the shoreline therefore indicates the direction of net sediment movement. However, as noted by Chadwick & Morfett (1998), neighbouring bays may not necessarily be similarly aligned, as local refraction effects can alter the direction of incoming wave crests significantly.

Littoral drift will cause beach accretion or depletion when the supply of material between two points along a coast is not balanced. This may be due to a breakwater or groyne which has cut off the supply of material to down-drift parts of the coast, and so could eventually lead to part of the coast reaching static equilibrium. Static equilibrium is reached once there is no further input or removal of sediment to the section of coast, which is possible when for example; a deep channel is dredged across the shelf upcoast (Hsu and Evans 1989). As there is no further supply of material to the beach, static equilibrium means that the indentation of the coastline has reached a limit which is in equilibrium with the incoming wave climate, the littoral drift is zero and no further changes in shoreline position occur.

Coastal Management 2003, Thomas Telford, London, 2003.

The plan shape of bays depends on the orientation of the coastline to the predominant wave direction, with the down-drift end of the bay tending to have a straight coastline parallel to the predominant inshore waves, whilst the up-drift end curves sharply, shaped by the diffraction of waves around a headland, (Muir Wood and Fleming, 1981).

Two equations have been suggested over the past 30 years, in order to describe this shape that beaches form once they have reached static equilibrium (Silvester and Hsu 1997): a logarithmic spiral and a parabolic curve. The logarithmic spiral formula has several weaknesses however. The main problem according to Hsu and Evans (1989), is that the spiral shape produced is not true to the downcoast part of the bay. The spirals did not fit the beach shapes if the spiral's centre was on an upcoast headland point, where diffraction had taken place. This is more noticeable for typical, small values of β. It is also a difficult method to use, as many templates must be used for checking.

Parabolic bay theory

The logarithmic spiral has since been dropped in preference to a parabolic equation. The idea of describing a bay shape as parabola was first suggested by Mashima in Japan (Silvester and Hsu 1997). A curved coastline section, but not the headlands and points of diffraction upcoast, could be decribed by the parabola

$$y = px^2 - b \qquad \text{Equation (1)}$$

Since further examination of data from both models and equilibrium prototype bays, opinions have changed again. The change has been to a method that involves construction of a control line from the point of diffraction to the downcoast limit of the beach. This arc has a radius R_0 and is formed at an angle β to the wave crest line. The point of diffraction will be the lee of the upcoast headland. It may be part of the mainland, but Hsu and Evans (1989) identify other possibilities such as a rocky island or a reef. The wave crest line is normal to the wave orthogonal but parallel to the downcoast tangent of the stable bay and forms an angle θ with the radii. The tangential section downcoast being parallel to the wave crests in a bay occurs when the bay has reached static equilibrium, and in this condition the incoming waves will refract and diffract into the bay, breaking simultaneously along the whole periphery (Hsu et al. 1989a).

This equation relates the ratio of arc lengths (other R) to R_0 with β and the corresponding angles between the various other radii and wave crest line θ.

$$R / R_0 = C_0 + C_1(\beta / \theta) + C_2(\beta / \theta)^2 \qquad \text{Equation (2)}$$

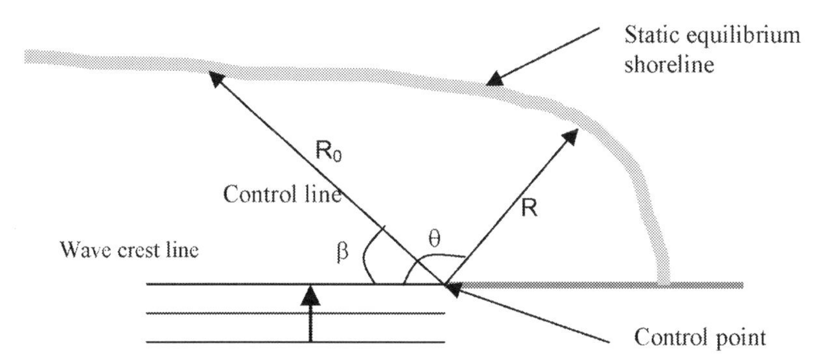

Figure 1: Definition sketch of variables employed in parabolic solution of bays (adapted from Hsu and Evans 1989)

It was found that the order of this equation was sufficient in order to predict bay shapes that matched real bays very well for the complete periphery (Silvester and Hsu 1997). The coefficients C_0, C_1 and C_2 are defined for β values of between 0^0 and 80^0, after having been extrapolated from the original test data of between 22^0 and 75^0. The values of these coefficients vary uniformly with β.

This method of the parabolic equation does not take into account other variables such as beach profile and characteristics of the waves (wave period, height etc.). However, Tan & Chiew (1994) undertook a detailed investigation of the importance of these parameters and concluded that the most important parameter is β. They gave the following explanation for this observation: the shape and size of a bay in the lee of a breakwater that has reached static equilibrium is the result of the redistribution of wave energy in this location and the reaction of the beach material to it. If the wave direction remains constant, but there is an increase or decrease in the size of the waves, a larger or smaller bay will be formed, respectively. However the pattern of wave energy redistribution on the beach in both the larger and smaller bays are similar. This means that both bays will have similar geometric shapes.

Application

The concept of static equilibrium can be of practical use in coastal defence. The idea is to construct artificial headlands along a coastline so that bays are created or to maintain or change the shape of existing bays by reinforcing or extending the natural headlands. If conditions are such that static equilibrium can be reached the beaches will evolve towards this equilibrium and form the shape predicted by the above parabolic equation.

The longshore and offshore positioning of new headlands can be designed so as to prevent large areas of erosion. For example, a large bay could have two new headlands added at intervals along the bay, so as to create three crenulate shaped beaches, which would bring the line of static equilibrium seawards, and so the retention of more land. Another advantage of headland control, as discussed by Hsu et al. (1989b), is that the shape and orientation of headlands can be chosen to reduce the cost of artificial headland structures by designing the equilibrium shape so that beaches form in front of the structures.

As with all coastal constructions a possible problem could be the erosion downcoast of the scheme due to accretion in the new bays, (see eg. Kamphuis 2000). During the evolutionary phase supplementary beach nourishment may be necessary until the beach adopts a configuration that is close to its equilibrium shape.

Project site

Beadnell lies on the northeast coast of England. The study area ran from the rocky outcrop of Comley Carr, a little further upcoast from Beadnell Harbour, to Snook Point, the southerly headland of Beadnell Bay. The harbour influences the shape of the sandy bay, which is sensitive to any changes due to its soft dune backing. The harbour piers and some other buildings are listed structures and are important for the area's tourism. The harbour itself is used by the local fishing fleet. The dunes are used for recreation and behind them, in the northerly part of the bay, are a caravan park and housing on low-lying land.

The beaches to the south of the harbour are partially protected by the rocky outcrops know as Burn Carrs. Salients in the low water line behind the rocks are evident, indicating that Burn Carrs are acting in a manner similar to detached breakwaters. South of the Burn Carrs the Long Nanny Burn cuts across the beach to outfall in Beadnell Bay. Running south the beach is backed by sandy dunes until the rocky outcrops at Snook are reached, and then lead on to Snook Point.

Coastal defences were already in place where required to the north of the harbour, and so the main purpose of the study was to examine the economic damage that could occur if a "Do Nothing" policy was adopted and therefore the harbour was abandoned. A Shoreline Management Plan incorporating Beadnell Bay had shown that without the harbour there would be significant reorientation of the dunes, resulting in the loss of harbours and the possibility of flooding to Beadnell village.

In Autumn 1997 the harbour wall at Beadnell, near Berwick-upon-Tweed, was breached during a severe storm and further damage occurred during severe storms in 1998. Berwick-upon-Tweed Borough Council, the Coastal Protection Authority of the Beadnell area commissioned Posford Haskoning to carry out a strategy study and project appraisal of the Beadnell area. This was completed in 1999. The study investigated the importance of the harbour as a coastal defence, as it was felt that the harbour provided protection to the dunes in the bay area.

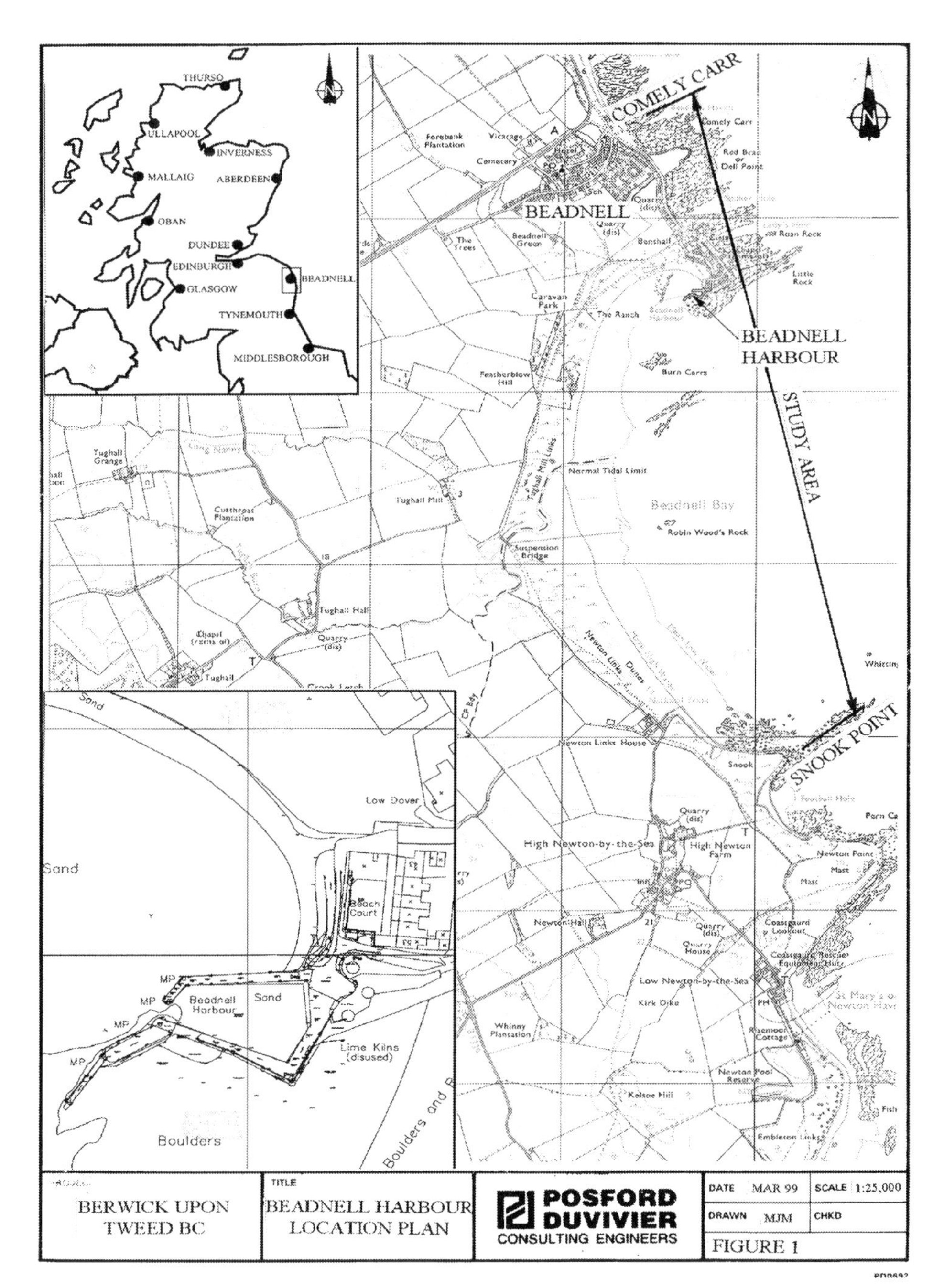

Figure 2: Site map of Beadnell Bay and surroundings

The project work involved mathematical modelling, including the prediction of the long-term beach response to possible changes in the harbour structure. Parabolic bay theory was used to examine the effect of different options on the crenulated shape of the beach. This confirmed the importance of the harbour in maintaining the line of the bay. The study showed that "without the harbour, the northern section of Beadnell Bay will, within approximately five years, retreat in places by up to 50 to 80m." The consequential damage: the loss of 51 residential properties and estimated total discounted damages over a period of fifty years of nearly £4 million would have been unacceptable.

The options looked at included doing nothing, only carrying out maintenance, rebuilding the existing harbour wall, building a new harbour wall and allowing the harbour structures to fail, whilst constructing a new breakwater or providing a rock revetment along the dune frontage. After careful consideration and examination of these possibilities, it was recommended that rebuilding the outer harbour breakwater by carrying out substantial restoration of the harbour structures would be the best solution, bringing great benefits, both economically and for safeguarding the environment.

Wave climate

The original study established the annual wave climate for the Beadnell area. The most frequently occurring waves, in the nearshore zone were from the North East. However, due to the orientation of the coastline and the position of the harbour (this is the only west opening harbour on the east coast), the harbour protects the bay from waves not merely from this North East direction but also from a wide spread of other directions. With the harbour in place, the dominant wave energy impinging on the northern section of the bay comes from the East.

Water levels

Tide levels for North Sunderland were taken from the Shoreline Management Plan and are shown in Table 1. At North Sunderland, Chart Datum is 2.4m below Ordnance Datum Newlyn (ODN). The extreme water levels are shown in Table 2.

Table 1. Tide levels at North Sunderland (Posford Duvivier, 1998, St. Abb's Head to River Tyne Shoreline Management Plan)

Condition	Level (m ODN)
Mean High Water Springs	+2.4m
Mean High Water Neaps	+1.3m
Mean Sea Level	0.0m
Mean Low Water Neaps	-0.8m
Mean Low Water Springs	-1.7m

Table 2. Extreme Still Water Levels at Berwick upon Tweed (Posford Duvivier, 1998)

Return Period (years)	Level (m ODN)
1	+2.84m
10	+3.04m
50	+3.30m
100	+3.38m
500	+3.49m
1000	+3.54m

Bay morphology

Beadnell Bay is in a state close to equilibrium, though there is some transfer of sediment between sections of the bay due to the variability in within the wave climate. A future change in the direction of the net wave energy could cause a realignment of the shore. Future management strategy, therefore, recommended that the bay's behaviour should be monitored.

Application of stable bay concepts

It was always recognised that in developing a strategy for the frontage and in evaluating options, applying the concept of stable bays would have to be based on some quite broad assumptions because of the existing limitations associated with the theory. This would lead in turn to uncertainty in results and a risk approach had therefore, to be taken. This was addressed through sensitivity analysis.

In particular, the key uncertainties arose from:

- Limitations of theory relating to tidal range and to the relative level of the assumed headlands;
- The difficulties in defining precisely where headlands should be fixed, taking into account the shape of the harbour in relation to the coast to the north and recognising the problems of defining a clear down drift headland;
- The broad spread of incident energies around the net energy direction determined from available records;

As part of the project the relative difference between the bay shapes assuming a with- and without-harbour condition was analysed. Two points were analysed: one at the end of the harbour's breakwater and the other adjacent to the lime kilns. A wave direction of 60^0N was used for both points. This was tested, in the case of the existing harbour acting as the headland against the actual shape of the bay, and gave a good correlation between the existing bay shape and that predicted by the equilibrium theory.

Further work has since been undertaken at the University of Nottingham. This considered other critical variables including: the choice of the downcoast point; the specification of dominant wave direction, and the position of the upcoast headland point for the case that the harbour was removed and the wave direction was so that the point of diffraction was further round the coast to the North, in the direction of Little Rock. The following sections cover these issues.

Upcoast headland position

Results of three cases are shown in Figure 3: the current situation at MHWS; the current situation at low water; the current wave climate and MHWS with the harbour removed.

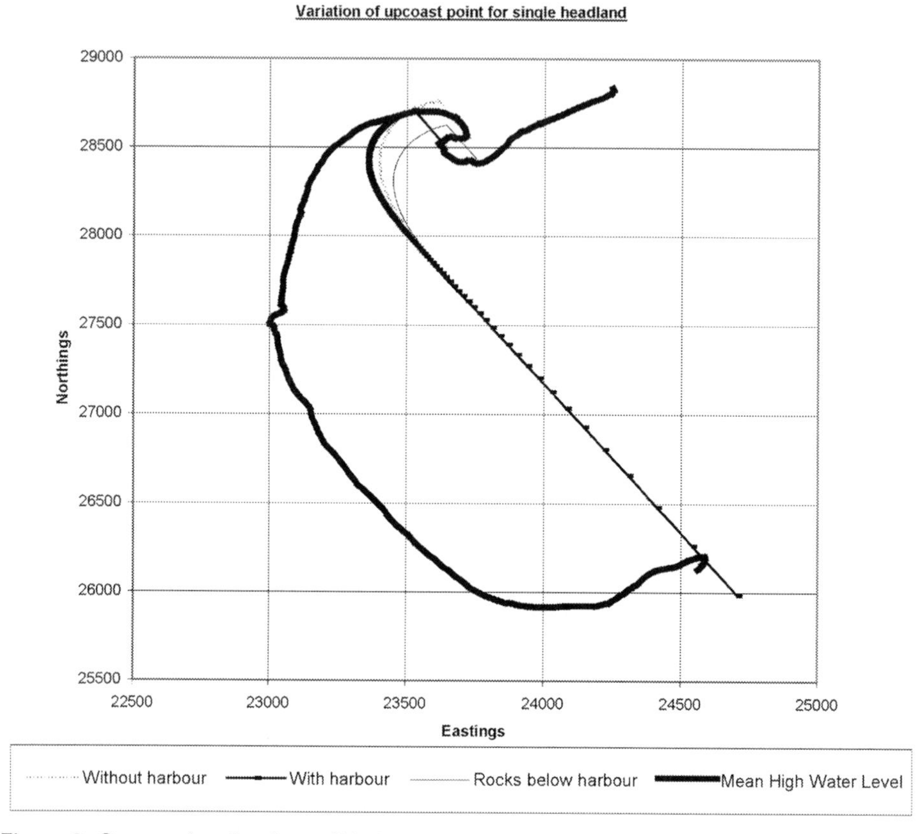

Figure 3. Curves showing the validation against the existing coastline, the result for low water and result for mid-tide level with the harbour removed.

In applying the static equilibrium method there often appears to be little reason to question the choice of the upcoast point. However, at a tidal site consideration needs to be taken of the effects of changing water level. This is particularly so in a case such as Beadnell Bay where there is steep relief, and relatively small changes in water level can lead to significant changes in the shoreline topography. The upcoast point for the without harbour scenario is placed on the tip of the harbour, situated on the 5m ODN contour. At low water the rock to the southeast of the harbour emerges and is the natural choice for upcoast point for the calculations. The choice of upcoast point can have a noticeable effect, particularly in the immediate lee of the headland, which is significant in this case.

Choice of downcoast point

Hsu et al (1989a) suggest taking the downcoast point as the point where the beach is normal to the orthogonal of the predominant wave direction. Even so, in general there can be difficulties in identifying this point, as noted by Silvester et al (1980). The most appropriate downcoast point may not necessarily be the main downcoast headland, even though this may be implied from some existing literature.

Figure 4 shows the results of taking the downcoast point as either the original point used for calibration or half way down the bay with the harbour in place. Both curves were calculated using a wave direction of 95^0. Also shown is the mean high water line.

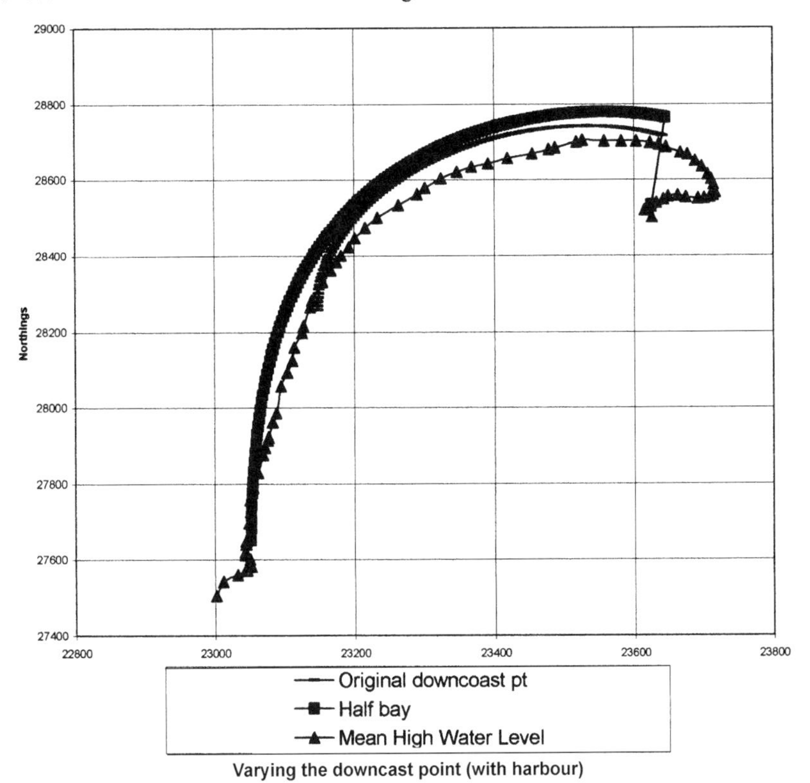

Figure 4. Effect of choice of downcoast point.

The point close to the midpoint of the bay corresponds to the position that arises from applying the method proposed by Gonzalez & Medina (2001). In this case the original downcoast point provides a better fit with the existing shoreline close to the headland while the alternate downcoast point provides a better broadscale fit.

Predominant wave direction

The static equilibrium bay shape is strongly dependent on wave direction, as might be surmised from the extremely sensitive dependence of littoral drift on wave angle. Tests were carried out for several wave directions and different downcoast points. The reasoning behind this was to provide as much information as possible, particularly as it was thought that the sensitivity could be quite different for the area of bay examined (determined by the choice of the downcoast point). Figure 5 shows the results for the original downcoast point. Figure 5 suggests that the current bay is not quite in equilibrium but is relatively insensitive to the changes in wave angle of about 10^0. Some erosive tendency in the lee of the harbour could be expected. Figure 6 shows the corresponding curves for the mid-bay downcoast point. These results suggest the bay is further from equilibrium, which does not concur with the observed behaviour of the beach. The choice of downcoast point, as calculated using the method of Gonzalez & Medina (2001), does not appear to give good results in this case.

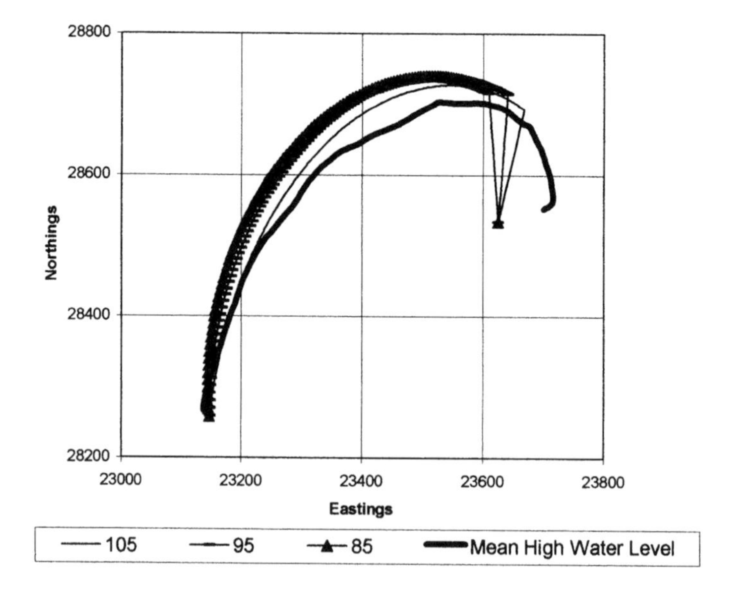

Figure 5. Sensitivity to wave direction for original downcoast point (with harbour)

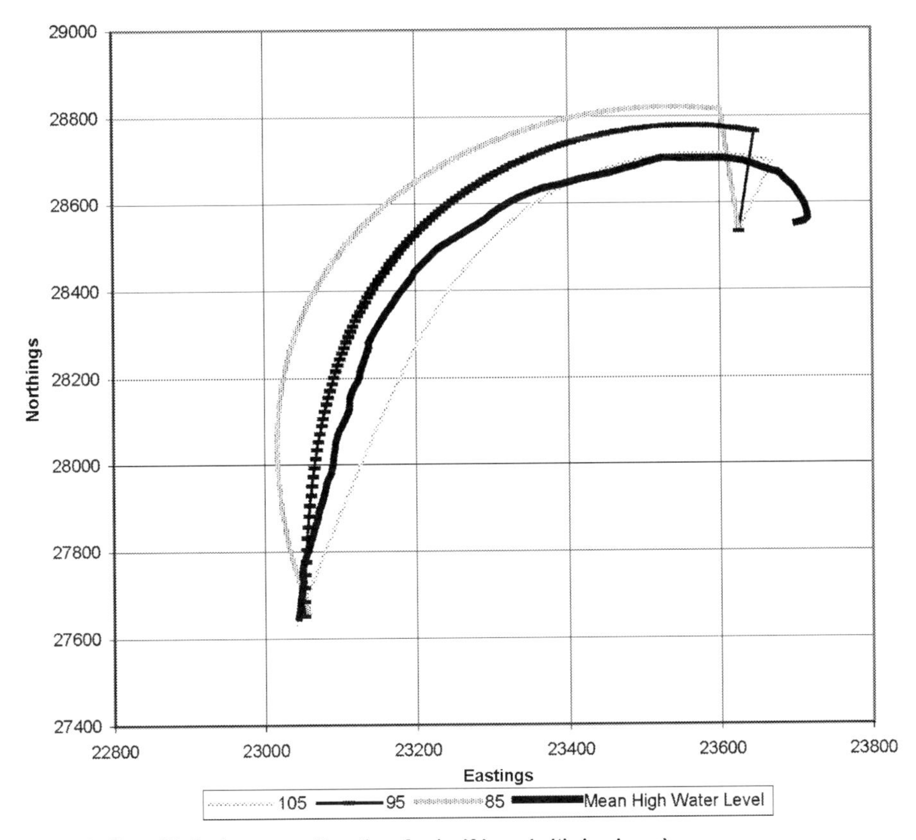

Figure 6. Sensitivity to wave direction for half bay (with harbour).

Single v double headland

One reason for this problem could be the presence of another headland at the south of Beadnell Bay. At a broadscale the bay could be considered as being governed by a double headland. More significantly the equilibrium beach lines in the northern and southern parts of the bay appear to be responding to slightly different predominant wave climates. Figure 7 shows the equilibrium shorelines calculated as two joining segments using the mid-bay 'downcoast point' for Beadnell harbour and the southern headland separately. The closest fit with the current bay shape is obtained with a wave direction of 110^0 in the northern part of the bay and a wave direction of ~80^0 in the southern part of the bay. The fit in the southern part of the bay is not as good as in the northern part, suggesting it is further from equilibrium.

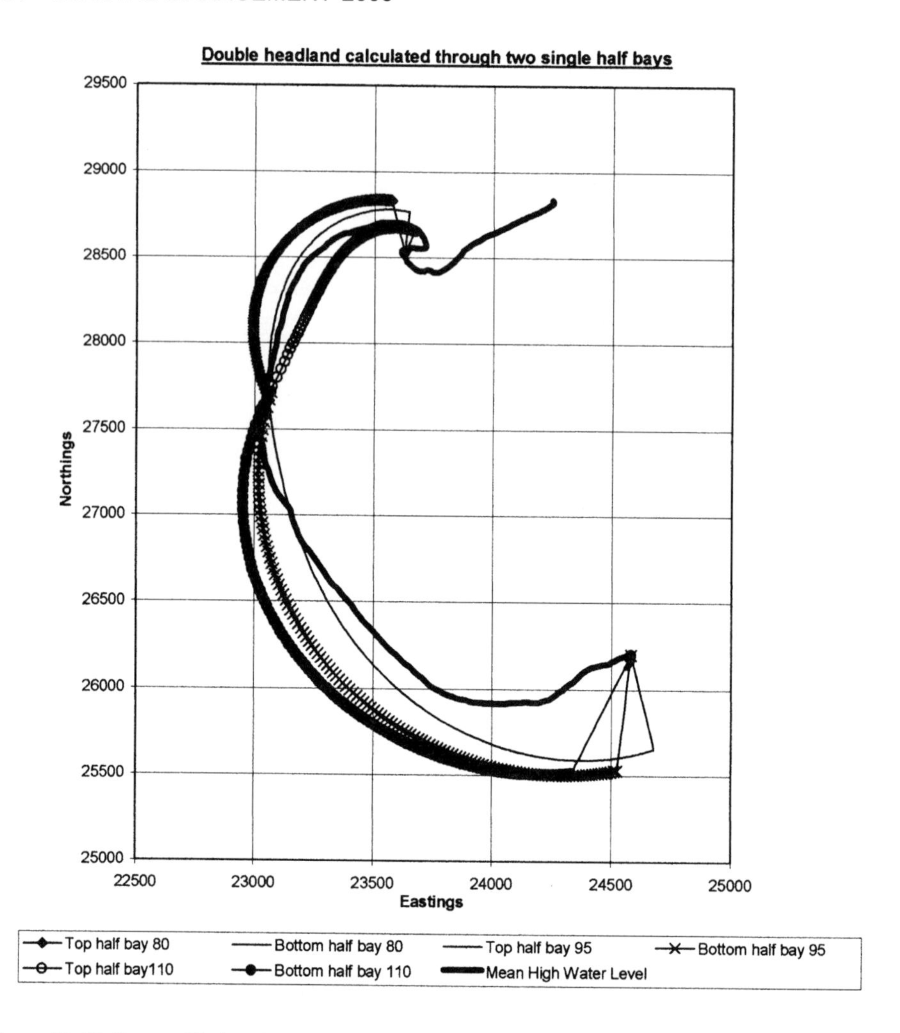

Figure 7. Static equilibrium bay shapes for a range of predominant wave directions.

Conclusions

In this study we have investigated the sensitivity of the results obtained using static equilibrium bay shape theory to the choice of model parameters and environmental variables. This included the examination of how sensitive the shapes predicted are to slight changes in the variables that might accompany climate change. The results of the sensitivity analysis showed that the existing configuration of the bay near Beadnell harbour is close to equilibrium. However, significant adjustment in shoreline position may be expected should the predominant wave direction change from its current sector by a little as 10^0.

Agreement between the current MHWL and the position suggested by equilibrium theory in the southern part of the bay is not as good, but is best towards the centre of Beadnell Bay, where there is plenty of sand. Near Snook Point there are large rocky outcrops that are not adjusting to equilibrium as fast as the sandy beach.

We have also investigated the sensitivity of the equilibrium bay to the choice of down coast point. This is normally taken to be a point where the shoreline is parallel to the incoming wave crests. From an inspection of the MHWL in Beadnell Bay, (see Figure 2), it is evident that there is a sharp change in alignment of the MHWL where Long Nanny Burn reaches the sea. The current bay shape appears to be the result of shoreline response to different predominant wave directions reaching the northern and southern parts of the bay. However, one would expect to see a smooth change in shoreline orientation in the area where the two different wave climates meet, but such a region is not observed. This suggests that the Long Nanny Burn plays an important morphological role in sustaining a longshore gradient of sediment transport by transporting sediment in the upstream/cross-shore direction.

It must be remembered that the theory for static equilibrium bays considers the predominant wave climate for a particular bay. Fluctuations in the variables over the timescale of tides, storms and seasons can be anticipated to lead to significant variation from static equilibrium. In this case, equilibrium theory provides a useful first order estimate of the likely range of coastal response to fluctuations in wave climate.

References

Chadwick, A.J. and Morfett, J., 1998. *Hydraulics in Civil and Environmental Engineering*, 3rd Edition, E & FN Spon, London, 600pp.

Gonzalez, M. and Medina, R., 2001, On the application of static equilibrium bay formulations to natural and man-made beaches, Coastal Engineering, Vol. 43, p209-225

Hsu, J. R. C. and Evans, C., 1989. Parabolic bay shapes and applications, Proc. Inst. Civil Engrs., 87, p557-570.

Hsu et al.,1989a, Static equilibrium bays: new relationships, Journal of Waterway, Port, Coastal and Ocean Engineering, May 1989, Vol. 115, No. 3, p285-298

Hsu et al.,1989b, Applications of headland control, Journal of Waterway, Port, Coastal and Ocean Engineering, May 1989, Vol. 115, No. 3, p299-310

Kamphuis, J. W., 2000, *Introduction to Coastal Engineering and Management*, Singapore: World Scientific, p534

Muir Wood, A. M. and Fleming, C. A., 1981, *Coastal Hydraulics*, 2nd ed., London: Macmillan and Co Ltd

Posford Duvivier, 1999. Beadnell Harbour Strategy Study and Project Appraisal, for Berwick-upon-Tweed Borough Council, May 1999.

Silvester, R. and Ho, S. K., 1972. Use of crenulated shaped bays to stabilise coasts, Proc. 13th Intl. Conf. Coastal Engrg., ASCE 2, p1347-1365.

Silvester, R., Tsuchiya, Y. & Shibano, T., 1980. Zeta bays, pocket beaches and headland control, Proc. 17th ICCE, ASCE 2, p1306-1319.

Silvester, R. and Hsu, J. R. C., 1997. *Coastal stabilisation*, Advanced Series on Ocean Engineering, Vol. 14, World Scientific, pp578.

Tan, S-K. and Chiew, Y-M., 1994, Analysis of bayed beached in static equilibrium, Journal of Waterway, Port, Coastal and Ocean Engineering, March/ April 1994, Vol. 120, No.2, p145-153.

An Awareness of Geomorphology for Coastal Defence Planning

Helen Jay, Kevin Burgess, & Adam Hosking, Halcrow Group Ltd, Swindon, Wiltshire, UK

Introduction

There have been considerable advances in coastal engineering and coastal defence management over the past two decades, which has included great progress in the ability to model coastal processes and analyse coastal change in response to man's intervention. The work of geomorphologists has, in the past, often been regarded by many in the coastal defence industry as more academic than practical for present day defence management; perceived as concentrating on behaviour and evolution over hundreds and thousands of years.

In more recent years coastal engineers and managers have sought a strategic approach to planning for future generations, and through this to think about coastlines over longer time periods and over larger areas. As such, the need to appreciate the structure of the shoreline and its behaviour as a system comprising different elements functioning on different time and space scales has been recognised. Despite this recognition, the integration of a geomorphological understanding with other more commonly applied approaches was not widely taken up in some long term planning, this being a notable exception within the first generation Shoreline Management Plans (MAFF, 2000). The Futurecoast project (Halcrow, 2002) was undertaken to address this, by bringing an appreciation of coastal geomorphology and a better understanding of coastal behaviour to the coastal engineer and manager, at a level appropriate for its consideration and integration into future coastal defence planning.

This paper does not specifically discuss the development or content of the Futurecoast project, which is well documented elsewhere (e.g. Burgess et al, 2002a; Burgess et al., 2002b; Cooper & Jay, 2002), nor does it attempt to provide the definitive text on coastal geomorphology, a fuller understanding of which can be obtained from many excellent publications (e.g. Bird, 2000; Haslett, 2000; Carter, 1988; Pethick, 1984 etc.). Rather, through presenting some of the principles employed in making the Futurecoast assessments (Halcrow, 2002: Futurecoast\Thematic Reports), it aims to provide those involved in coastal defence planning with an improved awareness of coastal geomorphology and hopefully a better appreciation of the coastal environment.

Examining coastal environments

Within coastal engineering, there has traditionally been a strong focus upon littoral processes and modelling forcing parameters, with this approach frequently used as a basis for analysing coastal change and assessing future policy options and impacts. Historic data sets are also often extrapolated to predict the future evolution of the shoreline. Whilst modelling and knowledge of historic change are important components for understanding coastal behaviour,

Coastal Management 2003, Thomas Telford, London, 2003.

they have limitations (see Box 1) and other factors also need to be taken into account when assessing future shoreline evolution.

Box 1. Constraints on the use of certain approaches in understanding coastal evolution

Numerical Modelling	**Historical Data Analysis**
Numerical models often provide an excellent source of information on the exchange of sediments in response to forcing conditions, but they only provide part of the picture because:	Historical information is essential to understanding how a shoreline is behaving, although using this directly to predict future change can be misleading because:
▪ they are usually based upon short-term data-sets; ▪ there is an assumption of linear change; ▪ features other than the beach are not well represented; ▪ there is an inability to reproduce changes in landform behaviour; ▪ feedback mechanisms are poorly represented.	▪ inaccuracies in data sets can often exceed actual changes; ▪ data usually offers only a snapshot in time and no knowledge of interim states; ▪ extrapolation assumes a linear change in a system, ignoring potential step changes/ threshold exceedence; ▪ future evolution may not mirror past change, as controlling influences change, e.g. management practices & forcing.

Analysis of coastal dynamics and evolution is especially difficult due to both the range of spatial and temporal scales over which coastal changes occur, and the complex interactions that result in shoreline responses of varying, non-linear and often unpredictable nature (see Box 2). There is also inter-dependence between different geomorphic features that make up the natural system, such that the evolution of one particular element of the coast is influenced by evolution in adjacent areas. Often these influences extend in a number of directions, thereby further complicating the task of assessing change.

Box 2. Understanding the time frame (Halcrow, 2002: Futurecoast\Thematic reports\onshore reports)

Coastal change may be considered to be occurring over three characteristic timescales:

- Macro-scale secular change (centuries to millennia);
- Meso-scale cyclic or periodic change (years to century); and
- Micro-scale episodic or individual event change (days to years).

Care needs to be taken, particularly when using short-term data or knowledge, not to simplify the reasons for coastal behaviour, as coastal change can be operating at these three scales simultaneously. Not all processes are operating at the same intensity and they may also be operating in different phases, although it can be difficult to determine the presence and strength of the driving forces.

The coastal response to these drivers is also complex as there may be thresholds and lags to this response. The present coast is predominantly a function of variable timescale changes that have been operating over the last 10,000 years (the Holocene Period) and some of the trends in shoreline changes observed today are inherited from the early and pre-Holocene. Understanding the dominant controls is important in making predictions of future change.

The Futurecoast project concluded that the coast is best considered in terms of 'behavioural systems' (Halcrow, 2002: Futurecoast\ Framework reports\ General methodology). This approach involves the identification of the different elements that make up the coastal structure and developing an understanding of how these elements interact on a range of both temporal and spatial scales; this being central to determining behaviour.

In analysing shoreline response, each feature and its influence upon coastal evolution must be considered, including management practices. Whilst generic behaviour and response characteristics often has to be concluded at the individual feature level, an essential requirement is to consider the coast as a whole system, not to consider each feature in isolation. As such, the analysis must continually consider the following:

- What features are present and how are they organised?
- How is each feature reacting to circumstances around (i.e. forcing factors, controls, behaviour of other features)?
- Why is the feature reacting in this manner?
- What are the consequences elsewhere of this reaction (i.e. the impact and response of other features)?

Feedbacks invariably play an important role; these occurring as the result of change in a system, and can affect the net morphological response of the coastal zone. Changes in energy/sediment inputs that affect one feature can in turn affect other features, which themselves give rise to a change in the level of energy/sediment input. An example of a feedback is where cliff erosion releases sediment, which allows the beach to build up, thus reducing the amount of wave energy reaching the cliff toe and thereby reducing cliff erosion and supply to downdrift beaches.

Identifying influences upon coastal behaviour

Key to appreciating larger scale behaviour is the understanding of the main **controls** on the system. These may fundamentally influence both the present morphology and the nature/rate of future response to environmental forcing. Typical examples of such controls are presented in Table 1.

Table 1. Typical controls on coastal behaviour

Headlands	Promontories, often due to more resistant geology, can have a major influence upon the coastal processes and the resultant orientation of the shoreline, in particular the formation and evolution of embayments. These form due to wave diffraction around at least one fixed point, although often between two fixed points; a soft coast between two resistant points will readjust its orientation to minimise the wave-generated longshore energy. Examples at different scales include Flamborough Head and Porlock Bay, whilst the Dungeness foreland offers a good example of a non-geological and mobile headland.
Shore platforms	Hard rock or armoured foreshores will limit exposure of the backshore to wave activity and be resilient to shore lowering, restricting the rate of evolution of the shoreline. Examples of this type of control can be seen at various locations around the UK, including Northumbria, the south-west coast of the Isle of Wight and Devon.
Offshore banks and features	Accumulations of sand, or submerged headlands, can both shelter areas of the shoreline, and alter the wave regime inshore, which will be influential upon the rate and nature of sediment movement. Notable examples include the sand banks offshore between Gt. Yarmouth and Lowestoft, and the submerged headland off Selsey Bill.
Estuaries and tidal inlets	Flows from these water bodies can deflect sediment from the shoreline to offshore. The formation of deltas can act as "soft" shore platforms and headlands, limiting exposure of the backshore and diffracting waves.
Shoreline defences	The intervention by man, with some defence structures acting as new headlands and others causing an interruption to sediment movement. Impacts can be both local and regional.

Many changes tend to occur at scales that relate to long-term responses to past conditions. Often the underlying pressures for shoreline change are related to large-scale reorientation of the coast, therefore an understanding of **past evolution** is important. Knowing how a feature formed or a particular coastal alignment developed can assist in determining how it will respond to future changes in the forcing parameters.

Coastal morphology also changes due to the **forcing** that acts upon it, i.e. waves and currents. Although much of our knowledge regarding processes is contemporary, an understanding the coastal response to present conditions can assist us in predicting future response.

The formation and maintenance/growth of some geomorphological features is dependent upon a supply of sediment of an appropriate size-grade. This therefore requires a suitable source and a transport pathway, so an understanding of the sediment **linkages** is important. At the local-scale, response of the geomorphological elements is key to the predictions of future coastal evolution. Understanding why the feature is where it is and its particular **morphological character** is essential to determining future behaviour.

Understanding the impacts of these factors on behaviour of the local-scale geomorphology, requires the following points to be considered:

- changes in foreshore response to wider-scale and local factors;
- assessment of the implications of foreshore response on backshore features;
- wider-scale geomorphological assessment of this coastal response (e.g. feedback interactions); and
- identification of any potential changes in geomorphological form (e.g. breakdown of gravel barriers).

Appreciation of these factors helps enable the long-term and large-scale evolutionary tendencies to be broadly established and in particular identify where a change from past evolution may be expected. To ensure that relevant factors are addressed it is useful to have a framework of generic questions, examples of which have been developed and are outlined in Table 2. These questions detail some of the main issues that need to be addressed when assessing future geomorphological behaviour and coastal evolution.

Shoreline types

Although the identification of large-scale behavioural systems and units is possible, generic categorisation of these is exceptionally difficult due to the different organisation of the individual features within them. Knowledge of shoreline type can, however, be useful in identifying likely behavioural tendencies and some broad characteristic types of shoreline are described below.

The evolution of hard coasts is almost exclusively a function of the resistant nature of the geology, with the influence of prevailing coastal forcing on the orientation of these shorelines only occurring over very long timescales (millennia), whereas the evolution of softer coasts is more strongly influenced by coastal forcing, although geology continues to play a significant role in both influencing this forcing (e.g. diffraction of waves around headlands) and determining the resistance of the shoreline to change. The plan-form of these shorelines will, over timescales of decades to centuries, tend towards a shape whose orientation is in balance with both the sediment supply and the capacity of the forcing parameters to transport available sediment. In general, soft shorelines have already undergone considerable evolution; some shorelines may have reached their equilibrium plan-form in response to prevailing conditions, whilst others have not and will continue to change.

Table 2. Key Questions (Halcrow, 2002: Futurecoast\ Data & supporting information\ Coastal Geomorphology Reference Manual)

Controls and Influences	Linkages
• What are the key geological controls, e.g. are there predominant headlands, and are these composed of hard or soft geology? How is the geology changing over time, i.e. what is the resistance to erosion and what is the main failure mechanism? • Are there any offshore or nearshore controls e.g. banks or islands? Are these changing and/or is there potential for them to change in the future? What control do they have on the shoreline, e.g. are they providing shelter or causing wave focussing? • What estuarine or inlet controls are present? Is there a delta that is an influence on the shoreline, e.g. by providing protection? Are there spits, and how are these behaving? • What tidal processes operate? Are coastal processes effectively tidal-driven? Is the shoreline subject to storm surges? What has been the past response to such events? • What wave processes operate? Are coastal processes effectively wave-driven? What are the predominant directions of wave approach? Are there differences in wave energy along the shoreline, e.g. due to wave diffraction?	• What are the key sources of sediment within the system? What sizes of sediments are released? How does this compare with the composition of the depositional features, e.g. dunes or beach ridges, present? • Have previous sources of sediment now been exhausted or removed from the system, e.g. due to rising sea levels? • Are there key sinks of sediment? Can these be considered permanent or temporary stores? If temporary, are these volumes likely to be released in the future and under what processes, e.g. cannibalisation of a barrier as it migrates landwards? • What are the key mechanisms of sediment transport, e.g. suspended or bedload, onshore or longshore? • What are the interactions between features? Over what temporal and spatial scales are linkages evident? What is the relative strength/importance of these linkages?
Past Evolution	**Morphology**
• How and why has the feature formed, and over what timescales? Have some features disappeared and what are the possible reasons for this? • What has been the historic behaviour of the feature at millennial, centennial and decadal timescales, e.g. has the volume held within a dune system changed, or has there been a change in position? • Are the processes that caused the features to form still occurring today, or can the features be considered relict? • How does the evolution of a certain feature, or geomorphological element, fit into a larger-scale pattern of change? Is the contemporary landscape a product of a previous different landscape form?	• What are the key internal physical controls on the behaviour of the feature, e.g. geology/composition, resistance to erosion, height, width, position etc.? • What are the key external physical controls on the behaviour of the feature, e.g. is it a wind-created feature, wave-dominated feature etc.? • Does it depend on a sediment supply and if so what are the key sources? Is the source of sediment contemporary or relict? • What are its links with neighbouring geomorphic units, e.g. does is depend upon another feature for its sediment supply, or is it a source for other features? • Does its evolution fit into a larger-scale pattern of change?

Reasons why such shorelines have not reached a dynamic equilibrium, and may still be adjusting in orientation, include constraints upon the rate of change (e.g. the level of resistance of the geology) and changes in conditions (e.g. sediment availability, emergence of new controls, breakdown of older features, changes in offshore topography). It should be recognised that we are presently in a generally transgressive phase (i.e. a period of rising relative sea levels) and shorelines are still adjusting to this.

Table 3 lists some of the shoreline types that are found around the UK, although this is only a broad, generic characterisation.

Table 3. Typical shoreline types

Geologically-dominated	Rocky shore	Hard coast, relatively independent of wave action. Evolution will be dictated by resistance to erosion of formative material. Examples include many areas of South-west England and Wales.
	Embayment	Strongly headland controlled shore, shaped by wave action. Within this system are various bay types. Examples at different scales include the Holderness shoreline and Porlock Bay.
Wave-dominated	Drift aligned shoreline	Dissipative shore with sediment moved alongshore by wave action. This form will include features such as spits and nesses. Evolution of this coast is largely dependant on wave direction and sediment supply. The Suffolk shoreline is a good example of this shoreline type.
	Swash aligned shoreline	Reflective shore, dominant waves normal to beach, formed in homogenous material, with the form including features such as barrier beaches and barrier islands. This coast is largely dependant on wave direction. Examples include Chesil Beach.
Tide-dominated	Tide-dominated shoreline	Mostly estuarine environments, such as the Humber.

Geologically-dominated shorelines

Hard rock coasts are resilient to significant changes in orientation over decadal to century timescales. Differential erosion may occur where there is an area of softer geology, or faulting, which can be exploited by wave activity. In many cases embayments may have been formed, with the geology having a major influence upon the coastal processes and the resultant orientation of the shoreline, in particular the formation and evolution of deeper embayments. Differential resistance of the backing geology will also influence the position of the shoreline, which in some cases may produce secondary embayments as new headlands emerge. A further influence is the response of the different geomorphological elements that comprise the backshore. This could, for example, create floodplains or inlets that alter the hydrodynamics operating within the bay and thus the alignment tendencies at the shoreline.

Wave-dominated shorelines

The natural tendency for most shorelines is to become orientated to the predominant wave direction, although clearly there are many constraints and influences upon this. This concept applies equally to the shoreface, foreshore and backshore, although is perhaps best illustrated by beach behaviour; the shoreline adjusts in form because sediment is moved, giving rise to areas of erosion and deposition.

Beaches can be defined as swash-aligned (or swash-dominated) or drift-aligned (or drift-dominated). Swash-aligned coasts build parallel to incoming wave crests, with little or no longshore transport but cross-shore movement. Drift-aligned coasts are generated by obliquely-incident (but not necessarily uni-directional) waves. Due to variability in the wave climate, few beaches are entirely swash- or drift-aligned, but identification of the predominant characteristic can help in predicting likely future evolution.

Where shorelines have become adjusted to the prevailing pattern of the waves, i.e. are in 'dynamic equilibrium', they reach a state of relative stability. Where changes are made to the shoreline controls, whether natural or anthropogenic (for example removal of defences), there may be tendency towards greater drift-alignment, with increased mobility of foreshore sediments and backshore erosion.

Tide-dominated shorelines

All coasts are affected by tides, but only a few types of coastal environments can be considered to be tide-dominated. Tide-dominated coasts generally occur in more sheltered areas where wave action is largely removed, e.g. due to shoaling or by direct shelter, such as by a spit at a river mouth, and are therefore most commonly associated with estuaries, although there are parts of the open coast around England and Wales where tidal influences are most dominant upon the shore plan-form. The features present reflect the change in dominant influence from waves to tides and sediments tend to be characterised by silts and muds due to the lower energy levels. These shorelines are generally low-lying and the shoreline plan-form arises from the deposition of fine sediments, which creates large intertidal flats.

The future plan-form evolution of tide-dominated shorelines is perhaps the most difficult to predict accurately due to the complex interactions within these environments. One of the key influences on the evolution of tide-dominated coasts is the change in tidal currents. This may occur for a number of reasons, but one of the main causes is due to changing tidal prisms, i.e. the amount of water that enters and exits an estuary every ebb-flood tidal cycle. Another influence on evolution is the configuration of ebb and flood channels, which affects the pattern of erosion, transport and deposition both across the intertidal zone and at the shoreline. In many estuaries, changes in the position of these channels have had a significant impact upon the adjacent shorelines (e.g. Morecambe Bay); where a major channel lies close to the shoreline it allows larger, higher energy waves to attack the marsh cliff, whereas where there are sandbanks adjacent to the marsh wave energy is attenuated. It is often not clear what causes a channel to meander because there are a number of interacting factors involved.

Landform types

Whilst appreciating the interactions between different features is crucial to understanding shoreline behaviour and response, in terms of actual coastal defence management it is usually individual landforms that will tend to be the focus of intervention. Therefore knowledge of landform types and characteristics is important.

Despite the complications associated with understanding coastal systems, their constituent components, and predicting or estimating their future evolutionary trends and tendency, sufficient information is known about certain coastal geomorphological features to enable generic statements to be made describing how, in theory, they form and evolve. However, there must be recognition that there can be significant variations in the type and response of similarly classed landforms to either ongoing or changing forcing. For example, a variety of cliff types exist which due to their composition have radically different failure mechanisms that vary dramatically in both magnitude and time; e.g. some cliffs will be subject to ongoing linear erosion whilst others are only periodically active albeit resulting in massive landslides (Defra, 2002). Whilst this is a commonly recognised variation other features, such as dunes and saltmarshes, can also exist in a variety of forms and exhibit different behavioural characteristics.

Tables 4 to 10 provide a brief summary of the main shoreline geomorphological features, together with their morphology, types and influence. Reference should be made to Halcrow, 2002; Futurecoast\Data & supporting information\Coastal Geomorphology Reference Manual, for more information.

Table 4. Dunes

Definition	Accumulations of sand blown from the foreshore to the backshore by the wind; the sediment accumulates above the mean high water mark where it becomes vegetated. Further sediment is trapped by the presence of vegetation and deposition accelerates.
Morphology	The formation of sand dunes is dependent upon two main factors: ▪ an abundant supply of sand-sized sediment; and ▪ a strong onshore wind to enable entrainment and transportation of sand from the beach to the dunes. Backshore dune development is aided by a low gradient sandy beach, which provides a large expanse of beach sand exposed at low tide. Although establishment of colonising vegetation can influence dune morphology, it is not essential for their formation (Pye & Tsoar, 1990). Coastal dune mobility, in general terms, is controlled by (1) the rate of sand supply, (2) wind energy and (3) vegetation cover (Pye, 1983). The marine erosion of dunes is more complex than that of cliffs because of the close interaction between the beach and the dune, which means that dunes can both accrete and retreat. Erosion rates of dunes can be very high and rapid because they are composed of unconsolidated sands. The type of dune failure varies due to exposure, dune morphology and vegetation cover. Three main modes of failure have been recognised: avalanche (sudden fall of sand), tabular (whereby dune fails in large blocks) and rotational (when small 'terraces' are formed) (Carter & Stone, 1989).
Types	The characteristic behaviour of a dune will depend upon its stage of evolution. A number of broad types can be recognised: ▪ **Embryonic dunes** - represent the first stage in the development of dune ridges and are formed by the deposition of sand along the high tide mark. They are low-lying mounds of sand and are often vegetated by salt-tolerant species; they are easily overwashed and removed during storms, releasing sand back to the beach. ▪ **Foredunes** - continuous or semi-continuous ridges of sand, often vegetated, which lie at the back of the beach and parallel to the shoreline. Parallel dunes can be modified during storm surges when wave overwash and breach may occur resulting in sand being swept landward in the form of fans or sheets. The height of foredunes is dependent upon the wind strength and sediment supply. Foredunes may become cliffed at their seaward margin during storms and undercutting at the dune toe can cause collapse and failure of the dune cliffs. Foredunes are also vulnerable to overwash (depending upon height) and breaching, particularly where the ridge is narrow and/or characterised by a series of blow-outs. ▪ **Climbing dunes** - occur on some cliffed coasts there are sand dunes either piled against the cliff, forming climbing dunes, or at the top of the cliff. Where dunes have spilled inland and become separated from any source of sand, they have become relict cliff-top dunes, such as observed at locations along the Cornish coast. ▪ **Relict dunes** - are also present where there are no contemporary sources of sand or where the link between the beach and dunes has been broken e.g. on shingle beach ridges such as at Blakeney Point, Norfolk (Steers, 1964). ▪ **Blow-outs and parabolic dunes** - generally form where dunes are unstable, possibly due to a lack of stabilising vegetation cover. There are two main ways in which they form: (1) where natural gaps or storm-damaged cliffs in the foredune ridge are exploited by winds and (2) by erosion processes, e.g. the deflation of a poorly vegetated terrain. The movement of blow-outs and parabolic dunes is dependent upon the direction, frequency and strength of the onshore winds. Parabolic dunes may form close to the shoreline or by disruption of vegetated older dunes inland (Pye, 1983). They represent vulnerable sections of shoreline in terms of coastal defences. ▪ **Transgressive dunes** - mobile dune forms which develop where sand blown inland from a beach has been retained by vegetation or where previously vegetated dunes become unstable and the numerous blow-outs merge to form an elongate dune (Bird, 2000). If vegetation along the trailing arms of a parabolic dune is destroyed then the dune form evolves into transgressive sand sheets (Hesp and Thom, 1990). Along the coastal margin, periodic wave erosion may play an important part in transgressive sand sheet development by removing any incipient foredunes (Pye, 1990).
Influence	Dunes perform two functions in terms of defence: (1) they provide a temporary store of sediment to allow short-term adjustment of the beach during storms and (2) they provide a protective barrier to the hinterland.

Table 5. Coastal cliffs and slopes

Definition	Defined as areas of elevated relief often forming a distinct break in slope between the hinterland and the shoreline.
Morphology	Cliffs are vertical or steeply sloped faces cut by marine action, whereas coastal slopes lie at a slightly gentler gradient. Coastal slopes may represent older, naturally degraded cliff lines where changes of coastal orientation over centuries has led to the building of beach width in front of the former cliff. Alternatively they may form where wave action is too limited to cause scarping and cliff formation. Also, in many areas coastal slopes represent artificially landscaped cliffs, which may now be protected by sea walls and promenades.
Types	A range of forms exists, dependent upon an infinite combination of subtle differences in both geology and processes (marine, slope and sub-aerial). A number of characteristic cliff forms have been defined (MAFF, 1996; Defra, 2002): ▪ **Simple cliff face systems** - generally characterised by a steep cliff face, narrow foreshore zone and rapid removal of toe debris. Erosion typically occurs as rock falls, topples or slides from which material is deposited directly on the foreshore. ▪ **Simple landslide systems** - first time failures in previously un-sheared ground, or repeated failures in recently sheared ground. Toe erosion of cliff debris leads to oversteepening of the cliff face and a deep-seated rotational slide develops. ▪ **Composite systems** - typically comprise inter-bedded hard and soft rocks; this can generally be as either soft rock caps resting on hard rock or as hard rock caps resting on softer rock. ▪ **Complex systems** - comprise a series of sub-systems, such as scarp and bench features, within the cliff. Each sub-system has its own input, storage and output of material, whereby the output from one sub-system forms a cascading input to the next. ▪ **Relict systems** - comprise sequences of pre-existing landslides, which could be susceptible to re-activation, e.g. due to debris removal, foreshore lowering or increasing porewater pressure.
Influence	Cliffs may be a significant source of material for the coastal system. They are also a key control on wider-scale coastal orientation, dependent upon their composition.

Table 6. Tidal flats and marshes

Definition	Formed by an accumulation of fine sediments, such as sands, silts and clays, at the shoreline. They are usually formed in areas with a relatively large tidal range and a degree of shelter against direct action from ocean-generated waves. Tidal flats are often characterised by sandflats and/or mudflats and vegetated saltmarshes.
Morphology	Deposition of sediment flocs (mass of mud particles) occurs when the shear velocity of the water flow is at, or close to a minimum, therefore an important factor in determining the rate of accumulation of cohesive sediments is the duration of slack water periods. A second factor is the suspended sediment concentration, which is a further control on the rate of accumulation. Erosion of a salt marsh may occur due to: ▪ sea level change; ▪ sediment starvation; or ▪ an increase in marine forcing conditions.
Types	• **Sandflats** – generally wave-dominated features, but tidal currents may also play a role in their morphology (Bird, 2000). Deflation processes depend upon several factors including moisture, wind strength and sediment size. • **Mudflats** - mudflat deposition is primarily governed by tidal processes with mudflat erosion primarily governed by wave processes (NRA, 1994). • **Pioneer saltmarshes** – when the tidal flat is high enough to result in a decrease in the frequency and duration of tidal inundations of the upper sections of the profile, vegetation that is tolerant to high salinity levels begins to colonise the surface. This results in reduced tidal velocities and increased sediment deposition (Wayne, 1976). • **Saltmarshes** - as the increased sedimentation within pioneer saltmarshes occurs, so the frequency and duration of tidal inundation decreases further, leading to the colonisation of the sediment by many varied salt-tolerant plant species.
Influence	Tidal flats and saltmarshes are extremely efficient dissipaters of wave and tidal energy (Möller et al., 1999; Cooper, 2001) and as such are vitally important for reducing the risk of flooding to low-lying hinterland. They may also be a sink of sediment.

Table 7. Shore platforms

Definition	Relatively flat bedforms extending across, and sometimes seaward of, the intertidal zone (Carter, 1988). They can be formed in various ways, but are generally the remains of a coastal slope that has been progressively eroded over time (Pethick, 1984).
Morphology	The morphology of a shore platform (and its rate of lowering) depends upon: ▪ the hardness of the material; ▪ structural weaknesses within the material; ▪ the solubility of the material; ▪ the nature of wave attack.
Types	Bird (2000) discusses three generic types of shore platform: ▪ **Horizontal surface lying at high tide level** - tend to be well-developed in micro-tidal, low wave energy environments and are typically composed of sandstones and mudstones; ▪ **Horizontal surface lying at low tide level** - best developed on micro-tidal, limestone coasts; ▪ **Sloping surface between high and low tide levels** - can be cut into a variety of rock types and are the most common type around the coastline of England and Wales.
Influence	The shore platform helps to dissipate and limit incident wave energy before it reaches the backshore and also provides a minor source of sediment to the coastal system. Shore platforms can also act as a control on coastal orientation, similar to a cliffed headland.

Table 8. Beaches

Definition	An accumulation of unconsolidated deposits of sand and gravel (shingle) on the shoreline, representing a store of littoral sediments.
Morphology	Beach morphology is primarily dependent upon: ▪ Tidal range; ▪ Wave energy; ▪ Sediment size; ▪ Sediment supply; ▪ Underlying geology.
Types	▪ **Sand beaches** - commonly form a low gradient dissipative profile, with transport (predominately saltation and suspended) occurring during most wave conditions. During storms, erosion of the upper beach occurs, with newly-released material being deposited on the lower profile, often beyond mean low water. This results in a wider, flatter dissipative profile. Onshore transport operates during calmer conditions to return sand to the upper profile. ▪ **Shingle beaches** - commonly form the steepest possible beach profile gradients with a berm present at the upper beach. Sediment transport occurs primarily as bedload, although the sediment exhibits low to medium mobility, even under storm conditions when, typically, sediment remains on the upper and mid beach. ▪ **Mixed beaches** - comprise a poorly sorted mixture of sediment types (sand, pebbles and cobbles), which, typically, are relatively coarse and steep in profile at the upper beach and increasingly sandier and flatter further seaward. Due to the relatively high percentage presence of sandy material, the upper beach berm is much less permeable than a pure gravel counterpart. This results in backwash processes serving to remove sand-sized material and transport it seaward, resulting in a cliffed beach profile. ▪ **Composite beaches** - comprise a well-sorted mixture of sand and gravel with, effectively, a standard gravel beach overlying a base sand beach. Each of the two separate components tends to behave like its corresponding pure sand or pure gravel beach counterpart. The upper gravel beach may be classed as a barrier beach (see separate section on Barriers and Spits). ▪ **Boulder beaches** - formed of large boulders. The boulders are immobile under most marine forcing conditions and may remain in-situ for a relatively long period, forming a robust apron which affords a great deal of protection to the backshore.
Influence	They act to dissipate wave energy and, as such, afford considerable protection against direct marine forcing to backshore features. Beach response is dependent upon the sediment composition of the beach; sand beaches tend to respond more rapidly than beaches composed of coarser sediment and are therefore more mobile than shingle beaches.

Table 9. Barriers and spits (detached beaches)

Definition	Depositional features, formed of sand or gravel (shingle), which are closely related to beaches, but may be completely or partly separated from other shoreline features. Many of the shingle features around the coastline of England and Wales are essentially relict and there is finite amount of shingle within the system that is continually being reworked.
Morphology	Morphology depends upon: ▪ Hinterland morphology; ▪ Sediment size; ▪ Sediment supply; ▪ Marine forcing parameters.
Types	They can be broadly classified into: ▪ **Barrier islands** - completely detached from the shoreline (e.g. Scolt Head Island, North Norfolk) and commonly are associated with tidal deltas. Generally formed in a high-energy environment, where ebb and flow currents are strong enough to prevent deposition of sand and shingle that would seal the intervening inlets. ▪ **Barrier beaches** - attached to the shoreline at both ends, but may leave the shoreline in their centre. They are not restrained in landward movement by backshore topography and tend to be mobile features. Some beaches of this type extend from the main shoreline to an offshore island (e.g. Chesil Beach in Dorset). ▪ **Spits** - beaches built up above high water, which leave the main coastline, usually at estuary mouths or bay, and project into deeper water before terminating, e.g. Hurst Castle Spit and Spurn Head. They often exhibit a cyclic behaviour, whereby there is a period of elongation, before the spit becomes breached, e.g. by river flows, and material from the cut-off segment is dispersed and then the elongation process may start again. ▪ **Cuspate forelands** (nesses or salients) - depositional features that are low coastal forms with a triangular plan-form projecting as a promontory. They are characterised by a series of parallel, or sub-parallel, beach ridges, e.g. Benacre Ness in Suffolk. Some are migratory.
Influence	Protect low-lying land by forming a barrier and also are sediment stores.

Table 10. Deltas

Definition	Deltas are accumulations of sediment, which form at the mouth of rivers, where the rate of sedimentation exceeds the rate of removal under wave and current action.
Morphology	The shape and size depends upon: ▪ rate of sediment input; ▪ coastal configuration; ▪ nearshore bathymetry; ▪ relative important of fluvial, tidal and wave processes.
Types	Deltas can be subdivided into: ▪ **River deltas** - tend to be long, narrow features that protrude far into the sea; ▪ **Wave-dominated deltas** – tend to be wider in form than river deltas; ▪ **Tide-dominated deltas** - comprise both flood tide deposits (flood delta) inside the estuary mouth and ebb tide deposits (ebb delta) slightly seaward of the estuary mouth.
Influence	Deltas influence coastal processes by: deflecting sediment transport; dissipating wave energy; acting as a 'soft shore platform' and providing a coastal control and acting as a sediment store/ sink.

Future behaviour

Shorelines of any form have the potential to evolve in three ways: continuation of present form; breakdown of present form, or transition to a different form. Influential upon these changes will be the long-term, large-scale, adjustment of the shoreline in response to the controls and influences previously described. However, it is also important to look for situations where the system response might be to switch to a different state, for example, the catastrophic failure of a spit, or the switching of channels as a consequence of episodic storm events. Future large-scale reorientation of the coast may also be influenced by the emergence of new features; Box 3 provides some examples of possible changing situations.

The process of evaluation is necessarily an iterative one to investigate response of the coastal morphology to changes in energy and sediment exchange as a result of feedbacks within the system. The consequence of potential changes in controls and influences may be an alteration in the large-scale alignment tendency for the shoreline, which results in modification in sediment exchange, for example:

- a change in sediment input from the backshore or from offshore;
- a change in sediment outputs to offshore or to the backshore;
- a change in longshore transport, (both volume and direction).

This may then modify backshore response and system feedbacks, which in turn may influence the larger-scale behaviour.

Box 3. Examples of changing controls and influences

Examples of changing controls and influences over time include:

- changes in **geological controls** (e.g. emergence of headlands in eroding cliffs, variations in backshore geology);
- alteration to **hydrodynamic forcing** (e.g. increased or decreased wave diffraction processes around headlands or over offshore banks, interruption of drift by newly created tidal inlets, development of tidal deltas);
- changes in **sediment regime** (e.g. exhaustion of relict sediment sources, shorelines switching from drift- to swash-alignment); and
- **human intervention** (e.g. cessation of sediment supply due to cliff protection, the introduction of new hard points, or improvement of sediment supply due to defence removal).

Climate change influences

The behaviour of coastal systems at any time has been described as a function of the processes acting upon it, the response of the constituent landforms, and the resultant feedbacks. It is reasonable, therefore, to assume that any changes in the forcing agents resulting from climate change could result in a change in coastal behaviour.

Based upon our understanding of landform behaviour, it is possible to identify theoretical responses of various shoreline features to changes in certain controlling parameters, such as key climate variables. Table 11 identifies the generic sensitivities of different landforms to the forcing parameters, and largely relates to some moderation of the existing functioning of the landform. However, it is also possible that climate change could promote conditions whereby the generic type of coastal form and/or behaviour may alter at specific locations. These are referred to as 'step changes'. The point at which these changes are likely depends on the sensitivity of the system as well as the magnitude of the forcing, i.e. the exceedence of thresholds. Significant and partly unpredictable changes in the nature and magnitude of behaviour are likely to accompany these transitions. Examples of the most likely situations where this might occur are provided in Box 4.

Whilst there is clearly the potential for future climate change to increase the likelihood of behavioural thresholds within these sensitive coastal units being exceeded, it must be noted that these processes could also occur under current conditions.

There are two important points to make with regard to the potential implications of climate change for coastal defence management (Halcrow et al., 2001):

- Climate change will not result in any new physical process, and hence hazards, on the coastline; increasing sea levels, storminess, etc will bring about changed patterns and intensities of existing hazards, but new hazards will not be created. Therefore we know what the physical impacts of climate change will be; however, it is their magnitude that remains uncertain.
- Climate change is just one variable in the future coastal management framework. The changes in society and the political economy that will inevitably occur over the next 50 years will be as significant as climate change and sea-level rise in determining how coastal risks are accepted, tolerated and managed. So regardless of climate, managers must be prepared for future change.

Table 11: Relative sensitivity of landforms to climate change

Landform Type	Climate Change Sensitivity			
	Sea level Rise	Storm Surge	Precipitation	Wave Direction
Simple Cliff	High	Moderate	Moderate	Low
Simple landslide	High	Low	High	Low
Composite cliff	Moderate	Low	Moderate	Low
Complex cliff	Moderate	Low	High	Low
Relict cliff	High	Low	High	Low
Embryonic dunes	High	High	Low	Low[1]
Foredunes	High	High	Moderate, impacts on vegetation	
Climbing dunes	Moderate	Moderate		
Relict dunes	Low	Low		
Parabolic dunes	Moderate	High	Low	
Transgressive dunes	Moderate	Moderate	Low	
River deltas	High	High	Moderate	Moderate
Tide dominated deltas	High	High	Low	Moderate
Wave dominated Deltas	High	High	Low	High
Shore Platforms	High	Moderate	Low	Low
Sandflats	High	High	Low	Moderate
Mudflats	High	High	Low	Low
Pioneer Saltmarsh	High	High	Moderate, impacts on vegetation	Low
Saltmarsh	High	High		Low
Sand beach	Moderate	Moderate	Low	High
Shingle beach	Moderate	Moderate	Low	Moderate
Mixed beach	Moderate	Moderate	Low	Moderate
Composite beach	Moderate	Moderate	Low	Moderate
Boulder beach	Low	Low	Low	Low
Barrier island	High	High	Low	High
Barrier beach	High	High	Low	High
Spit	High	High	Low	High
Cuspate foreland	Low	Low	Low	Low

Notes: Sea level rise = sensitivity to accelerations in the rate of sea level rise.
Storm surge = sensitivity to changes in intensity/frequency of storm surges (will depend upon exposure of site).
Precipitation = sensitivity to changes in pattern/intensity of precipitation.
Wave direction = sensitivity to changes in wave direction (e.g. changed sediment transport patterns).
[1] Dunes are likely to be sensitive to changes in wind direction.

Box 4. Examples of Step Changes

1. An eroding beach or marsh fronting coastal lowland may alter to form a soft cliff or slope where recession cuts into any abrupt rise in backshore gradient that is formed in a soft lithology. Local shorelines will receive additional sediments from the eroding cliff. In the long term, the recession rate is likely to reduce as a result of the occurrence of this transition occurring on an undefended coast.
2. A barrier beach may alter to form a tidal inlet, with spits and tidal deltas, through the permanent breaching or breakdown of the barrier beach due to sea-level rise. The inundated backshore could become an estuarine environment. This might generate significant flooding hazard during the initial breakdown or breach. This type of transition may also occur due to failure, removal, or abandonment of defences.
3. A tidal inlet could alter to a barrier beach through sealing of the inlet as a result of changed patterns of longshore drift or increased sediment supply. Although estuarine flooding will be reduced, the new behaviour would involve significant flood and erosion probabilities on the new barrier, and further episodes of breaching and inundation might be expected.

Summary

A knowledge and understanding of the physical structure of the coast is important as this provides the foundation for effective shoreline management and defence planning. This paper has illustrated some of the characteristics of the shorelines that we are attempting to manage, and indicated the complexity and variability that exists.

This understanding has already been applied in developing the Futurecoast project (Halcrow, 2002) to provide an assessment of coastal behaviour and a vision of shoreline evolution throughout England and Wales over the next century. This provides an informed starting point for those managing shorelines. However, in planning future coastal defence management policy, coastal engineers and managers should have an appreciation of coastal geomorphology. The objective of this paper has been to help facilitate this through delivering an awareness of the subject at an appropriate level.

References and further reading

Bird E C F (2000). *Coastal geomorphology: an introduction*. John Wiley & Sons Ltd, Chichester.

Burgess K A, Jay H & Hosking A S (2002a). Futurecoast: Predicting the future coastal evolution of England and Wales. *Littoral 2002*, 22-26 September. Porto, Portugal.

Burgess K A, Balson P, Dyer K R, Orford J & Townend I H (2002b). Futurecoast – the integration of knowledge to assess future coastal evolution at a national scale, *28th International Conference on Coastal Engineering*, ASCE, New York.

Carter R W G (1988). *Coastal Environments*, London, Academic Press

Carter R W G & Stone, G W (1989). Mechanism associated with the erosion of sand dune cliffs, Magilligan, Northern Ireland. *Earth Surface Processes and Landforms* 14, p1-10.

Cooper N J (2001). *Coastal data analysis: The Wash. Study 2: Wave attenuation over inter-tidal surfaces*. Report to Environment Agency (Anglian Region), January 2001.

Cooper N J & Jay H (2002). Predictions of large-scale coastal tendency: development and application of a qualitative behaviour-based methodology. Proceedings of the International Coastal Symposium (ICS 2002). *Journal of Coastal Research Special Issue* 36, p173-181.

Department for Environment, Food and Rural Affairs (2002). *Soft Cliffs Study: prediction of recession rates and erosion control techniques.*

Halcrow Group Ltd, University of Portsmouth, University of Newcastle and the Met. Office, (2001). *Preparing for the Impacts of Climate Change.* Report to SCOPAC.

Halcrow (2002). *Futurecoast.* Defra.

Haslett S K (2000). *Coastal Systems.* Routledge Introductions to Environment Series, Routledge, London.

Hesp P A & Thom B G (1990). Geomorphology and evolution of active transgressive dunefields. In: Nordstrom K F, Psuty, N P and Carter, R W G (Eds) *Coastal Dunes: Form and Process.* John Wiley & Sons, Chichester, 253-288

MAFF (1996). *Soft cliffs: prediction of recession rates and erosion control techniques: Literature review and project definition study.* Report by High Point Rendel, HR Wallingford, Cambridge Coastal Research Unit, River and Coastal Environments Research Group to MAFF. March 1996.

MAFF (2000). Review of existing Shoreline Management Plans around the coastline of England and Wales. R & D report by Universities of Newcastle and Portsmouth to MAFF.

Möller I, Spencer T, French J R, Leggett D J and Dixon M, (1999). Wave transformation over salt marshes: a field and numerical modelling study from North Norfolk, England. *Estuarine, Coastal and Shelf Science* 49, 411-426.

NRA - National Rivers Authority (1994). *A guide to the understanding and management of saltmarshes.* R&D Note 324.

Pethick J S (1984). *An Introduction to Coastal Geomorphology.* London, Edward Arnold.

Pye K (1983). Coastal Dunes. *Progress in Physical Geography* 7, 531-557.

Pye K (1990). Physical and human influences on coastal dune development between the Ribble and Mersey Estuaries, northwest England. In: Nordstrom K F, Psuty, N P and Carter, R W G (Eds) *Coastal Dunes: Form and Process.* John Wiley & Sons, Chichester, 339-359.

Pye K & Tsoar H (1990). *Aeolian Sand and Sand Dunes.* Hyman Ltd., London.

Steers J A (1964). *The Coastline of England and Wales.* Cambridge University Press, Cambridge

Wayne C J (1976). The effects of sea and marsh grass on wave energy. *Coastal Research Notes*, 14, 6-8.

EUROSION: Primary results of the review of experience in erosion management across European coasts.

Dr J. Serra, Eurosion WP Leader. University of Barcelona (UB), Barcelona, Spain
Dr C. Montori, Eurosion WP Coordinator. Autonomous University of Barcelona (UAB), Barcelona, Spain
P.K. Schoeman MSc Eurosion WP4 Collaborator,
O. Gelizo, BSc Eurosion WP Co-Worker. Autonomous University of Barcelona (UAB), Barcelona, Spain
E. Roca, BSc Eurosion WP Col-Worker. Autonomous University of Barcelona (UAB), Barcelona, Spain

Background

One quarter of the European Union's coast – 16,000 km – is currently eroding despite the development of a wide range of measures to protect shorelines from eroding and flooding. The prospect of further sea level rise due to climate change and the heritage of mismanagement in the past imply that coastal erosion will be a growing concern in the future. Recently the European Council of ministers has adopted a recommendation for the implementation of integrated coastal zone management (ICZM), where sustainable coastal defence and transparent information management are key issues. In this way the DG Environment has commissioned a service contract called EUROSION, regarding policy advice on coastal erosion to a consortium lead by the Dutch Institute for Coastal and Marine Management (RIKZ), and includes the largest European network of coastal practitioners "EUCC-The Coastal Union", the international branch of the French Geographic Institute (IGN France International) specializing in GIS and mapping engineering, the French Environment Institute (IFEN), the Autonomous University of Barcelona (UAB), the European Information Technology EADS SD&E (ex-MATRA S&I), and the French Institute for Geological and Mining Research (BRGM), (EUROSION WP 5 Trend Analysis, 2002).

The overall objective of EUROSION is to provide the European Commission with a package of recommendations for policy-making and information management practices to address coastal erosion in Europe, after thorough assessment of knowledge gained from past experiences and of the current status and trends of European coasts. However the project also aims at producing results of immediate value for policy makers and managers on other administrative levels. The project is divided in seven work packages (WP):

WP1 – Project Management
WP2 – European Level Database
WP3 – Guidelines for Developing Local Information Systems
WP4 – Review of Experience in Erosion Management
WP5 – Formulation of Policy Recommendations
WP6 – Dissemination and Networking
WP7 – Defining User Requirements and Feedback

Coastal Management 2003, Thomas Telford, London, 2003.

Objectives of WP4

The task of WP4 will be to analyse the state-of-the-art of current practices in coastal erosion prevention and management at various levels. This state of the art is being based on a Europe wide review of successful and unsuccessful strategies, measures and experiments to prevent or manage erosion for different types of coast. The immediate result of this undertaking will be an on-line "shoreline management" database of practical examples from all over Europe, highlighting the weaknesses and strengths from technical, economical, and social points of view. The aim of the *Shoreline Management Guide* (SMG) is to facilitate embedding of lessons learnt from existing local shoreline management practices into decision-making processes and make this knowledge easily accessible to a wide range of local stakeholders. The information obtained will be placed in the context of the entire spectrum of coastal defence measures and highlight the relationship with ICZM, also containing technical documentation on the methods adopted for the maintenance and protection of eroding coasts. The next flow-chart (Figure 1) show the WP4 relationships within Eurosion project.

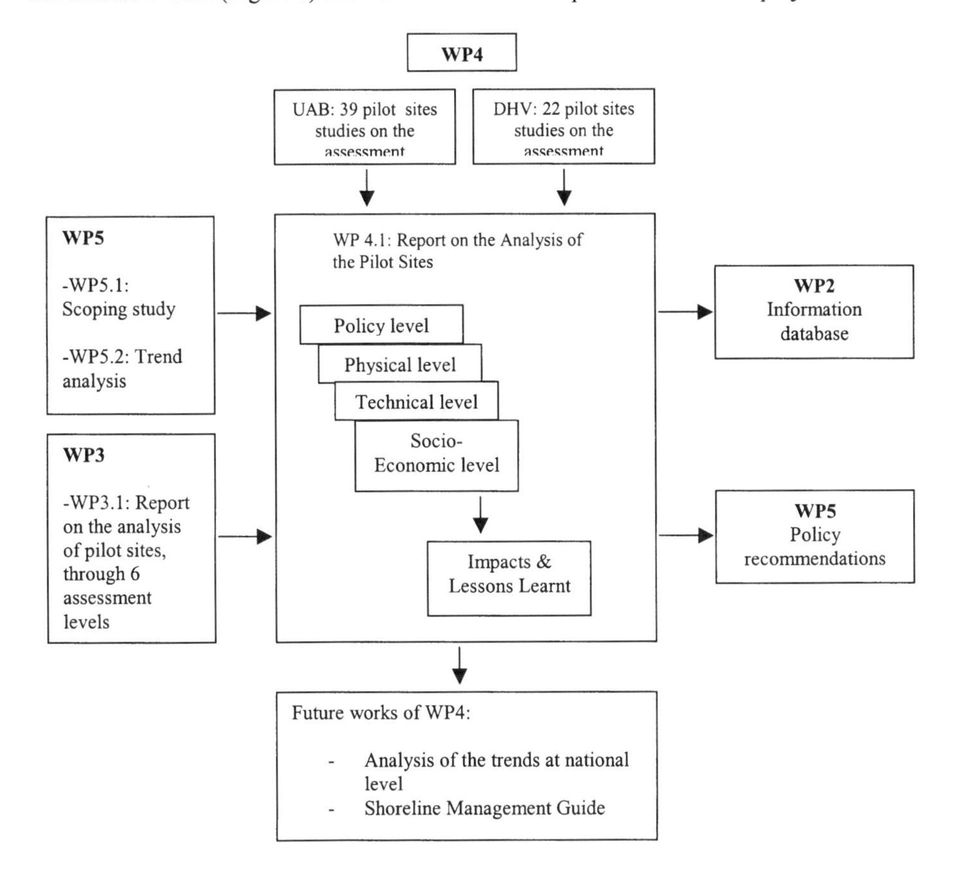

Figure 1. Flow-chart of WP4 within the framework of EUROSION project.

Methodology of reviewing cases

The study is based on **61 cases** selected throughout European Union and presently accessing countries (see Table 1 & Figure 2), twenty-two of which are provided by RIKZ (assisted by DHV)and thirty-nine cases by UAB. The analysis of the sites is divided in two major areas of study: the Baltic Sea – North Sea (RIKZ) and the Mediterranean Sea – Black Sea (UAB). The Atlantic coast is analysed by both institutions.

The way for reviewing case studies is by documenting the geomorphological and hydro-dynamic patterns and the socio-economical functions fulfilled by the coastal zone. For each case study, it will also reviewed technical operations implemented at the local level and underline their strengths, weaknesses, opportunities and threats regarding both their assigned objectives and their undesirable short, medium and long term effects.

Table 1. The WP4 selected sites for reviewing study (In bold=UAB pilot sites).

	NATIONAL	REGIONAL	LOCAL	NAME OF THE SITE
1	Belgium	West-Vlaanderen	De Haan	De Haan beach
2		West-Vlaanderen	Zeebrugge	Zeebrugge beach
3	**Bulgaria**	**Dobrich**	**Shabla**	**Shabla - Krapetz**
4	**Cyprus**	**Larnaka**	**Cape Kiti to Zigi**	**Dolos-Kiti**
5	Denmark	Zealand	Roskilde	Koge Bay
6		Jutland	-	Western coast of Jutland
7		Zealand	Frederiksborg	Hyllingebjerg-Liseleje
8	Estonia	-	Tallinn county	Tallin
9	Finland	Varsinais-Suomi	Turku	Western coast
10	**France**	**Aquitaine**	**Lacanau to Cap Ferret**	**Côte Aquitaine**
11		**Provence-Alpes-Côte d'Azur and Languedoc-Roussillon**	**Camargue**	**Rhône delta**
12		**Haute-Normandie**	**Pays de Caux**	**Haute-Normandie**
13		**Poitou-Charentes**	**Chatelaillon**	**Plage de Chatelaillon**
14		**Pays de la Loire**	**Sables d'Olonne**	**Sables d'Olonne**
15		**French Guyana**	**Commune de Rémire-Montjoly**	**Rémire – Montjoly**
16	Germany	Hamburg	-	Elbe
17		Schleswig-Holstein	-	Isle of Sylt
18		Mecklenburg-West Pomerania	Rostock	Rostock beach
19	**Greece**	**Achaia (Peloponese)**	**Lakkopetra bay**	**Lakkopetra**
20		**West Greece**	**Mesollogi**	**Mesollogi lagoon area**
21	Ireland	Donegal	-	North Donegal
22		Waterford	Tramore	Tramore beach
23	**Italy**	**Veneto and Emilia Romagna**	**Po delta**	**Goro Po mouth**
24		**Emilia Romagna**	**Ravenna**	**Marina di Ravenna-Lido Adriano**
25		**Liguria**	**Sarzana**	**Marinella di Sarzana**
26		**Lazio**	**Ostia**	**Vecchia Pineta**
27		**Sardegna**	**Agligentu**	**Lu Littaroni La Liccia**
28		**Sicilia**	**Taormina and Giardini**	**Giardini-Naxos**
29		**Toscana**	**Regione Toscana**	**Marina di Massa Marina di Pisa**
30		**Campania**	**Isle of Procida**	**Cirqaccio- Ciracciello**

31	Latvia	Vidzeme	Riga	Gulf of Riga
32	Lithuania	Klaipeda province	-	Klaipeda beach
33	**Malta**	**Malta**	**San Pawl il-Bahar nad Mgarr municipalities**	**Xemxija - Ghajn Tuffieha**
34	The Netherlands	**North Holland**	**Scheveningen**	**Holland Coast**
35		Delta	-	-
36		Wadden	-	Wadden Sea islands
37	**Poland**	**Westpomerania**	**Swinoujscie, Pomorski, Dziwnow,... municipalities**	**West Polish coast**
38		**Pomorski**	**County Puck, Commune Wladyslawowo and Jastarnia**	**Hel Peninsula**
39	**Portugal**	**Algarve**	**Loulé municipality**	**Vale do Lobo**
40		**Aveiro**	**Costa Nova**	**Vagueira - Mira**
41		**Porto**	**Póvoa de Varzim**	**Estela**
42		**Setubal**	**Costa da Caparica**	**Cova do Vapor**
43		**Açores**	**-**	**Açores islands**
44	**Romania**	**Dobrogea**	**Constanta municipality**	**Mamaia**
45		**Constanta**	**-**	**Danube Delta**
46	**Slovenia**	**Slovenian coast**	**Piran and Izola municipalities**	**Slovenian coast**
47	**Spain**	**Cataluña**	**Garraf**	**Sitges**
48		**Cataluña**	**Amposta**	**Ebro delta**
49		**Illes Balears**	**Alcúdia**	**Can Picafort**
50		**Valencia**	**Almazora**	**Castellón**
51		**Vasque Country**	**San Sebastián**	**Playa de Gross**
52		**Murcia**	**San Pedro del Pinatar-San Javier-Cartegena**	**Mar Menor**
53		**Canarias**	**Granadilla de Abona**	**El Medano**
54	Sweden	Skane län county	Falsterbo	Falsterbo Peninsula
55		Skane län county	Ystad	Ystad beach
56	United Kingdom	**Essex (Walton-On-Naze)**	**-**	**Essex estuaries**
57		**Isle of wight**	**Ventnor**	**Luccombe to Blackgang**
58		East Sussex	-	-
59		Yorkshire	-	Coast of Holderness
60		Northumberland	-	Humber estuary
61		*To be selected*	*To be selected*	*To be selected*

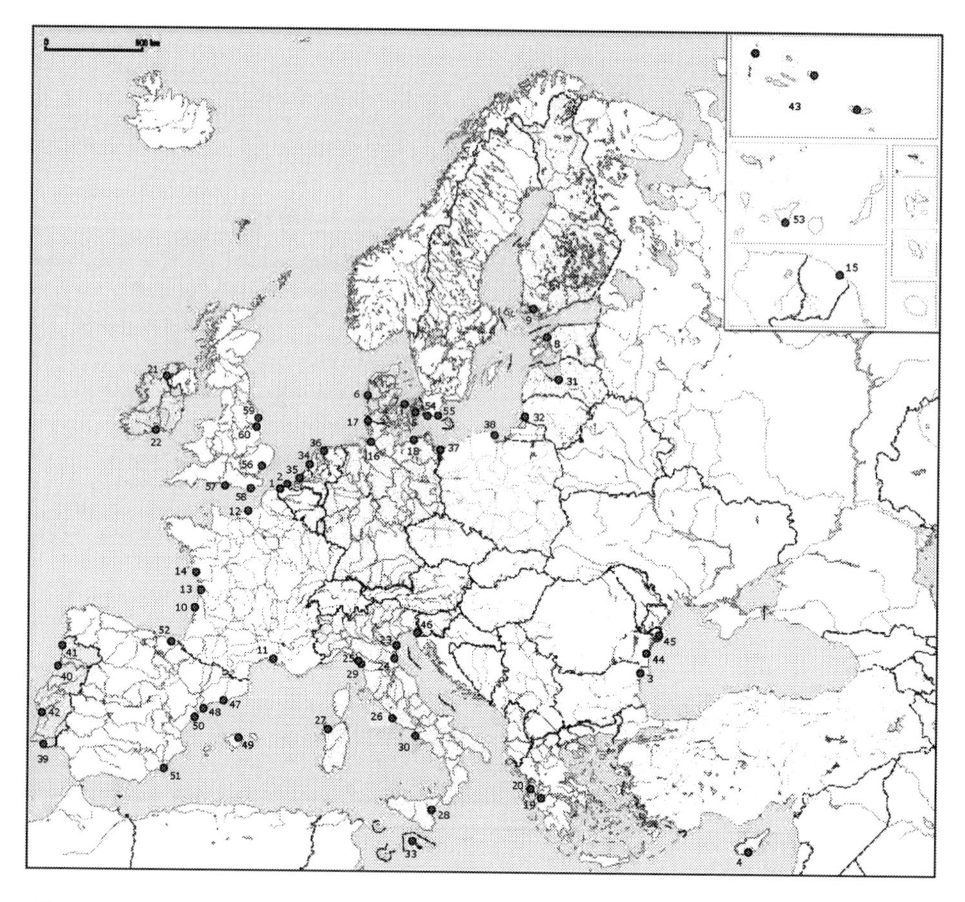

Figure 2. Location map of the selected sites.

The criteria for selection of the pilot sites are: erosion problem, physical type, policy options, socio-economic function, technical protection measures adopted (including innovative ones) and geographical distribution (Table 2). The analysis of the information will be focused on coastal types, characterised by distinct erosion behaviour, the responses to the driving forces, physical (climate, climate change) and human (constructions in the coastal zone) and the responses for the control and management of coastal erosion. Technical documentation on the methods adopted for the maintenance and protection of eroding coasts will be part of the information needed.

The objective of the first phase of the research is to review and analyse several aspects of these case studies, which are grouped in *5 assessment levels*: coastal typology, policy options, technical measures, socio-economic aspects and impacts. The analysis emphasizes in the technical measures, their effectiveness and the possible undesirable effects.

Table 2. Selection criteria for case studies.

Criteria	Goals Foreseen
Erosion problem	Selected sites have to face erosion.
Physical types	Selected sites have to be representative of the 5 EUROSION's coastal types: rocky coast, beaches, muddy coasts, artificial coasts and mouth (e.g. estuaries, rias, fjords,...).
Policy options	Selected sites have to be representative of the 5 policy options available to manage erosion (DEFRA, 2001): hold the line, limited intervention, move seaward, managed realignment and do nothing.
Socio-Economic functions	Selected sites have to be representative of the following socio-economic functions of the coast: agriculture, forestry, fishery & aquaculture, tourism, urbanisation, water management, transports, ships & ports, industry, energy, nature conservation, others (military defence...).
Technical measures	Selected sites have to be representative of existing shoreline management and coastal defence practices including pioneer and innovative technical solutions.
Geographical distribution	Geographically distribution of the selected sites has to cover all the European Union member states, overseas territories and accessing countries.

The information of *good and best practice examples* will be stored in the database without judging or giving advise, showing the measures taken and the consequences derived, and providing background information on erosion, causes, effects, possible counter measures and the way these elements find their respective places in management strategies. This needs to be done by the users because of two reasons:

- Coastal managers are most probably not willing to deliver information on coastal erosion if they are classified as bad example,
- Measures taken in one region may be good, while the same measure in a different region may not be the best one.

Results

After a primary analysis of the information provided by the different pilot sites in their reports, some results can be extracted. The results presented correspond to the thirty-nine UAB selected pilot sites (see Table 1 for location), following the 5 assessment levels described above.

Coastal typology

The pilot sites analysed show a wide diversity of coastal typology and a variety of origins for the erosion. The distribution of these types of coasts along the south Europe is as follows:

In one hand, *cliff landscape* type (rocky coasts) is widely present in the Atlantic Europe. The more exposed rocky and cliff coastlines extend along the northern France, the north west of Spain and south west coast of Portugal. The coastlines of northern France and northern Spain also include *estuaries* within the hard-rock dominated landscapes. Throughout much of this zone, the coast is characterised by a macro-tidal range which in the west of the France is the highest in the area.

On the other hand, along the exposed west-facing "Atlantic" shores, *coastal plains* (beaches) corresponding to sedimentary cells are also well represented (e.g. Aquitaine coast, Aveiro coast). Western France also has one of the largest expanses of dunes anywhere in Europe at "Les Landes", in the Aquitaine coast. Here their development involves the progressive stabilisation of dune forms as the sand is blown inland and vegetation helps to create the more typical undulating dune landscapes.

The two broad categories of coastal landscapes (high cliffs and low-lying flat land) are not mutually exclusive, nor restricted to particular geographical areas. There are, for example extensive cliffs in the Mediterranean, and areas of coast with little sediment input which correspond to the hard rock cliff landscape described for the north and west. These cliffs and more gently sloping rocky shores are often composed of various types of limestone which form the basis for the 'karst' landscapes of the hinterland (e.g. Sitges coast, Spain).

Deltas and narrow coastal plains, generally occupied by wetlands, lagoons and sand spits help to define the landscapes of the Mediterranean coasts. These are present throughout the region and are most extensive in areas backed by mountains where major eroding catchments deliver large quantities of sand and silt to the coast by torrents. This causes floods which also enhance sedimentary processes. This process combined with the small tidal range help to create some of the largest deltas in Europe: Danube, Ebro, Rhône and Po river deltas. All of these have been modified in some way by human activity whether through deforestation in the hinterland, damming of rivers delivering the sediment or other activities in the deltas themselves. The majority of the deltas are of considerable importance for both human use and wildlife, providing food and locations for urban and industrial development for the former, and migration, breeding and wintering areas for the latter. The *dunes* associated with these sedimentary areas may also be extensive. These often support a shrub vegetation similar to that of the hinterland. Many large afforestations (e.g. with pines and Australian Eucalyptus) have reduced the area of natural dune landscapes. The development of infrastructure associated with the growth of the tourist industry has also contributed to change many natural areas and has had a major impact on much of the landscape.

Offshore there are significant areas of relict sediments or with low terrigenous sedimentary input (mainly in Mediterranean insular environments), which support rich and varied benthic communities. However, one of the most significant marine landscapes derives from the presence of extensive sea grass communities, notably *Posidonia oceanica* prairies. These play a vital role not only in the biodiversity of the seas but also as nursery of many of the commercial fish stocks of the Mediterranean Sea, as long as the first source of beach sediments in many cases, derived from the high carbonate productivity of organisms (e.g. Balearic Islands, Spain).

The relative sea level rise can have important implications for the future of Mediterranean deltas, or Atlantic estuaries. However, the pattern of change is much more complicated with tectonic movements and natural or human-induced subsidence at different time scales. When this is coupled with human influences which exacerbate sea level rise, significant problems of erosion, salt water intrusion and flooding can occur. These effects are especially important in the major deltas where a decrease in sediment availability and a continuous natural subsidence coupled with water pumping or the sheer weight of infrastructure may be some of the factors which give rise to substantial problems of erosion and flooding as is being experienced in several of the major Mediterranean deltas.

Progressively, artificial (e.g. Piran harbour, Slovenia) or high protected shorelines (e.g. Toscana, Italy) are increasing their presence along the Europe. This tendency seems to be likely to continue in the future if other coastal protection policies don't take into account the diversity of coastal typologies and dynamic processes related to them.

Technical level

There is a wide range of engineering methods and techniques for coastal defence which operate in different parts of the shoreline. Offshore (nearshore) techniques (away from the shoreline) include *detached breakwaters*, stable bays, barrages and lately nourishment. Techniques operating on the lower shore between low and high tide include *beach nourishment, groins, revetments*, and sedimentation *polders*. *Seawalls*, revetments and flood embankments are located at or just above high tide on the upper shore as is the landward extent of the managed retreat technique. Finally, suprashore techniques which operate at or above spring tides include *dune building, cliff strengthening*, and *beach ridge restructuring* (Ministry of Agr. Fish. and Food, UK,1993).

The methods for addressing coastal erosion can be classified in two different approaches, (hard and soft engineering solutions).

- **Hard engineering**: establishment of structures which aim to resist the energy of the waves and tides. Such structures include; breakwaters and seawalls designed to oppose wave energy inputs, groins designed to increase sediment storage on the shore, and flood embankments and barrages designed as water tight barriers. It is important to notice here that the hard engineering structures modify strongly the natural hydrodynamic movements of the water and sediments of the area where they are applied. Therefore, in the most part of the cases the hard engineering structures can increase the erosion processes in neighbouring areas.

- **Soft engineering**: establishment of elements which aims to work with nature by manipulating natural systems which can adjust to the energy of the waves, tides, wind and especially the sediment transport. This approach has benefits while reducing environmental impact of traditional engineering structures. The methods which can be used include artificial beach and dune nourishment, the set back of wind sediment traps and plantations.

At present, in the Mediterranean countries of Europe most coastal defences include aspects of both these approaches. This method is known as "mixed" solutions for dealing the coastal erosion. Current trends in Europe focus on the concept of shoreline management, working with the dynamic nature of the coastal environment rather than fighting against the forces of the sea. This is best exemplified by the widespread move away from hard engineering methods of coastal defence which act to restrain coastal processes, towards soft engineering approaches which recognise the dynamic nature of the coastal environment by utilising these processes to take advantage of them. Soft engineering methods usually have a lesser impact on the environment and may require less maintenance.

Examples of "mixed" and soft solutions can be found in the Atlantic coast of France (e.g. Aquitaine region), where beach nourishment and dune regeneration were implanted as the best solution for stopping the erosion problems. In the north west of Italy (e.g. Marina di Massa, Italy) the construction of submerged groynes added to the beach nourishment of the area has been established as the best solution in order to deal the erosion issues. At present, in

Spain (e.g. Can Picafort and Sitges) and in Portugal (e.g. Vagueira) the mixture between hard and soft engineering solutions are adopted for dealing the erosion issues. It is important to point here, that not always the "mixed" and soft solutions are the best methods for protecting the coastline. Examples of this can be found in Spain (e.g. Castellón) and in France (e.g. Rhône delta), where the construction of different perpendicular groynes and coast parallel breakwaters have been established respectively as the best method for stopping the erosion of the coastline.

Other countries as Bulgaria, Romania, Malta, Slovenia, Greece and Cyprus where the defence of the coastline is a new concept, have adopted the construction of hard engineering structures in order to protect the coastline till latest 1990s. In most part of these countries the hard structures were built illegally or without any environmental studies, producing negative effects in the surrounded areas. For example in Dolos-Kity (Cyprus), different groins have been built by private owners to protect their properties without any environmental study.

The application of soft engineering solutions, taking into account the natural conditions of the environment, is a new or even non-existent concept in these countries. In Slovenia, for example, the most part of the coast is covered of hard structures (seawalls, submerged breakwaters and rock dikes) producing an aesthetic impact in the landscape. In Bulgaria, due to the lack of financial resources, only some hard measures as dikes and concrete walls have been built recently for fighting erosion problems of the coast. In Romania there have adopted different "mixed" measures. For example in Mamaia (Constanta), at the end of the communism period, different parallel breakwaters and beach nourishment were established for dealing the erosion problem. The beach nourishment was not effective in this area because of the restoration of the beach was done with less grain sized sediments.

These results show us a clear difference between west and east European countries. The former have started to fight coastal erosion since early the last century, while the latter ones have started in the 80s or even in the 90s. Besides, there is also a clear difference between the techniques adopted for dealing with the erosion of the coastline in the different countries. Portugal, France, Italy and Spain have developed, since the 90s, the mixed and soft techniques as the best methods for dealing with erosion problems and taking into account the natural environment. On the other hand, in countries like Cyprus, Malta, Bulgaria, Romania, Slovenia and Greece, soft and mixed techniques are new methods that are starting to be applied. Presently, all European countries are taking into account the principles of sustainable coastal management and different policies according to these.

Socio-Economic level

This level can be divided in three coastal domains following the main socio-economic aspects described in Table 2, as seen:

Atlantic Coast

Coastal erosion is an important socio-economic problem on all the Atlantic European coast, especially in France, Spain and Portugal, but also in the North Sea and Baltic countries. The rate of concentration of population in relation to the rest of its territory is high in the Portuguese and north Spanish coasts, being less than that of the Mediterranean sea. The main conflict in these regions is among all the urbanisation but also the occupation of agriculture or forestry areas, the over exploitation of natural resources, the occupation of risk areas with the consequent need for protection causing hydrodynamic unbalances. In certain areas of the

Atlantic French coast, like Aquitaine, forestry with the introduction of pine plantations to stabilise the dunes has been a major industry that at the same time has contributed to avoid erosion and to the stability of the coastal ecosystems. Tourism is an important sector in France, Portugal and Spain. In Portugal it accounts for approximately 5% of the employed population. In Spain and France Atlantic coasts the tourism is an important resource, but always in a less degree than in the Mediterranean coast. However, tourism in the Atlantic coasts and the associated industries has the problem of seasonality.

In those countries in which coastal management is centralised, like France, Portugal or Spain, the finance of coastal works to avoid erosion problems is mainly based on Government budget, although regions are progressively having a more important role. In Spain the responsibility belongs to the Ministry of Environment and the autonomies has some rather reduced competences and a consultive role. In Portugal the central level represented by the Ministry is organized in territorial demarcations.

Mediterranean Coast

Tourism and recreation are, without any doubt, the most important activities on the Mediterranean coast, where coastal tourism accounts for 30% of total world tourism (France, Spain and Italy are the 3 leading touristic countries) and in many places, and differently from the Atlantic coasts, it works most of the year round. Current trends show migration to coastal zones for work and welfare reasons. The majority of the Mediterranean regions show a high concentration of population on the coast. All this causes high pressure on the environment, but statistics show a continuous increase in this activity. Coastal erosion is a major concern for managers of the coastal zones for many of the touristic resources and infrastructures are highly dependent on the existence of beaches. Consequently, coastal dynamics and processes have an important role to maintain beach equilibrium. Examples of incorrect coastal management activities leading to the loss of beach systems are frequent (e.g. Can Picafort-Mallorca Island, Marina di Massa-Italy, Cyprus). In touristic localities it is often necessary to make a big effort in terms of maintenance by public services of the beaches and that implies an important economic and management effort for local authorities. When erosion is not very patent this often makes people forget the important measures to avoid or to prevent it.

Due to the negative hydrological balance in the Mediterranean region, water is one of the main socio-economic constraints for economic development. As urban environments grow with tourism it is necessary to improve the different uses of the hydrologic resources that can be an important source of socio-economic conflict.

Soft techniques, as beach nourishment, are being perceived more and more as an expensive, temporary and not always environmental friendly, but effective and a quick solution. In a long term it is necessary to achieve other kind of solutions. Dikes and other hard engineering solutions are normally also very expensive measure but in certain emplacements, when the changes in sediment dynamics have been very well foreseen, it can provide for the necessary stability to prevent erosive processes. Although much cheaper and often more useful than hard engineering solutions, soft or mixed solutions like the use of sand fences and plantations, combined with permeable groins, drainage systems or minor structures, can help the stability of beach systems in the Mediterranean coast.

Industry is another human use that takes place on the coast and puts the environment under high pressure. Inappropriate sitting of this activity has led to the loss of coastal ecosystems and incompatibility with other uses as it happens in Cyprus or in the region of Tuscany (Italy) where it is an important human activity specially in construction.

Black Sea Coast

The socio-economic conditions of the coast in Black Sea countries are generally characterised by current depressed economic conditions, structural problems in industry, unemployment, outdated equipment primarily in ports and lack of investments. Priority actions are set to achieve an economically self-sustainable system of land use and nature use, to develop a sustainable tourism strategy to attract investments for tourism and recreational destinations, to promote the efficient use of transportation, to adopt a better management of the agriculture and forestry and to develop aquaculture. The Strategic Action Plan for the Rehabilitation and Protection of the Black Sea represents an important step in the process towards attaining sustainable development in the Black Sea region.

Part of the current improvement of the coast is channelled through the touristic potential and its capitalisation. The most remarkable resort in the zone is the Danube Delta. Its favourable climate as well as to its rich fauna and flora, make up a consistent touristic potential for it can meet various internal and international touristic demands and even generate an special attractive for scientific and oriented tourism. Cultural and historical heritage of the Black sea coasts of Bulgaria and Romania are important mark-ups that have not yet reached its full potential.

Statistical indicators show a distinct trend of decreasing the cultivated areas, however the reed economy is still intensely used in the rural household economy as well as in a bigger scale. Fish farm breeding arrangements for consumption is revealing as an important resource for the capitalisation of the natural resources and is in the increase and is divided in the fresh water fishing resources (mainly in the Danube Delta) and the marine fishing resources (spread on about 113,000 ha of the Black Sea).

The industrial capitalisation of these resources has brought about the development of some specific industrial branches: oil processing, transports activities (ship building and repairing, prefabs, naval equipment), easier transport of raw materials like limestone mainly on the Danube or fishing. Thus, the active population employed in industry is an important percentage of the total employment.

No information available about cost of coastal protection carried out in the communism time. Romania and Bulgaria have a centralised model based on the main Ministries related to environment. The historical way of managing politic affairs in Romania and Bulgaria has the current consequence of not involving an active participation of stakeholders an local institutions.

Policy level

Coastal protection policies can be made for reasons of protecting financial goods (e.g. capital at risk), securing the local population, or protection of ecologically valuable areas. The review of policy options used in the various EU Member States provided a particularly useful approach of Generic Policies (DEFRA, 2001). The EUROSION project adopted this approach, analysing the following policy options: *hold the line, limited intervention, do nothing, managed realignment and move seaward.*

Broadly speaking, it can be said that coastal policies start out with massive tourism activity, being ca. 1970 the inflection point of this trend in Mediterranean countries. Beaches are ever since intensively used, buildings, roads, shopping centers, airports, marinas are set anywhere within this narrow strip of land. Thus, high valued properties pop up throughout the shore and hazards derived from coastal erosion becomes a problem. Before 1970, except for Greece, policies consist of an amount of individual isolated cases, most of them on sea-facing cities where ports and auxiliary industries (shipyards, fish food industry, etc.) exist, which were the most valuable assets at risk.

Then the policy can be easily traced through a rough legislation analysis. When a country feels itself threatened, particularly when human activities are considered to be of importance, a written piece of regulation is produced. Actually, speaking about coast (shore) legislation is to speak about land-use planning on a particular piece of land, a narrow strip where land and sea come together and, occasionally, it is difficult to separate the one from the other. However, just Romania has produced a land-use planning act for the whole country, distinguishing a separate policy for coasts within it, whereas the rest of them have produced specific texts for regulation of coasts, shores or the littoral area. In France, besides the preponderance of central management every coastal project involves a great diversity of institutions (local, regional and national) and it is, by far, the one that has more tradition in including stakeholders in the decision-making process) out of the three mentioned ones.

Apart for tourism, other policies might have affected the quality of the coast and littoral zone over the century; however no importance seem to have been put on them. *Fishing policies* near shore: dragnets leading to decay of *Posidonia oceanica* meadows; these ones contributing with sand on the beach system; *agricultural policies* demanding irrigation systems and electrical power generation, both driving to the building of dams, these trapping sediments to nourish beaches, and so on, are just a few examples.

Policies need agents and funds in orde to make an adequate progress. First public bodies dealing with coastal issues under the different laws are usually port or harbour authorities (e.g. Spain, Italy, Portugal) or mercantile marine (e.g. Greece) switching to specific departments under the public works sector through the 1970s and, more recently, to environment and nature protection bodies.

Troughout the 1980s and 1990s, policies were reviewed all over Europe. As a major consequence of hard defence structures (dikes, seawalls, breakwaters, etc.), commonly used during the 1970s, adverse results elsewhere (particularly down drift) ocurred often. Except for a small number of cases, these kind of *hold-the-line* projects are forgotten and the new policies of groyne structures and 'beach nourishment' continued until the very end of the 19th century, more on a *managed realignment* and *limited intervention* way. This last policy, however is brought under discussion as it is proved to be only a short-term and an extremely expensive solution. It was thus in the 1990s when the concepts of integration were first considered under what is called Integrated Coastal Zone Management (ICZM). All related policies (former tourism, land-use, transport but also agriculture, fishing and above all, nature conservation) are brought to the same decision-making process. So, sustainability and holistic approach are two milestones under which erosion control policies start this 21st century.

Previous considerations are generally valid for all different types of coast although it has been basically extracted from EU Mediterranean countries' experiences. The Atlantic coast of Portugal and north Spanish coast face two different variations to the problems identified for the Mediterranean Sea:

a) tourism pressure is lower, and
b) coast is adapted to a high energetic sea climate.

These different characteristics are still typical of the coastline today.

Finally, the Black Sea countries (Bulgaria and Romania) have fearfully started to define policies for coast management in the 90s, although major concerns are focused on water pollution. The Black Sea has been declared by UN as one of the most polluted seas in the world; the 'Bucharest Convention for the Protection of the Black Sea' has been signed by all the countries sharing their territories with it and is the sole expression of a political willingness to address coastal erosion (art. 69 of the Black Sea Strategic Action Plan).

Environmental & Socio-Economic Impacts

The major impacts on the environment are due to the presence of coastal defence structures and the hydraulic works that restrain sediment input to the coastal system. Sometimes the protection structures are noticed to have a positive effect on the coast they are protecting, but a negative one on the neighbouring regions. The impacts of hard structures are usually more severe than soft solutions, and they are irreversible. Hard engineering generally results in long term changes in coastal and near shore bottom morphology alongside protected areas. The construction of hard structures caused the destruction of most part of the habitats like dunes system and coastal and submarine vegetation represented by Mediterranean flora. The tourism sector, agriculture and the fishery activity have proven to be the most affected by erosion.

Soft measures are proven to be more effective than hard measures. They also are a more environmentally friendly approach which works toward providing a dynamic equilibrium at the coast whereby erosion and flooding are kept to a minimum. Only, a well knowledge of coastal and sedimentary processes, can help to take a correct solution. One typical example can be beach nourishment plans: the maintenance of beach amplitude for some years needs of periodical recharges, not previously taken into account in the scheme. The beach drainage systems used in Sables d'Olonne (France), Ebro delta (Spain), Isle of Procida (Italy) and Vecchia Pineta (Lido di Ostia, Italy) are proven to be effective to stabilise the shoreline when there is a good maintenance.

Discussion & Conclusions

Erosion is a natural process like the other natural coastal processes of transport and sedimentation, all of them originated by marine and wind energy. Broadly speaking, only that fraction of erosion induced by humans activities might be undesirable but not unavoidable.

As erosion processes depend on the sort of materials along the coastline (sedimentary or rocky coast) different types and rates of erosion can be considered. While in the Atlantic coast major erosion causes include 'relative sea level changes', sediment supply and anthropogenic causes, in the Mediterranean Sea erosion problems are presently often related to anthropogenic littoral modifications closely connected to touristic infrastructures which, basically, hinder the sedimentary drift, in addition to other natural processes.

A thorough legislation analysis often provides a good indicator to the necessities of a country. Laws, acts and other type of regulations on coast and shores are produced in a close relation to tourism development. Apart from Greece, where a Law "On the Foreshore and the Waterfront" (Law 2344/1940) appears as early as in 1940, coastal policies throughout Southern Europe countries initiated in the 70s. Until then, erosion problems were related to other coastal policies, especially, those concerning ports safety and flooding prevention in metropolitan areas.

As stated before, policies need agents and funds for an adequate progress. First public bodies dealing with coast issues under the different laws were usually port or harbour authorities (Spain, Italy, Portugal) or mercantile marine (Greece) switching to specific departments under the public works bodies during the 70s and, more recently, to environment and nature protection ones. As valuable properties increase, erosion and flooding control, especially following big storms and surges becomes a critical issue. So, a clear 'hold-the-line' policy is implemented, and sometimes overcome by a more aggressive 'move-seaward' one.

Eastern countries like Slovenia are now starting to consolidate tourist infrastructures and it would be wise if they could learn from the mistakes made in the past in Western Europe. On the other hand, land-use planning plays an important role in the Mediterranean countries. Forestry, agriculture or local industry are often important sources of conflict to coastal management policies. With the exception of France, centralist politics of Mediterranean countries is a major element influencing the public participation, which can be rated until now as poor. Centralised governments have a weak tradition in broadening participation processes. As a consequence stakeholders involved on coast defence have fewer opportunities to influence in political issues, thus management is subject to less control and transparency.

Apart from tourism, other policies might have affected to coast and littoral health along the century; however no importance seem to have been put on them. A few examples: Fishing policies near shore (dragnets leading to decay of prairies of *Posidonia oceanica*) agricultural policies demanding irrigation systems and electrical power generation, both pushing to the building of dams, these trapping sediments to nourish beaches, and so on.

Concerning technical aspects, at the beginning major coastal defences consisted of hard structures (dikes, seawalls and break-waters) in a very 'port-style'manner, perhaps because most of the managers came from these departments . Planning-use was mainly focused on urban development by the sea, thus seafront promenades, vehicle access to the beach and other capital-intensive structures popped up everywhere, particularly on the Mediterranean coast. Together with dwelling areas, a number of shopping centres, roads, airports and marinas completely covered the narrow 500 m-wide strip from the sea. This situation was seriously worsened by the massive building of dams throughout the continent.

Currently, there is a clear differentiation between hard (seawalls, groins, seawaters), soft (beach regeneration, dune revegetation, introduction of wetlands and buffer areas), and mixed (submerged groynes plus beach nourishment) techniques to solve coastal erosion problems. Broadly speaking, soft techniques are apparently preferred when conditions are suitable, because of the lower costs and environmental impacts, but not always. During the 80s coastal policies evolved from mentioned hard works of stone and concrete to a less capital-intensive projects (basically beach nourishment) which have lasted until very recently. Yet, these policies are currently under discussion as they prove to be only short-term and extremely expensive solutions.

It was during the 1990s when the concepts of synergy and integration are considered under what is called Integrated Coastal Zone Management (ICZM). All involved policies (tourism, land-use, transport but also agriculture, fishing and above all, nature conservation) are brought to the same decision-making process. So, sustainability and holisticapproach are two milestones under which erosion control policies start with this 21st century. Conservationist groups are becoming each time more influential on politics and ICZM policies are progressing slower than in Northern Europe.

References

Bolsius, E et al. (1999). A Coastal Zone Perspective. Preparatpry study. Interdepartmental Vision for the Coast, Dutch Gov, Den Haag, The Netherlands.

EUCC (1987). Threats and Opportunities in the Coastal Areas of the European Union. National Spatial Planning Agency of the Ministry for Housing, Spatial Planning and Environment. The Netherlands.

DEFRA (2001). Shoreline Management Plans: a Guide for Coastal defence Authorities. London.

EUROSION (2002). Trends of coastal erosion in Europe. September 2002.

EUROSION (2003). WP4.1 D.2 Report of UAB Pilot Sites. Internal Report.

EUROSION (2003). WP4.1 Review of existing practical experiences and experiments in erosion management and control at national level. Analysis Report (Internal).

Hanson, H., Brampton, A. et al. (1999). Beach nourishment projects, practices and objectives. A European overview, SAFE, European Commission.

Lechuga, A. (1994). Littoral dynamics and shoreline erosion: selected Spanish cases. U.S. Spain Workshop on Natural Hazard

Muñoz-Pérez, J.J. (2001). Cost of beach maintenance in the Gulf of Cadiz (SW Spain), Coastal Engineering 42 p. 143-153

Segar, D.A. (1998). Introduction to Ocean Sciences, Wadsworth Publishing Co., New York.

Climate change and the coastal challenge: innovative means of stakeholder involvement

Naylor, Larissa A. [1]Environmental Consultant, Komex, 129 Cumberland Road, Bristol, BS1 6UY, email: lnaylor@bristol.komex.com and [2]Visiting Fellow, Tyndall Centre for Climate Change Research, University of East Anglia, Norwich, NR4 7TJ
O'Riordan, Timothy [2]Tyndall Centre for Climate Change Research, University of East Anglia, Norwich, NR4 7TJ
Gill, Jennifer A. [2]Tyndall Centre for Climate Change Research, University of East Anglia, Norwich, NR4 7TJ
Watkinson, Andrew R. [2]Tyndall Centre for Climate Change Research, University of East Anglia, Norwich, NR4 7TJ

Abstract

A major research theme at the Tyndall Centre for Climate Change Research, UK, is concerned with 'Sustaining the Coastal Zone'. Research projects within this theme are focussed on examining the effects of different sea-level rise and adaptation scenarios on coastal habitats and communities in East Anglia. A key aspect of this work is to involve stakeholders throughout the research process – by encouraging local citizens to help participate in shaping the future of coastal communities and habitats. Part of this research has involved a series of agenda setting workshops entitled, 'Redesigning the Coast' which has led to some very fruitful discussions about new forms of institutional governance in the coastal zone and innovative means of stakeholder dialogue. The series has progressed from engaging with scientists, government and non-government organisations, to the residents of Norwich and, lastly, to include 67 representatives from local parishes along the coast of East Anglia. This paper details how the 'Redesigning the coast' workshops have been used to improve the dialogue between scientists and stakeholders, by raising local awareness of coastal responses to climate change and helping feed local knowledge into research projects.

I. Introduction

The soft coast of Britain is facing unprecedented change in character and governance. The coastline in the east is gently dipping seawards, due to iso-static adjustment following the ice-age. Sea level rise prompted by climate change and global warming invites a prospect of stormy conditions and ever increasing high tides (IPCC, 2001a, 2001b). The current line of defence cannot be guaranteed under such conditions, not at least without huge expenditures, much disruption to the natural character of the littoral zone, and uncertain outcomes for human livelihoods and well-being. Coastal and river flooding has become a major issue of testing competence in government. The widespread floods of 2000 and 2001, plus the prospect of more stormy and turbulent seas, have resulted in the financing and strategic management of flooding becoming high priority issues in national and local political debates. A series of reviews by the House of Commons Committee on Agriculture (1998), the Department of Environment, Food and Rural Affairs (2001a-c, 2002) and the Environment Agency (2001) have begun a serious public examination of just how well current administrative and financing arrangements are meeting the long term and reliability tests of sustainable flood and coastal zone management. These issues, coupled with the complex natural functioning of coastal systems, mean that coastal zone management is increasingly politicised, procedurally complicated and fraught with communication and administrative challenges, because many different bodies manage different components of the coastal zone (Turner, 2000; Turner, 2002).

As an introduction to examining the coastal challenge and as a mechanism to effectively illustrate the coastal arena, its inherent natural complexities, the current pressures on the coastal zone and how climate change maps onto this intricate web of disciplines, processes and policies, a theatre production will be used. It is this arena in which the Tyndall Centre for Climate Change Research is working, and where new methods of stakeholder involvement are being explored.

II. The Theatre (e.g. the coastal arena)

The Theatre Company is an amalgam of vested interests shaping the coastal zone. The stage will be the world as backdrop and, UK coastal systems, on the boards, facing current pressures on coastal systems. The shifting global backdrop will illustrate how the climate envelope has the potential to affect coastal systems while revisions to the plot will highlight ways of grappling with the climate change and the coastal challenge. The actors represent all of the different vested interests in the coastal zone such as tourism, the local economy, biodiversity, fisheries, local communities and international obligations. Meanwhile, the directors and financiers are the users – the governments that administer, fund and monitor coastal zone management initiatives carried out by regional – local authorities and non-governmental associations, such as the National Trust. The audience represents the local stakeholders – those who live, work or depend on coastal areas for their livelihoods and homes. Lastly, the critics of the production represent the media, which play a considerable role in shaping public perception of coastal issues and the prospect of climate change. The production is called:

'Climate change and the coastal challenge: a United Kingdom perspective'.

Acts I and II: Setting the scene and current pressures

The stage: UK coastline

Coastal zones are dynamic, complex systems that are subject to natural variation and change; they support more than a third of the global population, yet they represent only 18% of the world's land area. Britain well illustrates these global processes and trends. It is a low-lying, island nation which is geographically situated at the margins of major northern and southern climate fronts where the Atlantic Ocean mixes with several seas and estuaries. Britain's land mass is also prone to change, with much of it, and southern and eastern England in particular, slowly sinking into the sea, as a result of isostatic rebound. Britain also has the second longest coastline relative to the total land area of the country; second only to Japan. Like Japan, Britain has historically had a strong affinity and dependence on coastal resource. During the 20th century great efforts were made by both countries to defend the coastline from erosion and flood risk. Britain responded to the catastrophic 1953 floods with generous government funding and hard engineering solutions, designed to defend the coastline and estuaries from a repeat of the 1953 floods. For two subsequent decades, this engineering biased approach has dominated coastal management arenas and has affected the public perception of risk and coastal change. The hard engineering structures have impeded the natural cycling of sediments in coastal systems, thereby causing accelerated erosion down the coast, as these areas become depleted of sediment. Meanwhile, the public has largely grown accustomed to a static coastline, with the expectation that the coastline is a relatively permanent feature and that the government has a responsibility to defend houses from coastal erosion. As evidence for this, successive reports of parliamentary committees reinforce a view that flood protection is a virtual "right" and that the job of government is to safeguard property for coastal and river flood risk.

Although there is a recent policy shift away from hard engineering approaches, to develop a defence system that works with the sea, the challenge lies in creating a shift in public perception from a static, defendable coastline to one that is dynamic and flexible. This deep-seated expectation is also evident in the first generation of shoreline management plans, when little change to existing coastal defences was accepted through public consultation. Indeed a study by Halcrow engineering (2002) has shown that only a dozen cases of coastal realignment have been permitted in UK shoreline management plans and that most of these are extremely small-scale.

The backdrop: current coastal pressures

Not only does the UK have a geographical position and form that is strongly affected by coastal processes, there is also a strong human dimension which adds pressure to coastal systems. The UK has one of the highest population densities in Europe with the majority of the UK population residing in coastal areas or along major riverine and estuarine systems. For example, important population centres in England are centred on soft estuaries, such as the Thames, Blackwater and Medway, all of which are vulnerable to increased coastal flood hazard.

Thus, the major pressures on the UK coastline are population and development pressures – which result in increased development in coastal areas, such as the flagship 'Thames Gateway' development, which aims to build a town the size of Cardiff in the outer Thames by 2016 (ODPM, 2003). Population, tourism and housing development growth has repercussions for many aspects of coastal systems – such as reducing the amount of habitat, which is commonly called 'coastal squeeze', overexploitation of coastal resources such as shellfish, fisheries and increased pollution.

Act III: the climate change dimension

The shifting backdrop: The Climate Change Envelope

When the risks of climate change are added to the mix, the challenges for the coastal zone are even more apparent. As the climate warms and sea level rises, the heavily populated low-lying areas of Britain will be at ever increasing risk of sea level rise associated with climate change. It is anticipated that 1:1000 year storm events will occur with a frequency of 1:100 years by 2030 (Woodworth et al., 1999). Coupled with this is the possibility of increased wave heights and incidence of storms, as well as the combined risk of high rainfall causing river flooding in winter months coinciding with high spring tides, thereby causing high intensity storms (Hulme et al., 2002). These changes have potentially serious ramifications for the low-lying and geologically soft areas such as East Anglia and the Thames Estuary. Meanwhile, the general complexity and uncertainty about coastal systems is compounded when climate change estimates are considered. For example, attempting to predict where species may move is challenging enough for vegetation specialists in sparsely populated areas, let alone in the coastal zone where development and coastal defence infrastructure may further limit where species can disperse. Furthermore, it is possible that the cost of defending the coastline may become so large, that it may become economically unfeasible to continue to defend large sections of the UK coastline.

Act IV: A Revised Script

The previous sections provide a simplified overview of the complex issues and challenges in the coastal zone. The present social, administrative and legislative structures are not equipped to cope with the additional pressures faced by climate change. Moreover, there is a need to model the complex interactions between the various factors influencing coastal change, to help understand the key drivers and controls on coastal environments – and to predict how climate change affects these processes (Rotmans et al., 1996; Holman et al., 2002; Schuchardt and Schirmer, 2002).

Current research at the Tyndall Centre for Climate Change Research is focussing specifically on the coastal zone, with nearly 1 million pounds of research funding being invested in examining the potential transformations of coastal dynamics as well as the economic and socio-cultural repercussions for coastal habitats and communities to climate change. Much of this research is focussed on three key aspects of the coastal challenge: (1) improving our scientific understanding of how climate change predictions will affect coastal processes such as flood risk, wave climate and sea level rise; (2) examining how coastal communities might adapt to climate change and; (3) most importantly, integrating the two spheres of research to examine the interrelationships between the two spheres of research and their combined affect on coastal sustainability in light of climate change.

It is hoped that this research can help de-couple many of the complexities of coastal systems, by pinpointing key factors influencing coastal communities and habitats and providing concrete measures which may help alleviate some of the climate change pressures on coasts. Possibilities include new forms of coastal governance, development controls and/or building regulations that may make coastal communities more resilient and coastal ecosystems more responsive to climate change.

Act V: Looking Forward: Successful coastal adaptation to climate change

As part of the Tyndall Centre's research portfolio on coastal systems, a series of workshops was initiated by the Tyndall Centre for Climate Change Research and the Centre for Social and Economic Research on the Global Environment to help explore the key scientific, policy and social issues surrounding the coastal zone. These workshops also examined different public perceptions of the coastal zone and coastal protection in the UK and sought to help engage and encourage local citizens and community groups to become actively involved in coastal management. The workshops, entitled, 'Redesigning the Coast' were co-organised, covered:

1. State of Scientific Knowledge: Coasts and Climate Change, October 2001
2. Policy Workshop, on new forms of coastal governance March 2002
3. Public Workshop, on coastal futures May 2002
4. Parish Workshop, on parish involvement September 2002

The concept of 'Redesigning the Coast' refers to the need to explore innovative and perhaps radically different means of managing coastal systems in light of climate change predictions. It is a world apart from the notion of 'reengineering' and/or 'controlling coastal processes through hard engineering options'. Redesigning the coast is aimed at achieving a conceptual shift in coastal management practice, where new forms of coastal governance are actively pursued and developed. The concept of 'Redesigning the coast' is underpinned by two important premises: (1) English Nature's Learning to live with the Sea campaign (see http://www.english-nature.org.uk/livingwiththesea/) and (2) the need to develop mechanisms to enable successful coastal adaptation[1] to climate change. As such, redesigning the coast is primarily concerned with developing new institutional frameworks and mechanisms to enable responsive, flexible and sustainable coastal management arrangements to grapple with the challenges of climate change.

Sea level is predicted to rise for at least the next 50 years, even if all carbon dioxide emissions were ceased tomorrow – and as a consequence, the most important driving factor influencing coastal change is how we choose to manage the coast. Consequently, an overarching principle of 'Redesigning the Coast' is that coastal responses to climate change, at the local to regional scale, are primarily dependent on the decisions local communities and governments have made. Where we build houses, how we choose to defend our coastline and whether we allow managed realignment (controlled coastal flooding of reclaimed or low-lying land, such as freshwater marshes) will strongly influence the impacts of climate change on coastal communities. Likewise, the behaviour of individuals and institutions in their production of greenhouse gases, from motor vehicles or houses, will also have a contributory influence on the rate of climate change. For example, the recent change in planning guidelines to reduce the potential for new building to be built on floodplains or known areas of cliff erosion, has the potential to greatly reduce the need for more flood defences and the damage caused by any floods which do happen (ODPM, 2001). As the decisions we make are strongly related to public perceptions of coastal systems and because the dominant public view is that 'the government is here to defend the coast for future generations', this paper focuses on methods and mechanisms to encourage greater public understanding of and interest in coastal processes and decision-making.

III. Redesigning the Coast Workshop Series

1. The Science Workshop

The Science workshop assessed the state of scientific knowledge over coastal oceanography, sedimentology, geomorphology, ecology and ecological economics to understand better the scope of strategic coastal redesign. The findings of this workshop have been published as a Tyndall Centre working paper (Gill et al., 2002). A number of eminent scientists came together to share research findings and identify the most likely changes to coastal systems in response to climate predictions and to highlight key gaps in our current understanding of how climate change may affect coastal systems.

The main findings of the workshop were that:

- There is likely to be increasing storminess, with higher sea levels over the next 50 years and beyond (Woodworth et al., 1999; Flather et al., 2001; Lowe et al., 2001);

[1] Adaptations are actions taken to reduce the magnitude of the adverse impacts of climate change (or to exploit the beneficial impacts). Adaptive Capacity is the ability of a system to adjust to climate change, including variability and extremes, to moderate potential damages, to take advantages of opportunities or to cope with the consequences, IPPC, TAR, 2001.

- Coastal geomorphology, such as sediment transport, erosion and deposition, will increase and become more noticeable (Nicholls and Wilson, 2002; Hosking et al., 2001; Lee, 2001);
- Coastal tidal mudflats and creeks will continue to evolve in a manner that may not be predictable (Pasternack and Brusch, 2002; Schuchardt and Schirmer, 2002);
- Coastal salt marshes are eroding as a consequence of various processes, including coastal squeeze (see Figure 1). Habitat restoration through coastal habitat recreation programs provides a basis for continued salt marsh habitats in the UK (Dixon, 2002). However, the results of habitat recreation trials and the natural conversion of farmland back into salt marsh after breach events are variable and unpredictable (Hughes and Paramor, 2002). Due to land use change, intensive agricultural production, changing water levels and climate-change related events (warming of surfaces, variable water quality and levels, altered biological activity), there is a large variability in the response of ecosystems and species to such changes and importantly, their resilience and response rates are not well understood (Huntley and Webb, 1989; Berry et al, 2002; Hawkins et al, 2002).

The main research needs were identified as:

- Systematic monitoring of coastal processes, including tides, wave heights, sediment transport and coastal ecology, including nearshore marine biodiversity change, is needed on a grand scale. This monitoring should be continuous, well networked across the UK and the results of such activities should be incorporated into experimental management schemes. Such monitoring and data gathering needs to be well managed and funded in a sustainable fashion, to deliver reliable, quantitative results.
- The various systems (e.g. geomorphological, ecological, oceanographic and coastal protection) affected by climate change need to be examined together, and subjected to rigorous evaluation of biocomplex[2] interactions between these systems and importantly, the cumulative effects of climate change on coastal biocomplexity.

---

[2] Biocomplexity is primarily concerned with exploring between the margins of systems, to understand and begin to decouple the complex interdependencies between organisms and the environments which sustain their populations, affect them and/or are modified by the organisms themselves (www.hsf.gov/od/lpa/news/media/99/fsbioenv.htm). Biocomplex systems are typically non-linear and chaotic which makes prediction difficult; the interactions within a given system can often span multiple spatial and temporal scales and biocomplex responses are found in extreme to benign environments, in microbial to mammalian communities (esa.sdsc.edu/factsheetbiocomplexity.htm). It is a fundamentally interdisciplinary area where biological, chemical and physical sciences work alongside social scientists, economists and computer modellers, to grapple with the complex interactions between different components of natural systems and/or natural systems and the human environment (Mervis, 1999).

Figure 1. Illustration of coastal squeeze, in Richmond, near Vancouver, Canada.

2. The Policy Workshop

The second workshop was designed to build on the scientific basis of the first workshop and bring coastal planners, managers and decision-makers together with coastal protection officers, scientists and researchers. It sought to examine how far current arrangements for managing, financing and constructing coastal change are suitable for changing circumstances and planning approaches, and to review the scope for redesigning these institutional arrangements. Potential shifts in coastal management – the redesign element – range from an evaluation of whether the current financial and economic-ecological appraisal mechanisms are appropriate to how current land ownership issues are dealt with in coastal defence planning.

The principal findings for this workshop were:

- the administrative effort to change the current coastline is increasingly demanding in officer time, documentation costs and participatory effort with landowners and other interests. Even small-scale elements of coastal realignment can take four years to complete. Major policy shifts for shoreline management could take ten years;

- the formal regulatory requirements for undertaking environmental risk and economic assessments for coastal change decisions are hugely demanding in paper work, inter-agency discussions and stakeholder dialogue. All the signs are that these aspects are becoming even more onerous;

- the uneasy mix of coastal erosion protection and sea flood management currently run by two different sets of bodies with hugely different financing and planning powers mean that complete co-ordination for changing coastal defence is deeply problematic;

- there is a clear need for a detailed assessment of fundamentally fresh approaches to coastal governance. This is now a matter for research priority that is being funded by the Tyndall Centre;

- public private partnerships for long term coastal management are not, as presently devised, suitable for proactive coastal governance. They will need to be amended if they are to become suitable and effective.

3. The Public Participation Workshop

The third workshop in the series was designed to engage with the public – to share the knowledge obtained at the previous two workshops and glean an understanding of public perceptions of coastal systems and climate change in Norfolk. The meeting explored both scientific explanations and social visions of coastal processes and delved into less tangible notions of the sea and coast, such as the spirituality of the coast. This workshop was attended by about 50 members of the local population and was a lively and engaging session.

The outcome from that event was a genuine interest in new forms of political and administration approaches to coastal management, and more comprehensive and co-ordinated involvement from local people. This latter aspect became the focus for attention from the final workshop, which, like the third event, was hosted by the diocese of Norfolk.

4. Local Coastal Parish Involvement

The final 'Redesigning the Coast' workshop sought to bring to together many of the ideas put forth in the early workshops and with the assistance of an important local community group in rural, coastal Norfolk, generate ideas for action. The importance of this is twofold. On the one hand, coastal parishes can work to inform, educate and encourage the local population to take a greater interest in coastal processes and change. On the other hand, the local community can be encouraged to work with researchers to help contribute to our understanding of coastal processes, thereby actively engaging in the research process. These aims can hopefully be implemented to help foster a greater appreciation of coastal change and the importance of local decisions in securing a sustainable coastal future. At the same time, by encouraging local residents to get involved and become part of the coastal research process, as 'the ears and eyes on the coast', the large-scale shifts in public perception of coastal systems that can facilitate redesigned coastal governance, can hopefully be realised. Numerous ideas to facilitate local involvement in coastal processes and the Tyndall Centre's research program were generated at this workshop.

Several positive suggestions for how local parishes can become involved in helping shape the future of their coastal communities and work with researchers at the Tyndall Centre were made. They are grouped into the following categories:

1. Education

- Inviting speakers to local parishes to help improve local citizens understanding of climate change and coastal processes; using churches as the focal point such activities.
- Primary Schools – developing an education package for the 117 Church of England primary schools dotted along the Norfolk Coastline.
- Adult Education – developing a video outlining the key concepts of dynamic coastal systems, historical coastal changes and the likely effects of climate change on coastal habitats and communities.
- Local Councils – suggestions were made about how to encourage different teams within local councils to become more aware of the risks of climate change.

2. Helping researchers collect data on coastal change
   - *Oral Histories of Coastal Change* – A useful input to research on coastal change would be to gain an understanding of how the coastal line and/or livelihood of coastal communities have changed over the past century. This can be achieved by developing a community activity to encourage long-term residents of coastal communities to talk about and record any changes they have noticed in their local area. Examples of topics would include changes in: the coastline, direction of wind or waves, amount of storms, fishing patterns, economic interests in the community (e.g. fisheries to tourists) and/or changes in saltmarshes.
   - *Observations of current coastal change* – as the 'Eyes and Ears' on the coast helping to record coastal changes that local people regularly observe along the coastal paths people regularly walk, including:
     i. Establishing a recording system similar to that used to monitor bird population arrivals and departures, perhaps with logbooks or a recording book in local churches.
     ii. Potographic recording of coastal change from the same location at regular intervals and sending the results to be compiled and used by researchers at the Tyndall Centre.
     iii. Establishing local groups who can record changes before and after winter storms.
     iv. Liaising with researchers about observations made and developing straighforward, easy mechanisms to do this, such as a web address where a simple observation form can be downloaded and submitted.

3. Informal Local Activities
   - Being the **'Eyes and Ears'** on the coast. Observing changes in your local area such as arrival of birds, changes in the beach profile, breakdown of coastal defences, local flooding after heavy rainfalls
   - Coordinating educational walks along the coast.

4. Contributing to more formal local processes
   - The Shoreline Management Plans (SMP) process includes an element of public consultation. Parishes could encourage local people to take a more active role in managing their local environment by providing a forum to review and provide feedback on the second phase of SMP guidelines which will be developed over the next few years.
   - Planning proposals and decisions in local councils. Local people can be encouraged to scrutinise planning proposals more actively, ask whether flood risk has been

considered, how sustainable/energy efficient the proposals are and how close the buildings are to recessing cliff lines.

- Local Coastal Fora; these groups exist around the UK and are a good opportunity to become involved in helping manage the coastline and are a good source of information on other local coastal management initiatives in the UK.

Of the suggestions generated at the final ‘Redesigning the Coast’ workshop in September 2002, one idea has been developed into a product for the local community. An education booklet has been produced jointly between the Tyndall Centre, University of East Anglia School of Environmental Sciences and the East Anglian Business Environment Club. It has been delivered to each member of the coastal clergy for distribution to primary schools. The booklet is designed to provide some background on coastal systems and how and why climate changes, as well as some simple experiments for children to do, to learn about climate change (Figure 2, Naylor and Torok, 2003).

CLIMATE CHANGE
COUNTING THE COST
AT THE COAST

Figure 2. An education booklet on coasts and climate change for primary schoolchildren in East Anglia.

IV. Conclusions

Several new ideas have been generated to help bridge the gap between scientific awareness of natural coastal change and the potential for accelerated impacts under climate change and policy responses to facilitate adaptation to climate change. Moreover, the importance of bringing the local public on board and finding mechanisms to encourage the public to take an active role in coastal decision-making and coastal research are proposed. What is now needed is the energy, financial backing and time for dynamic and engaging researchers and community leaders to initiate the ideas put forth. This, together with a concerted effort to re-examine the institutional arrangements for making coastal defence decisions in the UK, will enable us to 'Redesign the Coast'. In doing do, there is a strong possibility that successful, cost effective and socially acceptable measures can be made to ensure sustainable adaptation to climate change in coastal communities.

This is a timely proposal. Currently the European Commission is demanding that every member state undertakes a "stock take" of its integrated coastal zone management procedures and outcomes. This is a major exercise and is revealing many of the imperfections in management and financing opened up by these seminars. At the same time there is a significant shift in policy in favour of managed realignment of rivers and waste rather than the maintenance of "hard" and costly defences other than from commercial and residential property. In any case the advantages of a more integrated natural defence approach of coasts and rivers is seen as more environmentally robust and socially acceptable. So the time is ripe for a kind of root and branch institutional review advocated in this paper.

References

Berry, P. M., Harrison, P. A., Dawson, T. E. and Pearson, R. (2002). Integrated impacts on biodiversity. In I. P. Holman & P. J. Loveland *REGIS: Regional Climate Change Impact Response Studies in East Anglia and North West England.* DEFRA, London.

Department for the Environment, Food and Rural Affairs (2001a). *Managed Realignment: Land Purchase, Compensation and Payment of Alternative Beneficial Land Use.* DEFRA, London.

Department for the Environment, Food and Rural Affairs (2001b). *Final Consultation Review: Revised Scheme Prioritisation System.* DEFRA, London.

Department for the Environment, Food and Rural Affairs (2001c). *Flood and Coastal Defence Funding Review: Report to Ministers.* DEFRA, London.

Department for the Environment, Food and Rural Affairs (2002). *The Flood and Coastal Defence Funding Review: A Consultation Document.* DEFRA, London.

Dixon, N. (2002). Recreation of Tidal Wetlands: opportunities at Abbot's Hall and Salcott Creek, Blackwater Estuary Essex. Environment Agency, Colchester.

Environment Agency (2001). *Review of the Appraisal Framework.* Environment Agency, Bristol.

Flather, R., Baker, T., Woodworth, P., Vassie, I., and Blackman, D. (2001). *Integrated effects of climate change on coastal extreme sea levels.* Proudman Oceanographic Laboratory, Internal Document No. 140.

Gill, J., O'Riordan, T. and Watkinson, A. (2002). *Redesigning the coast.* Tyndall Centre and CSERGE, University of East Anglia, Norwich.

Halcrow Engineering (2002) *Coastal realignment study: a review of shoreline management plans.* Halcrow, Swindon.

Hawkins, S. J., Southward, A. J., Kendall, M. A., Burrows, M. T., Thompson, R. C. and O'Riordan, R. (2002). (MarClim) Marine biodiversity and climate change. In J. Gill, T. O'Riordan and A. Watkinson (eds) *Redesigning the coast.* Tyndall Centre and CSERGE, University of East Anglia, Norwich, 45-50.

Holman, I. P., Loveland, P. J., Nicholls, R. J., Shackley, S., Berry, P. M., Rounsevell, M. D. A., Audsley, E., Harrison, P. A. and Wood, R. (2002). *REGIS - Regional Climate Change Impact Response Studies in East Anglia and North West England.* DEFRA, London.

Hosking, A., Moore, R., Parsons, A., Bray, M., Lee, M. and McInnes, R. (2001). *Assessing the impacts of climate change on the SCOPAC coast.* Proceedings of the 36th MAFF conference of coastal engineers.

Hughes, R. G. and Paramor, O. A. L. (2002). The effects of biological and physical processes on saltmarsh erosion and restoration in SE England. In J. Gill, T. O'Riordan and A. Watkinson (eds) *Redesigning the coast.* Tyndall Centre and CSERGE, University of East Anglia, Norwich, 16-25.

Hughes, R.G. (1999). Saltmarsh erosion and management of saltmarsh restoration; the effects of infaunal invertebrates. *Aquatic Conservation: Mar. Freshw. Ecosyst.* 9, 83-95.

Hulme, M., Jenkins, G., Lu, X., Turnpenny, J.R., Mitchell, T.D., Jones, R.G., Lowe, J., Murphy, J.M., Hassell, D., Boorman, P., McDonald, R. and Hill, S. (2002). *Climate Change Scenarios for the United Kingdom: The UKCIP02 Scientific Report.* Tyndall Centre for Climate Change Research, University of East Anglia, Norwich, UK.

Huntley, B. and Webb, T. (1989). Migration: species' response to climate variations caused by changes in the earth's orbit. *Journal of Biogeography*, 16, 5-19.

IPCC (2001a). *Climate Change 2001: Impacts, Adaptation, and Vulnerability.* Cambridge University Press, Cambridge.

IPCC (2001b). *Climate Change 2001: The Scientific Basis.* Cambridge University Press, Cambridge.

Lee, M. (2001). Coastal defence and the Habitats Directive: predictions of habitat change in England and Wales. *The Geographical Journal* 167(1): 39-56.

Lowe, J.A., Gregory, J.M. and Flather, R.A. (2001). Changes in the occurrence of storm surges around the United Kingdom under a future climate scenario using a dynamic storm surge model driven by Hadley Centre climate models. *Climate Dynamics* 18: 179-188.

Mervis, J., 1999. Biocomplexity blooms in NSF's research garden. *Science* 286, 2068-2069.

Ministry of Agriculture, Fisheries and Food (1999). *Flood and Coastal Defence: Project Appraisal Guidance.* MAFF, London.

Naylor, L.A. and Torok, S. (2003). Climate Change: Counting the Costs at the Coast. School of Environmental Sciences, University of East Anglia.

Nicholls, R.J. and Wilson, T. (2002). Integrated impacts on coastal areas and river flooding. In I.P. Holman and P.J. Loveland (eds.) *REGIS: Regional Climate Change Impact Response Studies in East Anglia and North West England.* DEFRA, London.

O'Connor, B.A. (2002). Redesigning the coast: engineering and modelling considerations. In J. Gill, T. O'Riordan and A. Watkinson (eds) *Redesigning the coast.* Tyndall Centre and CSERGE, University of East Anglia, Norwich, 13-15.

Office of the Deputy Prime Minister (ODPM). (2001). Planning Policy Guidance Note 25: Development and Flood Risk.

Office of the Deputy Prime Minister (2003) *Sustainable Communities*, ODPM, London.

O'Riordan, T. (Ed). (2002). *Redesigning the Coast: a report of a Workshop.* CSERGE Working Paper PA 02-01, University of East Anglia, Norwich.

Pasternack, G.B. and Brusch, G.S. (2002). Biogeomorphic controls on sedimentation and substrate on a vegetated tidal freshwater delta in upper Chesapeake Bay. *Geomorphology* 43: 293-311.

Rotmans, J., Asselt, M.B.A. van., Bruin A.J. de., Elzen M.G.J. den., Greef J. de., Hilderink, H.B.M., Hoekstra A.Y., Janssen, M.A., Koster, H.W., Martens, W.J.M., Niesen, L.W., Vries H.J.M. de., (1996). *Global change and sustainable development: A modelling perspective for the next decade. RIVM report 461502004.* Bilthoven, The Netherlands.

Schuchardt, B. and Schirmer, M. (2002). Impact of a climate change scenario on the Weser estuary. In J. Gill, T. O'Riordan and A. Watkinson (eds) *Redesigning the coast.* Tyndall Centre and CSERGE, University of East Anglia, Norwich, 26-37.

Turner, R.K. (2000). Integrating natural and socio-economic science in coastal management. *Journal of Marine Systems*, 25, 447-460.

Turner, R.K. (2002). Concepts and methods for integrated coastal management. In J. Gill, T. O'Riordan and A. Watkinson (eds) *Redesigning the coast.* Tyndall Centre and CSERGE, University of East Anglia, Norwich, 51-60.

Woodworth, P.L., Tsimplis, M.N., Flather, R.A. and Shennan, I. (1999). A review of the trends observed in British Isles mean sea-level data measured by tide gauges. *Geophysical Journal International*, 136, 651-670.

Disseminating Coastal Zone Information through the Internet

B Tomlinson – Emu Ltd, Southampton, UK
C Hill – GeoData Institute, Southampton, UK
J Sadler – GeoData Institute, Southampton, UK

Executive Summary

Developments of legislative requirements on access to environmental information are providing an increased impetus to make public authority information accessible. The proactive release of data and information has significant benefits in coastal and marine operational applications. It also assists in delivering the objectives of the Aarhus Convention on Access to Environmental Information[1], in cost-effectively meeting the likely obligations of Environmental Information Regulations, in widening awareness and participation in decision-making.

Publication via the Internet appears to provide many advantages in addressing these requirements for integration with other proposed Directives, for integrated coastal zone management and for the removal of barriers to the exploitation of public sector coastal information.

Here, these international objectives are explored in relation to their relevance to UK legislation and how it is responding. The objectives are illustrated through their application to coastal information resources and programmes for coastal observatory development.

Introduction

Reading the extensive literature and Directives emerging from Europe, one could be forgiven for becoming confused as to the requirements, conditions and exemptions relating to the access to environmental information. Such confusion raises several questions:

- how does the Freedom of Information Act relate to the Environmental Information Regulations and relevant Directives;
- how will new Directives of public participation affect the access to information;
- how do all these regulations affect the development of a Information Age Society and the Exploitation of Public Sector information and;

[1] http://www.unece.org/env/pp/documents/cep43e.pdf UN/ECE Convention on Access to Information and Public Participation in Decision-making and Access to Justice in Environmental Matters 25 June 1998.

Coastal Management 2003, Thomas Telford, London, 2003.

- what is covered by information on the environment and which organisations are within the reach of these changes?

Despite these varied requirements the Directives need to be implemented in an integrated fashion to deliver their separate, yet complimentary objectives.

This paper explores how these regulations affect coastal information and how well their requirements match what is already happening in the development of online marine information systems. Of particular importance is identifying how these directives might make information more widely accessible and to what extent best practice can be turned into advice for coastal zone managers.

Access to Information - Legal Issues

Despite existing approaches to make coastal information and data accessible these approaches may need to change to come into line with the EU directives, and translations to regulations and guidance, especially where these are driven by public bodies with required compliance. The directives and legislation stemming from the adoption of the Aarhus Convention will update existing terms and introduce new requirements for access to environmental information. The Aarhus Convention provides an international framework for access to environmental information currently with 40 signatories and 25 ratified by member states of the UN/ECE. Many of the definitions and the scope of the Convention (what constitutes public authorities, environmental information, exceptions) are adopted in the subsequent national legislation and statutory instruments. The Convention covers public authorities, which are defined to include governmental bodies along with bodies performing public administrative functions from all sectors at national, regional and local levels.

This is not to say that there was no access to environmental information before Aarhus and there are various examples of existing and mature data dissemination policies (e.g. NERC 1999[2]). The Convention sets the broader framework and agenda for the presumption in favour of access to information, limits the scope of exceptions and, in due course, will set a minimum requirement for access across a wider range of organizations.

The Freedom of Information Act 2000[3] (FoIA) makes provision for the disclosure of information held by public authorities or by persons providing services for them. This general right provided by the FoIA 2000 makes specific reference to the terms of the Aarhus Convention (Section 74) and provides for exclusion of environmental information (Section 48) from the access rights where implementing information provision under the convention is set by regulations made by the Secretary of State. Defra, is now developing new regulations (Environmental Information Regulations (EIR)) under section 74, with the drafting process taking due account of the forthcoming EU Environmental Information Directive.

Legislative access to information on the environment has been available, within certain limits since 1992 (Statutory Instrument 1992 No 3240, EIR and guidance implementing the EC Directive 90/313/EEC). This Directive has been updated to address some of the experience

---

[2] Natural Environment Research Council 1999 NERC Data Policy Handbook ver 2.1 March 1999.

[3] http://www.hmso.gov.uk/acts/acts2000/00036--n.htm#74 Freedom of Information Act 2000

and shortcomings of operation of the existing regulations and to reflect the changes to access rights and specifically to reflect changes in both generation and storage technologies. The new EU directive (EC2003/4/EC Freedom of Access to Information on the Environment[4]) launched in Jan 2003 translates a number of the Aarhus Convention objectives.

There are important differences between provisions in the FoIA and the new directive. For example, the directive contains no exemptions for information that can reasonably be obtained from elsewhere, whether or not a charge is made for the information. These issues will affect how the EIR facilitates access and the scope for proactive delivery of information.

The Convention and translation to the Directive and the draft EIR has both active and responsive access proposals. Responsive access is based on requests which the earlier EIR required in written format but which the new proposals for EIR update to include electronic requests. Active information provision includes the duties of public authorities to make information 'effectively accessible' by providing details of the scope of the information holdings and the public acquisition process. These requirements are generally being met by the development of Publication Schemes providing details and terms of access to environmental information.

Other important provisions in relation to integrated coastal information are the form and format in which the information is to be made available and the implications and implementation of access through the Internet. The draft EIR also propose that environmental information should be supplied in the form requested - with potential implications for conversions and translation. There are important clauses within the Convention that limit the need to have open-ended formats where information is already publicly available or where it would be unreasonable for the authority to make it available in another form. Obviously the test for what is and is not reasonable is important but this also stresses the advantage of proactive publication and dissemination.

Access through the Internet is also a target of the Convention to make information available through electronic databases. Although the Convention stresses particular categories for Internet availability (state of environment reporting etc)[5] there is essentially little limitation to the scope of data and information for such dissemination approaches. Internet-based access also forms part of the UK Government e-Gov initiatives. The e-Government Interoperability Framework (e-GIF)[6] defines the essential pre-requisites for a web-enabled government, with the adoption of Extensible Markup Language (XML) allowing interoperability and cross-platform data sharing.

---

[4] http://europa.eu.int/eur-lex/pri/en/oj/dat/2003/l_041/l_04120030214en00260032.pdf DIRECTIVE 2003/4/EC OF THE EUROPEAN PARLIAMENT AND OF THE COUNCIL of 28 January 2003 on public access to environmental information and repealing Council Directive 90/313/EEC

[5] http://www.unece.org/env/pp/contentofaarhus.htm

[6] http://www.govtalk.gov.uk/documents/e-gif_v5_part1_2003-04-25.pdf e-Government Interoperability Framework Version 5 April 2003.

Coastal Information Resources

Coastal information collected and maintained by public authorities will become subject to the new EIR and thereby public access. However, information clearly has wider roles in facilitating coastal and marine management practice. Information provision is seen as a key integrator within ICZM and the implications of publicly accessible data are also made within the proposals for the ICZM Directive[7]. "*National strategies should ...include adequate systems for monitoring and disseminating information to the public about their coastal zone. These systems should collect and provide information in appropriate and compatible formats to decision makers at national, regional and local levels to facilitate integrated management.*" [8]

The development of the Shoreline Management Planning (SMP) process in England and Wales illustrates the advantages for data access as part of the public participation and the development of the management options. The first round of SMPs did little to publish their information more widely. Only latterly did paper-based plan formats get translated to the Internet either as .pdf documents or interactive databases. This was often a response to the large number of requests for access to documentation that was expensive and time consuming to duplicate and distribute. (e.g. Southeast Coastal Group[9]).

Data access driven legislation and initiatives are also strengthened by other coastal programmes and proposed directives such as Integrated Coastal Zone Management and Public Participation in Decision Making for projects and plans. Both these areas are the focus of proposed EU Directives. Proliferation of initiatives for coastal and marine data makes the integrated elements in ICZM harder to appreciate - with multiple metadata systems, geographic information systems and data portals (EDMED, CoastMap, ICZMap etc). Nevertheless these data and information driven approaches are seen as key solutions for delivering the integration within coastal zone management.

The scope of environmental information is very wide (Table 1) and is likely to include raw and processed data along with policy-based, programme and project decision information relevant at local and regional levels. Existing coastal resource databases tend to take a thematic approach with holdings of metadata, oceanographic data etc. It is usually the coastal fora or similar regional programmes (e.g. Solent Forum[10]) that seek to draw the varied thematic or sectorial information together within the context of integrated information for a coastal zone. This often includes socio-economic information and, although this may not be seen as explicitly 'environmental', it may act as an indicator of state of the environment.

Not all the datasets for coastal zones are generated by those organizations currently developing or maintaining the systems for national and international marine and coastal data

---

[7] http://ww2.unime.it/scienze-terra/aarhusconv-ICZM.pdf Application of the Aarhus Convention in the Integrated Coastal Zone Management (ICZM) Process, Mercardie, A. (1999).
[8] January 2002, Common Position (EC) No 13/2002 Concerning the Implementation of Integrated Coastal Zone Management in Europe
[9] http//:www.se-coastalgroup.org.uk Southeast Coastal Group Shoreline Management Plan
[10] http://www.solentforum.hants.org.uk/ Solent Forum homepage

centres (BODC, Sea Search, WaveNET etc) and more local and regional systems for dissemination will be essential to meet EIR requirements.

Table 1 Environmental Data Types

Environmental data types	**Characteristics of data types**
Real-time (raw and processed)	Real-time and near real-time data feed, with potential 'in-field' data processing
Delayed	Data collected and subsequently processed and displayed
Archive	Past records (of existing monitoring programmes) time series data and historic records (e.g. former surveys, historic maps/aerial photographs etc)
Metadata	Data about the data and may include usage, storage metadata (e.g. BODC, EDMED)
Analysis and Visualisation	Processed and analyzed data, information - this may be seen as 'value-add' information as distinct from 'raw' data.
Summary Information	Awareness and summary of datasets, descriptive information
Characterization	Background, ecological, socioeconomic and physical spatial / geographic datasets etc
Policy	Policies and plans for the coastal zone, e.g. defence options and strategy from SMPs, protective designations and development allocations etc.
Decisions	Decisions - planning decisions

There are already examples of the broadening of coastal environmental information systems. Driven by the development of shoreline management plans and coastal defence strategies, coastal groups (coastal defence authorities and Environment Agency) are collecting much information which does not fit the data types or data structure of the World Data Centre system. For example, the data for the annual beach monitoring surveys undertaken by the Environment Agency, UK are not drawn together within a single data access point and the historic information and aerial photographs from which the data are derived are not readily accessible.

The important lesson for these approaches is that of integration of access from distributed data sources and cross sector information. Through the adoption of standards the regional portals that facilitate access within a coastal management zone (sediment cell, regional seas etc) can still draw together the data and information to present a national picture. The web, as a dissemination medium, offers a range of technologies that facilitates sharing, integration of information and standardization. The development of procedural guidance for the second generation of SMPs (Defra 2003[11]) has recognized both the advantages of standardized data acquisition, coordinated management through GIS and the role of dissemination through the Internet.

[11] Department of Environment, Food and Rural Affairs (2003 – in press) Draft Procedural Guidance Defra/NAW.

Similarly, the present metadata systems and repositories for coastal data are largely targeted at marine information and offshore resources rather than coastal resource databases. Whilst public authorities may not have a statutory duty to maintain coastal data portals the access to environmental information regulations may help promote maintenance of these systems. It is this mix of data and information themes that are relevant to the coastal zone reflecting a local and regional perspective on decision-making, policy and planning related information.

Coastal Data Delivery - Technical Options

The technical requirements of the Internet-based coastal and marine information systems can be illustrated by a range of national and international web sites supporting metadata (EDMED, the European Directory of Marine Environmental Data), data access and data exchange initiatives (Wave Data Index, WaveNET, INSPIRE) and data gateways (SeaSearch, OceanNET).

There are also initiatives to promote discipline specific 'standardized' coastal spatial data models (ESRI 2002[12]) which cover the range of data types (time duration and instantaneous point, vector and area data and change data). Use of XML as a universal format for structured documents and data on the web provides the basis for non-proprietary storage and management of data and metadata allowing effective interoperability and exchange of information. Industry specific schemas for XML are also being developed to facilitate interoperability and exchange of data (Marine XML[13]). The Marine XML project is being developed within the global observing systems, with similar initiatives to develop schemas for other coastal data supply. Whilst many of these models and schemas are relevant the horizontal integration of varied disciplines (and hence specific data models and schema) within the coastal zone makes implementation more involved.

Regional consortia, such as the coastal and marine observation systems, offer a further cost-saving over more locally organized data collection and distribution. Recent development of the marine observing systems both for the UK and overseas illustrates the benefits of regional sea parameter monitoring and coastal data management (e.g. Channel Coastal Observatory and Dubai Coastal Observatory, Irish Sea and Liverpool Bay Coastal Observatory[14]) that bring together previously distributed data collection and management programmes. A selection of examples is presented in Table 2.

---

[12] Introduction to ESRI data models
http://support.esri.com/index.cfm?fa=downloads.dataModels.intro
[13] EU Marine XML Project http://www.marineXML.net/ XML - the extensible Markup language is a language for providing web based data content.
[14] http://cobs.pol.ac.uk/ Irish Sea and Liverpool Bay Coastal Observatory Proudman Oceanographic Laboratory.

Table 2 – Examples of Marine Observing Systems Presenting Access to Information on the Internet.

Name	Website Address
Dubai Coastal Observatory (Marine Works Unit, Dubai Municipality)	www.dubaicoast.org
Channel Coastal Observatory	www.channelcoast.org
Scarborough Coastal Observatory	www.geodata.co.uk/sbc
Coastal Observatory Irish Sea and Liverpool Bay (Proudman Oceanographic Laboratory)	http://cobs.pol.ac.uk/
Australian Coastal Data Centre	http://www.coastaldata.transport.wa.gov.au
WaveNET (CEFAS)	http://map.cefasdirect.co.uk/

Typical coastal observatories coordinate and enhance the recording and record management and dissemination through the Internet of real-time, time series, delayed, archive and policy data.

Figure 1 presents a broad insight into the structure and facilities under which coastal observatories may operate.

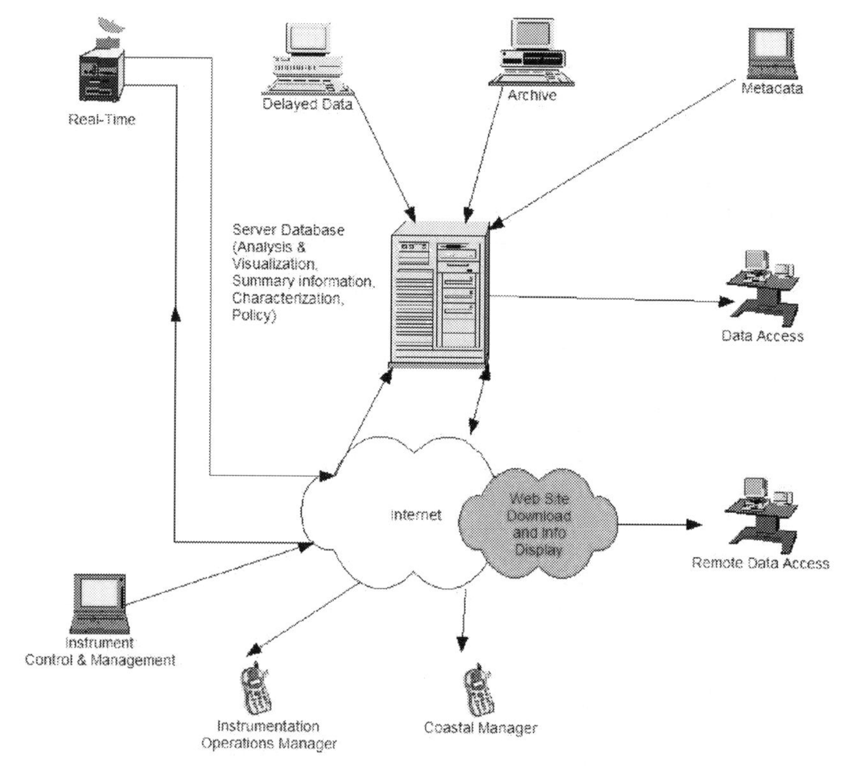

Figure 1 Typical Coastal Observatory Data Management and Dissemination System

These typical coastal observatory systems provide a browser-based interactive data access tool for marine metocean parameters with user-defined selection criteria and data download. Open source solutions are being promoted within both Europe and the UK[15] with eEurope[16] and eGIF policies to avoid proprietary lock-in, to advance interoperability and to reduce costs. Options for technical implementation are numerous. One approach adopted in Dubai (UAE) and Scarborough (UK) (Table 2) is based on an Open Source model, comprising PHP[17] scripts linked to a back-end MySQL[18] relational database storing data for a range of outputs (plots and data tables). Charts are generated using the JpGraph[19] PHP graph drawing library.

In a similar development (i-MARQ[20], Information System for Marine Aquatic Resource Quality supported under the EU Information Society Technology" Programme), web-based GIS functions are added for spatial query and display. The PostgreSQL[21] relational database is used for storing (NGDF[22]/XML compliant) metadata, while the PostGIS[23] extension enables direct storage and query of geographic objects within PostgreSQL. MapServer[24] and MapScript[25] are used for displaying spatial data and developing GIS interfaces within a PHP framework.

All applications are managed within a Content Management System developed in-house, allowing separation of content, navigation and design and enabling efficient compliance with web standards.

Users and Applications

Facilitating access to information and accessibility is established within the flexibility of adopting the web development standards (e.g. W3C[26], eGIF and OGC[27]) by divorcing the data content from the interface design through the use of style sheets. Such flexibility provides the potential to support a wide range of coastal data users:

- Operators and regulators (coastal zone managers, public authorities, environment protection and regulatory agencies)
- Stakeholders (wider public)

---

[15] http://www.govtalk.gov.uk/documents/oss_policydocument_2002-07-15.pdf Open Source software: Use within UK Government Office of e-Envoy 2002.
[16] http://europa.eu.int/comm/information_society/eeurope/documentation/index_en.htm eEurope Action Plan 2000.
[17] http://www.php.net/ PHP Homepage
[18] http://www.mysql.com/ MySQL Homepage
[19] http://www.aditus.nu/jpgraph/ JpGraph Homepage (Both Open Source and commercial licences have been deployed.)
[20] http://www.marinetech.co.uk/project_details/IMARQ/home.htm iMARQ homepage
[21] http://www.postgresql.org/ PostgreSQL Homepage
[22] http://www.ngdf.org.uk/ UK NGDF Homepage
23 http://postgis.refractions.net/ PostGIS Homepage
24 http://mapserver.gis.umn.edu/ MapServer Homepage
25 http://www2.dmsolutions.ca/webtools/php_mapscript/ MapScript Homepage
[26] W3C World Wide We Consortium
[27] OGC Open GIS Consortium

- Third party end users (research organizations, publishers etc)
- Emergency response (oil spill response, hazard response)
- Contractors, industrial users (consultants, commercial users)
- Educational users (schools, University, professional development)

Content management systems and user profiles offer a broad opportunity to customize the interfaces for these user groups. The Internet based system can deliver information through multiple channels with thick and thin client implementations. This may vary from map and graphical representation of data for thin clients to data download. Where 'value-added' data products form part of the cost recovery for data acquisition (e.g. NERC data) the ability to have registered users can be supported.

Additional services can include alerting users to data and data quality (for example industrial and regulatory users may require alerts of exceedence conditions (e.g. out of range or malfunctioning sensors or incidents) either as email or GSM calls.

Widening access is also about making the information available in the appropriate formats and in terms of widening participation in making data and information comprehensible and in notifying the potential users of the availability of information. Both the directives from the Aarhus Convention on public participation and the proposals for ICZM stress the wide participation (multiple users) which will require that information systems address these varied audiences.

Roles of Information Access

Development of Internet based dissemination of coastal data will clearly fall within the scope of Environmental Information Regulations and the Aarhus Convention given that it is generally public authorities (coastal defence authorities and government research organisations that collect and manage much of the information). Whilst meeting these legislative requirements there is a much wider role for access to information via the Internet (Figure 1).

These benefits include:

- Potential performance indicators for local authority in responding to data requests - (web tracking allows immediate reporting of requirements and delivery response rates).
- Potentially a way of making access to data exempt from EIR by virtue of information already available where the data is proactively published.
- Easing the tracking of information supply and profiling potential users of information, with eased management of licences and distribution.
- Offering the opportunity to automate supply and translation requirements (reasonable supply requirements)
- Supporting wider use of data, and with real-time data the opportunity to create web delivered exceedence messages and alerts (for data quality management and service), coupled modelling and forecasting.
- Obviating the need for users to have software - if the data is visualized on the web site.
- Supporting the public decision making - as a prelude to the EU Directive - but enshrined within the principles of ICZM and community decision-making.

- Supporting multiple user interfaces and output customization to different users (meeting 'format and form' requirements of EIR) and potentially meeting web accessibility requirements (W3C web content accessibility guidelines) and e-government initiatives (e-GIF).

Conclusions

This paper has particularly concentrated on emerging requirements for local dissemination of coastal information and therefore has not attempted to address many of the other issues surrounding the access to environmental information and access over the Internet, such as the control of copyright, intellectual property and the implications for charging for datasets. These are issues that will be affected by the new EIR and may have wide-ranging implications for the way that information is disseminated and potentially the ability to recoup costs of data acquisition by public bodies. Despite these concerns there is already widespread experience of the benefits accrued through making local and regional coastal information available through the Internet.

The legislative and best practice approaches clearly identify the advantage of considering the dissemination requirements within a project's specification. If these standards and requirements can be addressed at the data acquisition stage and proactive dissemination is facilitated through the Internet then it appears that there will be reduced costs and processing of responses to data requests under EIR. Wider benefits of this coordinated approach include a wide range of coastal management applications facilitating the effective use of the information. Whilst there is ample experience from coastal observatories focusing on (near) real time data delivery over the Internet the scope of environmental information for coastal zone management is more varied and will require more regional solutions and multiple user interfaces.